A Handbook for Sensory and Consumer-Driven New Product Development

T0329111

Related Titles

Emotion Measurement
(978-0-08-100508-8)

Multisensory Flavor Perception
(978-0-08-100350-3)

Developing Food Products for Consumers with Specific Dietary Needs
(978-0-08-100329-9)

Woodhead Publishing Series in Food Science, Technology and Nutrition

A Handbook for Sensory and Consumer-Driven New Product Development

Innovative Technologies for the Food and Beverage Industry

Maurice G. O'Sullivan

AMSTERDAM • BOSTON • CAMBRIDGE • HEIDELBERG
LONDON • NEW YORK • OXFORD • PARIS • SAN DIEGO
SAN FRANCISCO • SINGAPORE • SYDNEY • TOKYO

Woodhead Publishing is an imprint of Elsevier

WP
WOODHEAD
PUBLISHING

Woodhead Publishing is an imprint of Elsevier
The Officers' Mess Business Centre, Royston Road, Duxford, CB22 4QH, United Kingdom
50 Hampshire Street, 5th Floor, Cambridge, MA 02139, United States
The Boulevard, Langford Lane, Kidlington, OX5 1GB, United Kingdom

Notices
Knowledge and best practice in this field are constantly changing. As new research and experience
broaden our understanding, changes in research methods, professional practices, or medical
treatment may become necessary.

Practitioners and researchers must always rely on their own experience and knowledge in
evaluating and using any information, methods, compounds, or experiments described herein. In
using such information or methods they should be mindful of their own safety and the safety of
others, including parties for whom they have a professional responsibility.

To the fullest extent of the law, neither the Publisher nor the authors, contributors, or editors,
assume any liability for any injury and/or damage to persons or property as a matter of products
liability, negligence or otherwise, or from any use or operation of any methods, products,
instructions, or ideas contained in the material herein.

Library of Congress Cataloging-in-Publication Data
A catalog record for this book is available from the Library of Congress

British Library Cataloguing-in-Publication Data
A catalogue record for this book is available from the British Library

ISBN: 978-0-08-100352-7 (print)
ISBN: 978-0-08-100357-2 (online)

For information on all Woodhead Publishing publications
visit our website at https://www.elsevier.com/

Working together
to grow libraries in
developing countries

www.elsevier.com • www.bookaid.org

Publisher: Nikki Levy
Acquisition Editor: Rob Sykes
Editorial Project Manager: Karen Miller
Production Project Manager: Laura Jackson
Designer: Maria Inês Cruz

Typeset by TNQ Books and Journals

Contents

Part II
Product Quality, Development and Optimisation

Part III
Case Studies: Sensory and Consumer Driven NPD in Action

Woodhead Publishing Series in Food Science, Technology and Nutrition

Chilled foods: A comprehensive guide
Edited by C. Dennis and M. Stringer

Yoghurt: Science and technology
A. Y. Tamime and R. K. Robinson

Food processing technology: Principles and practice
P. J. Fellows

Bender's dictionary of nutrition and food technology Sixth edition
D. A. Bender

Determination of veterinary residues in food
Edited by N. T. Crosby

Food contaminants: Sources and surveillance
Edited by C. Creaser and R. Purchase

Nitrates and nitrites in food and water
Edited by M. J. Hill

Pesticide chemistry and bioscience: The food-environment challenge
Edited by G. T. Brooks and T. Roberts

Pesticides: Developments, impacts and controls
Edited by G. A. Best and A. D. Ruthven

Dietary fibre: Chemical and biological aspects
Edited by D. A. T. Southgate, K. W. Waldron, I. T. Johnson and G. R. Fenwick

Vitamins and minerals in health and nutrition
M. Tolonen

Technology of biscuits, crackers and cookies Second edition
D. Manley

Instrumentation and sensors for the food industry
Edited by E. Kress-Rogers

Food and cancer prevention: Chemical and biological aspects
Edited by K. W. Waldron, I. T. Johnson and G. R. Fenwick

Food colloids: Proteins, lipids and polysaccharides
Edited by E. Dickinson and B. Bergenstahl

Food emulsions and foams
Edited by E. Dickinson

Maillard reactions in chemistry, food and health
Edited by T. P. Labuza, V. Monnier, J. Baynes and J. O'Brien

The Maillard reaction in foods and medicine
Edited by J. O'Brien, H. E. Nursten, M. J. Crabbe and J. M. Ames

Encapsulation and controlled release
Edited by D. R. Karsa and R. A. Stephenson

Flavours and fragrances
Edited by A. D. Swift

Feta and related cheeses
Edited by A. Y. Tamime and R. K. Robinson

Biochemistry of milk products
Edited by A. T. Andrews and J. R. Varley

Physical properties of foods and food processing systems
M. J. Lewis

Food irradiation: A reference guide
V. M. Wilkinson and G. Gould

Kent's technology of cereals: An introduction for students of food science and agriculture Fourth edition
N. L. Kent and A. D. Evers

Biosensors for food analysis
Edited by A. O. Scott

Separation processes in the food and biotechnology industries: Principles and applications
Edited by A. S. Grandison and M. J. Lewis

Handbook of indices of food quality and authenticity
R. S. Singhal, P. K. Kulkarni and D. V. Rege

Principles and practices for the safe processing of foods
D. A. Shapton and N. F. Shapton

Biscuit, cookie and cracker manufacturing manuals Volume 1: Ingredients
D. Manley

Biscuit, cookie and cracker manufacturing manuals Volume 2: Biscuit doughs
D. Manley

Environmental assessment and management in the food industry
Edited by U. Sonesson, J. Berlin and F. Ziegler

Consumer-driven innovation in food and personal care products
Edited by S. R. Jaeger and H. MacFie

Tracing pathogens in the food chain
Edited by S. Brul, P. M. Fratamico and T. A. McMeekin

Case studies in novel food processing technologies: Innovations in processing, packaging, and predictive modelling
Edited by C. J. Doona, K. Kustin and F. E. Feeherry

Freeze-drying of pharmaceutical and food products
T.-C. Hua, B.-L. Liu and H. Zhang

Oxidation in foods and beverages and antioxidant applications Volume 1: Understanding mechanisms of oxidation and antioxidant activity
Edited by E. A. Decker, R. J. Elias and D. J. McClements

Oxidation in foods and beverages and antioxidant applications Volume 2: Management in different industry sectors
Edited by E. A. Decker, R. J. Elias and D. J. McClements

Protective cultures, antimicrobial metabolites and bacteriophages for food and beverage biopreservation
Edited by C. Lacroix

Separation, extraction and concentration processes in the food, beverage and nutraceutical industries
Edited by S. S. H. Rizvi

Determining mycotoxins and mycotoxigenic fungi in food and feed
Edited by S. De Saeger

Developing children's food products
Edited by D. Kilcast and F. Angus

Functional foods: Concept to product Second edition
Edited by M. Saarela

Postharvest biology and technology of tropical and subtropical fruits Volume 1: Fundamental issues
Edited by E. M. Yahia

Postharvest biology and technology of tropical and subtropical fruits Volume 2: Açai to citrus
Edited by E. M. Yahia

Postharvest biology and technology of tropical and subtropical fruits Volume 3: Cocona to mango
Edited by E. M. Yahia

Preface

In the mid-1990s, my first job, after completing my MSc in Food Science, was as a product and process development and optimisation scientist for a large blue chip meat ingredient company which supplied all the familiar pizza, chicken, Mexican and sandwich restaurant chains in Europe. I was tasked with a broad and diverse portfolio of New Product Development (NPD) projects. These ranged from bespoke projects for new meat-based pizza topping products to those designed to improve the yield of cooked meat toppings using efficient impingement cooking, with clean-label ingredient incorporation, to designing and commissioning of meat topping extrusion plates. Quite often, I found myself serving up the pizza topping (cooked or fermented meat) I had designed and optimised to a sensory panel made up of the different R&D teams and managers. Back then the demarcation line between hedonic and descriptive analysis was very clear. Quite simply we did not undertake any classical hedonic or affective analysis and typically used degree of difference from a standard (gold) scoring systems or weighting of differences from control tests using individual experts to evaluate our products. The hedonic side of the sensory coin was solely the realm of the external marketing company contracted to consumer test our optimised and developed products.

Even then as a novice sensory scientist, with no formal training, it occurred to me we were doing things backwards. We were developing products from inception and often from the idea of a senior manager, ingredient supplier or suggestion from a customer account manager without any consumer validation work being undertaken. The project was then implemented all the way through initial least cost formulation and regulatory compliance evaluation to prototyping, upscaling and ultimately to commercial batch production, safety and quality testing. Sensory quality was determined by in-house group discussions where a senior manager could either champion or kill the product based on individual enthusiasm or negative bias. The consumer was only asked to evaluate the product at the end of the chain when it was clear to me they should be brought on board at the beginning.

I knew then that I had to train properly as a sensory scientist and was extremely fortunate to have been offered the opportunity of studying for a PhD in sensory and consumer science with some of the best sensory scientists Europe then had to offer. I thus moved to Copenhagen, Denmark and commenced my studies, under the supervision of Professor Magni Martens, for the next 3 years at Sonsorik (Sensory) at the Royal Veterinary and Agricultural University (KVL,

Kongelige Veterinær-og Landbohøjskolen), now part of Copenhagen University. When I qualified with my PhD, it was then that I understood that product development needed to be holistic and multimodal in nature through the incorporation of the sensory (affective, descriptive, behavioural), instrumental and physicochemical, but it certainly had to be consumer-led. Harald Martens, a founding scientist for the application of sensory-based chemometrics (sensometrics) was Adjunct Professor during my time in Copenhagen and it was from his guidance and tutorship, and the opportunity to undertake research and publish with him that the secret to unlocking this complicated mass of multimodal data using multivariate data analysis (MVA) became my eureka moment.

I took this new found knowledge and launched back into industry-based research for the next few years with Diageo in Dublin, Ireland, successfully working on exciting product development and optimisation projects for famous cream liqueurs and beer brands.

In the present book, I will describe the sensory methods that can be broadly segmented into three different areas, Difference methods, Descriptive methods and Affective methods. Difference methods are the most basic form of sensory test and are relatively simple procedures whereby samples are compared directly to other samples and assessors are asked to determine if they are the same or different. Descriptive methods involve the training of panellists, which are then used to quantitatively determine the sensory attributes of samples. The assessors are trained and calibrated to produce a more standardised response in order to measure the attributes associated with the relevant sensory modalities of appearance, aroma, flavour, texture, taste and aftertaste. The language is descriptive and nonhedonic, in that assessors are not asked how much they rate or like the product being tested. On the other hand, 'Affective methods' use hedonics or ask untrained assessors their opinion of a product or products. Optimisation of food or beverage products are undertaken to maintain or improve the liking of that product with respect to consumer sensory quality. These affective or hedonic methods include rating their liking of appearance, aroma, flavour, texture and their overall impression or the overall acceptability of a product. Typical affective methods include the classical 'sensory acceptance test' and the 'consumer test'.

There have been a number of innovations in the last few years resulting in the development of some new sensory tools which are described here as Novel methods. These include methods such as 'napping' or 'preference mapping', 'flash profiling', 'ranking descriptive analysis (RDA)', 'ultraflash profiling' and 'check all that apply' methods. To a certain extent, these new novel methods have narrowed the divide between the rigid rules of classic descriptive profiling and the emotional responses involved with affective sensory methods.

Over the years, I have met NPD consultants and academic professors who have enthusiastically taught product development-related subjects without ever having worked a day in a factory, or R&D team within industry. Equally, I have frequently encountered industry professionals who work in NPD teams but do not fully understand the background sensory science or are aware of advances and innovations. As well as having an extensive NPD career in the food and beverage industries, for the last decade I have worked as a Sensory and Consumer Scientist at University College Cork, Ireland. As an academic, I have undertaken funded research, supervised PhD students, taught undergraduates and published prolifically, but I have also had the pleasure of applying this knowledge (current and new) with the collaboration of the food and beverage industry as part of the 'Sensory Unit' service. This sensory science-based service (at UCC) included testing the difference, affective and descriptive analysis of foods and beverages as well as assessing shelf life and comparing sensory results to microbiological and analytical (instrumental, physicochemical) data using MVA. We also routinely used the above described sensory methods as well as some of the new rapid methods, such as RDA, in our industry-based research and applications for NPD. Sensory testing is critical for NPD/optimisation, ingredient substitution and devising appropriate packaging and comparing foods or beverages to competitor's products. Again, the key to unlocking the hidden secrets of this multimodal data is with MVA.

This book describes these sensory methods, in detail, with case studies by way of detailed explanation where required and in conjunction with MVA. This book will appeal to all those interested in learning more about consumer-led product development, in an easy-to-read and user-friendly fashion, from the R&D scientist in the large multinational food or beverage company to their counterpart in a start-up or small enterprise to the undergraduate or postgraduate student and even the lay person.

Part I

Sensory Methods

Chapter 1

Difference Methods

INTRODUCTION

Sensory methods can be broadly segmented into three different areas: Difference methods, Descriptive methods and Affective methods.

Sensory perception is not a standard response in humans but is effected by age and gender (Michon et al., 2010a,b,c) as well as cultural influences (Yusop et al., 2009a,b) and many other factors. However, descriptive methods which involve the training of panellists to quantitatively determine the sensory attributes, in typically a selection of samples, can calibrate humans to produce a more standardised response. The assessors are trained to measure the attributes associated with the relevant sensory modalities of appearance, aroma, flavour, texture, taste and aftertaste. The language is descriptive and nonhedonic, in that assessors are not asked how much they rate or like the product being tested. The different methods for descriptive profiling include the flavour profile, QDA (quantitative descriptive analysis), spectrum, free choice profiling and texture profile methods. However, only some of these methods are widely used for product development and research purposes.

Affective methods use hedonics or ask untrained assessors their opinion of a product or products. These might include rating their liking of appearance, aroma, flavour, texture and their overall impression or the overall acceptability of a product. Typical affective methods include the classical 'sensory acceptance test' and the 'consumer test'.

There have been a number of innovations in the last few years resulting in the development of some new sensory tools which are described here as novel methods. These include methods such as 'napping' or 'preference mapping', 'flash profiling', 'ultraflash profiling' and 'check all that apply (CATA)' or rate all that apply (RATA) methods. These will be discussed in detail in Chapter 3.

Difference methods are the most basic form of sensory test and are relatively simple procedures whereby samples are compared directly to other samples and assessors are asked to determine if they are the same or different. They may also be asked to assign a score or grade to the sample or also even comment on why they think they are different. Generally, difference testing involves determining the difference between two (paired comparison), three (triangle) or four (tetrad) products. These tests can be categorised into overall

A Handbook for Sensory and Consumer-Driven New Product Development.
http://dx.doi.org/10.1016/B978-0-08-100352-7.00001-4

3

difference tests and attribute-specific directional difference tests. Difference tests are methods which determine if there is a detectable sensory difference between the samples, whereas attribute difference testing determines whether there is as a perceived specified attribute difference between samples. The most common difference tests are the duo-trio test, the triangle test, the simple same − difference test and the 'A' − 'not A' test (Lawless and Heymann, 1998a,b; Piggott et al., 1998). Attribute difference tests also include the simple ranking test and the alternative forced choice (AFC) test. These latter methods are more sensitive in the detection of sensory differences between samples. However, they are not practical for some complex products such as meat and some cheese studies because they require that only one sensory attribute varies independently of the other sensory properties (Byrne and Bredie, 2002). This is more difficult in these complex foods as typically there is an interaction between the sensory variables appearance, aroma, flavour, texture, taste and aftertaste.

A difference test can become an AFC test when specific differences are asked of assessors. For example, in a triangle test, three samples are given to the assessor, two are identical and the assessor is asked to pick the odd one out. This could become a 3-AFC test, if, for example, the assessor was asked to pick out the sweetest sample of the three presented. Each of these methods will be discussed below in detail. Similarly a paired comparison test becomes a 2-AFC test when assessors are given a criterion to differentiate between samples, i.e., sweetness, bitterness, etc. The panellist has to choose a sample in AFC tests even if they cannot differentiate between the samples. The tetrad test is more powerful than the triangle test, but the AFC tests are more powerful than either of these methods (Xia et al., 2015).

THE A − NOT A METHOD

This method is the simplest of difference methods. The panellists are presented with a series of coded samples, a minimum of two, which consist of sample 'A', while others are different from sample 'A'. The assessors are then asked whether or not the sample they are evaluating is identical to 'A'. The assessors evaluate sample A and become familiar with it prior to commencement of the test (Lawless and Heymann, 1998a,b; Piggott et al., 1998).

THE PAIRED COMPARISON TEST AND THE 2-AFC (ALTERNATIVE FORCED CHOICE) TEST

The paired comparison test is one of the simplest and fastest forms of difference test which can be used to differentiate between two samples based on assessor preference. In this case the assessor is asked which of two products is preferred. However, more usually assessors are presented with two coded products and are asked to differentiate them based on intensity differences between a specific attribute of interest such as saltiness, bitterness, sweetness, etc. The assessor may be asked which sample is more intense for a specific attribute and must give an answer. In this instance the paired comparison test becomes a 2-AFC test.

It is an easy and quick test used when the attribute that will change is known and it is thus often used to determine whether formulation changes to a product are detectable. However, it is not very statically powerful as there is a 50% chance of being right by simply guessing the answer. The coded samples should be presented twice in reverse order to assessors and should be randomised for presentation to larger assessor groups ($n > 30$) (Ennis and Jesionka, 2011) in order to obtain more reliable data. The paired comparison test is often used to detect differences in samples for quality control purposes, batch-to-batch variation, ingredient substitution trials, process changes, etc. It is also used in the screening of panellists for descriptive panels. See Table 1.1 for sample sizes required for significant differences.

THE DUO-TRIO TEST

Duo-trio test is a quick difference method where assessors are given three coded products with one of these identified as the control. They are then asked which of the products is the closest to the control. They are not asked about any specific attribute differences. Again, this test is not a very statistically powerful test as the probability of an assessor guessing the correct answer is 50%. For this reason, large assessor groups ($n > 32$) (Ennis and Jesionka, 2011) need to be used to increase confidence in the data. When undertaking the test the sample order should be randomised. The duo-trio test also is used to detect differences in samples for quality control purposes and also used in the screening of panellists for descriptive panels. See Fig. 1.1, for example, of panel scoring sheet and Table 1.1 for sample sizes required for significant differences.

THE TRIANGLE TEST AND THE 3-AFC TEST

Joseph E. Seagram and sons were the first to employ the triangle test in 1941 and since then it has been used in a variety of applications including product discrimination testing and panellist selection (Ennis and Rousseau, 2012). For the triangle test, assessors are presented with three blind coded samples. They are told two samples are the same, but are not told which two. They must determine which sample is the different one and are not asked any attribute-specific differences. This is also a quick method and is a little more statistically powerful than the duo-trio and paired comparison test as there now is only a 33% chance of an assessor guessing the correct answer. An advantage of the triangle test is that it does not require specification of the nature of the difference. Yet the triangle test requires large sample sizes to be effective (Ennis, 1993).

TABLE 1.1 Sample Sizes Required for Significant Differences for the 2-AFC, Duo-Trio, Triangle and Tetrad Difference Methods

Delta	Power = 0.3				Power = 0.85				Power = 0.9				Power = 0.95			
	2-AFC	DT	TET	TRI	2-AFC	DT	TET	TRI	2-AFC	DT	TET	TRI	2-AFC	DT	TET	TRI
0.30	237	23,293	5,311	20,704	273	27,010	6,159	24,047	319	32,171	7,307	28,625	394	40,563	9,229	36,148
0.35	173	12,687	2,919	11,314	201	14,709	3,389	13,125	237	17,491	4,022	15,603	296	22,080	5,066	19,692
0.40	137	7,508	1,753	6,709	156	8,713	2,025	7,792	184	10,365	2,397	9,247	226	13,088	3,019	11,676
0.45	111	4,753	1,116	4,246	126	5,522	1,293	4,933	143	6,561	1,540	5,843	182	8,253	1,934	7,369
0.50	89	3,160	752	2,825	102	3,661	870	3,286	122	4,350	1,029	3,901	150	5,473	1,293	4,898
0.55	76	2,203	531	1,972	87	2,553	611	2,271	100	3,018	723	2,696	124	3,799	902	3,398
0.60	65	1,579	386	1,415	74	1,828	440	1,642	87	2,162	528	1,940	104	2,713	657	2,435
0.65	56	1,170	290	1,049	65	1,348	332	1,206	74	1,591	389	1,427	89	2,003	491	1,800
0.70	49	891	223	792	56	1,020	251	919	65	1,213	301	1,084	78	1,524	372	1,360
0.75	42	689	173	614	49	792	198	706	56	936	234	841	69	1,174	293	1,055
0.80	37	542	140	488	42	627	162	565	49	734	187	660	60	928	234	827
0.85	33	436	113	389	40	496	129	451	47	596	151	531	56	736	187	663
0.90	30	357	94	318	35	409	110	366	40	484	124	434	49	602	154	545
0.95	28	294	78	262	33	338	89	304	35	394	105	355	44	494	129	443
1.00	26	241	65	220	28	279	78	251	33	334	89	301	40	413	110	369
1.05	23	205	57	184	26	237	65	212	30	279	78	251	37	346	94	310

1.10	21	173	47	154	26	203	57	179	28	237	65	212	35	296	81	265
1.15	21	152	42	135	23	173	47	154	26	203	57	184	33	254	70	231
1.20	18	135	39	116	21	152	42	135	26	175	47	157	28	220	60	198
1.25	18	113	34	102	21	135	39	116	23	154	42	140	28	190	52	173
1.30	16	100	29	89	18	115	34	102	21	137	39	121	26	169	47	151
1.35	16	89	27	81	16	102	29	94	21	122	34	110	23	150	42	132
1.40	16	78	22	70	16	91	29	81	18	109	32	94	23	135	39	116
1.45	13	74	22	65	16	80	23	73	18	96	29	86	21	117	34	105
1.50	13	65	20	57	16	74	22	65	16	87	25	78	21	109	32	94
1.55	13	58	16	52	13	67	20	60	16	78	22	70	18	98	29	86
1.60	11	51	15	47	13	60	18	52	16	69	22	60	18	87	27	78
1.65	11	49	15	42	13	56	16	47	13	65	20	57	16	78	23	70
1.70	11	47	15	39	13	49	15	42	13	58	18	52	16	74	22	65
1.75	11	40	13	37	11	47	15	42	13	56	16	47	16	67	22	60
1.80	11	37	13	34	11	42	15	39	13	49	15	42	16	60	18	53
1.85	11	35	11	30	11	40	13	34	11	47	15	40	13	56	18	50
1.90	11	33	11	29	11	35	13	34	11	42	15	39	13	51	16	47
1.95	11	30	9	29	11	35	11	29	11	40	13	34	13	49	15	42
2.00	8	28	9	23	11	33	11	29	11	37	13	34	13	47	15	39

From Ennis, J.M., Jesionka, V., 2011. The power of sensory discrimination methods revisited. Journal of Sensory Studies 26(5), 371–382. http://dx.doi.org/10.1111/j.1745-459X.2011.00353.x.

TRIANGULAR TEST WITH SCORING

Name:_____   Samples:_____   Date:_____

Instruction:
1. Taste these samples and determine if there is any notable difference between them.
2. Score and comment on findings/observations.
3. If '<u>no difference</u>' is detected, please tick the appropriate box, and record a score and comment for all three.

	Symbol	Score	Comment
Odd sample	☐	☐	_____
			_____
Paired sample	☐☐	☐	_____
			_____
No difference	☐	☐	_____
			_____

SCORING SCALE

10 = Perfect. 9 = Very good. 8 = Good. 7 = Borderline, 6 = Reject

Duo Trio TEST WITH SCORING

Name:_____   Samples:_____   Date:_____

Instruction:
1. Taste these samples and determine if there is any notable difference between them.
2. Score and comment on findings/observations.
3. If '<u>no difference</u>' is detected, please tick the appropriate box, and record a score and comment for all three.

	Symbol	Score	Comment
Control sample	☐	☐	_____
			_____
Odd sample	☐	☐	_____
			_____
No difference	☐	☐	_____
			_____

SCORING SCALE

10 = Perfect. 9 = Very good. 8 = Good. 7 = Borderline, 6 = Reject

FIGURE 1.1   Examples of Triangle and Duo-Trio test scoring sheets.

When undertaking the triangle test the sample order should be equally randomised across the three samples with a relatively large assessor group ($n > 30$) (Ennis and Jesionka, 2011). Assessors can also be presented with three coded products and are asked to differentiate them based on intensity differences between a specific attribute of interest such as saltiness, bitterness, sweetness, etc. The assessor may also be asked which sample is more intense for a specific attribute and must give an answer. In this instance the triangle test becomes a 3-AFC test. It is interesting to note that assessors choose the correct samples more often with the 3-AFC test compared to the triangle method even though the probability of guessing the right answer is one in three for both tests. The reason for this is that Thurstonian modelling of both methods shows that the 3-AFC test is more powerful than the triangle or duo-trio test (O'Mahony and Rousseau, 2003). This also means that fewer people are required, compared to a triangle test, to have the same degree of confidence in a result which is a major saving in time, effort and money. However, for the 3-AFC test to be affective, we need to know the sensory difference that is important before we start. This is often not the case as it can be hard to predict the effects of a change in formulation or process change and the subsequent consumer response (Ennis and Jesionka, 2011; Xia et al., 2015). See Fig. 1.1, for example, of panel scoring sheet and Table 1.1 for sample sizes required for significant differences.

THE RANKING TEST

In the Ranking test, assessors are asked to order a selection of coded samples (4−6) in increasing or decreasing intensity of a specific perceived sensory attribute. Samples are randomised across the assessors. For simple tests where fatigue is less likely, more samples may be used. It is often used in panel training or for screening samples prior to using other tests.

THE TETRAD TEST

The tetrad method is a difference test involving four samples where the assessor is presented with blind coded samples with two samples of one product and two samples of another product. The assessors must then group the products into two groups according to their similarity. Note that these instructions are different from asking the subjects to identify the two most similar samples (Ennis and Rousseau, 2012). The probability of guessing the right answer is similar to the triangle test (33%). The tetrad test has also received lots of interest due to its potential to provide increased power without specification of an attribute. This greater power means that for the same sample size, an existing difference is less likely to be missed (Ennis and Jesionka, 2011; Ennis and Rousseau, 2012). The tetrad method can thus reduce the likelihood of 'alpha' and 'beta' risk sensory testing errors. Alpha risk is the risk of making a wrong decision. If p is the decision point, then if $p <$ alpha, then the 'null hypothesis' is rejected. This is 'rejecting the null hypothesis', also called type I error and occurs when differences are found between samples when really there are not any. The opposite can also occur, by not rejecting the null hypothesis and is called beta risk, or type II error. Here, no differences are found between samples where differences really exist. Alpha and beta risks can be reduced by increasing the number of observations or the amount of data needed to make a decision. See Table 1.1 for sample sizes required for significant differences.

A considerable advantage of the tetrad test is that far fewer assessors are required compared to the triangle and duo-trio methods. According to Ennis and Jesionka (2011), p. 87, assessors would be required to achieve a significant (P < .05, 90% power, $d' = 1.5$) difference between samples for a duo-trio test, 78 for a triangle but only 25 for a tetrad panel. Greater power means that smaller sample sizes can be used to achieve the same performance as the triangle test as the sample sizes required by the tetrad test are theoretically only one-third that required by the triangle test (Ennis and Jesionka, 2011; Ennis and Rousseau, 2012). This could be of great commercial benefit in the saving of time, money and resources. However, the sensory scientist must determine through comparison which of the triangle or the tetrad method best suits their products and processes.

OTHER TESTS

Some other difference tests that can be used, particularly in the realm of quality assurance, are briefly described below and include the methods 'In/ Out', 'Ratings for degree of difference from a standard' and 'Weighting of differences from control' (Muñoz, Civille and Carr, 1992a,b; Lawless and Heymann, 1998a,b).

For the *'In/Out'* method, production batches can be evaluated by a trained panel as being either within or outside sensory specifications. The method can be limited though as it does not provide any descriptive information that can be used to amend problems (Muñoz, Civille and Carr, 1992b).

The relative deviation from a standard product is used in order to detect differences in production batches for the *'Ratings for degree of difference from a standard'* method. This method is also limited in that it does not provide any information regarding the source of differences compared to a control (Muñoz, Civille and Carr, 1992c). The degree of difference from a control or standard test is also quite often used in house for routine sensory testing, again using trained panellists (often external), particularly for monitoring of off-flavour defects and is widely used in the soft drink industry. Sources of off flavours could be flavour fading but more often involve cross-contamination issues due to inadequately CIP (cleaning in place) processes where residues in plant (pipework, containment vessels, balance tanks, etc.) from a previous production run contaminate the following run of a different product.

The final difference test '*Weighting of differences from control (Individual experts)*' is similar to the degree of difference from a standard method but involves an even more complex judgement procedure on the part of panellists. This is because it is not only the differences that matter, but also how they are weighted in determining product quality (Lawless and Heymann, 1998a; Muñoz, Civille and Carr, 1992b).

REFERENCES

Byrne, D.V., Bredie, W.L.P., 2002. Sensory meat quality and warmed-over flavour: a Review. In: Toldrá, F. (Ed.), Research Advances in the Quality of Meat and Meat Products. Volume Within Agriculture & Food Chemistry. Research Signpost, pp. 95—212.

Ennis, D.M., 1993. The power of sensory discrimination methods. Journal of Sensory Studies 8 (4), 353—370.

Ennis, J.M., Jesionka, V., 2011. The power of sensory discrimination methods revisited. Journal of Sensory Studies 26 (5), 371—382. http://dx.doi.org/10.1111/j.1745-459X.2011.00353.x.

Ennis, J.M., Rousseau, B., 2012. Reducing Costs with Tetrad Testing 15, 3—4.

Lawless, H.T., Heymann, H., 1998a. Sensory evaluation in quality control. In: Lawless, H.T., Heymann, H. (Eds.), Sensory Evaluation of Food, Principles and Practices. Chapman and Hall, New York, pp. 548—584.

Lawless, H.T., Heymann, H., 1998b. Descriptive analysis. In: Lawless, H.T., Heymann, H. (Eds.), Sensory Evaluation of Food, Principles and Practices. Chapman and Hall, New York (pp. 117—138, pp. 341—378).

Michon, C., O'Sullivan, M.G., Delahunty, C.M., Kerry, J.P., 2010a. Study on the influence of age, gender and familiarity with the product on the acceptance of vegetable soups. Food Quality and Preference 21, 478—488.

Michon, C., O'Sullivan, M.G., Delahunty, C.M., Kerry, J.P., 2010b. The investigation of gender related sensitivity differences in food perception. Journal of Sensory Studies 24, 922—937.

Michon, C., O'Sullivan, M.G., Sheehan, E., Delahunty, C.M., Kerry, J.P., 2010c. Investigation of the influence of age, gender and consumption habits on the liking for jam-filled cakes. Food Quality and Preference 21, 553 561.

Muñoz, A.M., Civille, G.V., Carr, B.T., 1992a. Comprehensive descriptive method. In: Sensory Evaluation in Quality Control. Van Nostrand Reinhold, New York, pp. 55—82.

Muñoz, A.M., Civille, G.V., Carr, B.T., 1992b. "In/Out" method. In: Sensory Evaluation in Quality Control. Van Nostrand Reinhold, New York, pp. 140—167.

Muñoz, A.M., Civille, G.V., Carr, B.T., 1992c. Difference-from-control method (Degree of difference). In: Sensory Evaluation in Quality Control. Van Nostrand Reinhold, New York, pp. 168—205.

O'Mahony, M., Rousseau, B., 2003. Discrimination Testing: a few ideas old and new. Food Quality & Preference 14, 157—164.

Piggott, J.R., Simpson, S.J., Williams, S.A.R., 1998. Sensory analysis International Journal of Food Science & Technology 33 (1), 7—12.

Xia, Y., Zhang, J., Zhang, X., Ishii, R., Zhong, F., O'Mahony, M., 2015. Tetrads, triads and pairs: experiments in self-specification. Food Quality and Preference 40, 97—105. http://dx.doi.org/10.1016/j.foodqual.2014.09.005.

Yusop, S.,M., O'Sullivan, M.G., Kerry, J.F., Kerry, J.P., 2009a. Sensory evaluation of Indian-style marinated chicken by Malaysian and European naïve assessors. Journal of Sensory Studies 24, 269–289.

Yusop, S.,M., O'Sullivan, M.G., Kerry, J.F., Kerry, J.P., 2009b. Sensory evaluation of Chinese-style marinated chicken by Chinese and European naïve assessors. Journal of Sensory Studies 24, 512–533.

Chapter 2

Descriptive Methods

INTRODUCTION

Sensory profiling methods first emerged in the 1950s and evolved from expert-based industry methods such as those employed by the established industry of winemakers, perfumers and brewmasters (Muñoz et al., 1992). Predominantly, these early methods involved the sensory detection of expected sensory problems and defects, which was only really effective for commodity-type products (Lawless and Heymann, 1998a). Sensory methods using trained panels evolved from these simple methods in the 1950s and 1960s (Muñoz, 2002). One of the most well-known contributions of this era was the 'the nine-point hedonic scale' or the 'degree of liking scale' in the 1940s invented by the US army (US Army Quarter-master Food and Container Institute, Chicago, Illinois) in the 1940s (Peryam and Pilgrim, 1957).

The next stage in the evolution of this process was the establishment of quality control programs in the food industry that included a more integrated sensory component (Muñoz, 2002).

Descriptive methods involve the training of panellists to quantitatively determine the sensory attributes in a sample or more usually a selection of samples. The assessors are trained to measure the attributes associated with the relevant sensory modalities of appearance, aroma, flavour, texture, taste and aftertaste. The language is descriptive and nonhedonic, in that assessors are not asked, for example, how much they rate or like the product being tested. Sensory perception is not a standardised response in humans and is effected by age and gender (Michon et al., 2010a,b,c) as many other factors including cultural influences (Yusop et al., 2009a,b). Descriptive analysis allows screened individuals to be trained to give a more standardised quantitative sensory response. The different methods for descriptive profiling include the flavour profile, texture profile, free choice profiling (FCP), spectrum and quantitative descriptive analysis (QDA) methods. Some of these are widely used for product development and research purposes, while others appear to be academic curiosities not widely used at all (Murray et al., 2001).

The very first descriptive method was the flavour profile method (FPM) (Caul, 1957) developed by the Arthur D Little Company in the late 1940s. FPM is a consensus technique, and vocabulary development and rating

A Handbook for Sensory and Consumer-Driven New Product Development.
http://dx.doi.org/10.1016/B978-0-08-100352-7.00002-6

13

sessions are carried out during group discussions, with panel members considering aspects of the overall flavour and the detectable flavour components of foods (Cairncross and Sjöstrom, 1950; Murray et al., 2001).

The texture profile method (TPM) evolved from FPM and was developed in the 1960s by General Foods. The technique aims to allow the description of only texture attributes from first-bite through complete mastication and also accounts for the temporal aspect of attributes (Murray et al., 2001). Sensory scales are anchored using sensory references for texture, but these references can now be quite hard to define either due to unavailability or reformulation. It is no surprise that this method has lost favour with sensory scientists.

FCP can also be used and this involves panellists developing their own descriptive terms (Williams and Arnold, 1985; Delahunty et al., 1997). The problem with this method is the subjective correlation of terms derived by different assessors may not, in reality, be related and for this reason, it has not been widely adopted.

Descriptive analysis is a method where defined sensory terms are quantified by sensory panellists. Detailed descriptions of sensory terminology and procedural guidelines for the identification and selection of descriptors for establishing a sensory profile by a multidimensional approach have been described in ISO (1992) and ISO (1994). A list of descriptive terms are determined initially and are referred to as a lexicon or descriptive vocabulary and describe the specific sensory attributes in the food or beverage sample and can be used to evaluate the changes in these attributes. These descriptive terms, determined from such lexicons of descriptive terms, have been developed and employed by a number of authors for the sensory evaluation of different products; Johnson and Civille (1986) for beef, Lyon (1987) and Byrne et al. (1999b) for chicken, and Byrne et al. (1999a, 2001b) for pork; Johnsen et al. (1998) for peanut flavour; Desai et al. (2013) for Greek yoghurt; Van Hekken et al. (2006) for Mexican cheese. The two most commonly used descriptive methods are the QDA and the spectrum methods.

The spectrum method developed in the 1970s (Civille and Szczesniak, 1973) is a descriptive profiling method which prescribes the use of a strict technical sensory vocabulary using reference materials. This method is pragmatic in that it provides the tools with which to design a descriptive procedure for a given product category. Its principal characteristic is that the panellist scores the perceived intensities with reference to prelearned 'absolute' intensity scales. The purpose is to make the resulting profiles universally understandable. The method provides for this purpose an array of standard attribute names ('lexicons'), each with its set of standards that define a scale of intensity (Muñoz and Civille, 1992; Meilgaard et al., 1999). With the spectrum method the scales are anchored using extensive reference points which may include a range of foods which correspond to food reference samples which apparently reduces panel variability. Panellists develop their list of attributes by evaluating a large array of products within the category. Products may be described in terms of only one attribute (e.g., appearance or aroma) or, they may be trained to evaluate all of them (Murray et al., 2001). One of the drawbacks is that extensive training of panellists is required when using the spectrum method. Also, similar to TPM, some of the reference materials are not available outside of America. Cultural differences of panels and the difficulty of quantifying an attribute over a range of different products are other problems of this method listed in the literature (Murray et al., 2001).

In the QDA method, experts with product knowledge can evaluate a sample set of the products to be profiled in the laboratory and suggest descriptive terms that specifically describe the product to be tested and the sensory dimension to be examined to produce an initial or 'meta' sensory term list. Also the vocabulary can be based on terms suggested by the panellists themselves in discussions under supervision of the panel leader. Sensory lexicons can also be provided which consist of lists of sensory terms describing appearance, aroma, flavour, texture, taste and aftertaste attributes. These lexicons are available for a vast array of food and beverage products with further examples including: Spanish Dry Cured Sausage (Pérez-Cacho et al., 2005), Honey (Galán-Soldevilla et al., 2005), Cheddar cheese (Drake et al., 2005), French cheese (Rétiveau et al., 2005). The QDA method was first proposed by Stone et al. (1974) and relies heavily on statistical analysis to determine the appropriate terms, procedures and panellists to be used for the analysis of a specific product. Initially, this statistical analysis is used in the sensory term reduction process during training and this training of the QDA panel requires the use of product references to stimulate the generation of terminology. These references help the panellist define and quantify the attribute they are assessing and greatly assist in the training process. The panel leader acts as moderator and facilitator, without directly influencing the group. Panellists cannot discuss data, terminology or samples after each taste session, but must rely on the discretion of the panel leader for any information on their performance. Feedback is provided by the facilitator based on the statistical analysis of the taste session data (Lawless and Heymann, 1998a,b; Meilgaard et al., 1999). From here, training involved sensory term reduction followed by sensory profiling which is described in detail in the case study section of this chapter below.

ENVIRONMENT AND PANEL CONDITIONS

For successful sensory evaluation the environment where the sensory analysis is conducted is of major importance. It is crucial to remove as many negative environmental factors as possible from the testing area so that the space does not interfere with the actual task of sensory judgement. This ensures that the panellist is not distracted or biased by their surroundings and that their assessment is purely on the product they are testing without any other external factors to influence them. This is achieved by designing the sensory testing area in such a way that these negative environmental factors are eliminated or at the very least minimised and is thus achieved by complying with the international standards and guidelines for the design of sensory testing facilities. ISO (2007) describes the requirements to set up a test room comprising a testing area, a preparation area and an office, specifying those that are essential or those that are merely desirable (ISO, 2007). The ASTM (2008) manual provides the latest guidelines for the sensory evaluation professionals who undertake the development of a new facility or remodelling of an existing sensory laboratory.

Sensory testing rooms should be separated from preparation areas to eliminate odour transfer. The test room must be divided into separate booths with adequate lighting and the area must have a suitable controlled atmosphere ISO (2007).

THE PANELLIST

The sensory panellist is the most important component of sensory testing and must be in optimal physiological condition when undertaking testing. Panellists must abstain from all food for a minimum of 1 h before undertaking an oral sensory evaluation. Muñoz et al. (1992) detail some practical guidelines for participation in sensory assessments. These guidelines include: being in sound mental and physical condition; knowing the scorecard; know relevant defects and intensities; evaluate aroma first (for relevant products where important); taste a sufficient quantity or volume of product; pay attention to the flavour sequence; rinse occasionally; concentrate; do not be too critical; be decisive; get feedback and check scoring; be honest; practice; be professional; do not eat, drink or smoke for at least 30 min prior to participating; do not wear perfume, cologne or aftershave or any fragranced cosmetics (Nelson and Trout, 1964).

SAMPLE PRESENTATION ORDER

In order to ensure that the true sensory response is obtained during sensory assessment, it is important that samples are prepared and presented appropriately so that potential sources of error are reduced. All methods of sample preparation must be standardised to ensure minimal influences on the sensory properties of the samples being tested. This is relatively easy with beverage products, with all samples served at the same temperature but can be more challenging for more complex products. This complexity is demonstrated very well when working with the very challenging samples that are fresh meat products. The mammalian carcass contains over 300 separate muscles which differ in their ratio to red and white muscle fibre types, with some muscles also being very heterogeneous. Scientists often standardise their sampling so that a standard thickness ($\sim$ 1 in., 2.54 cm) from the same parts of the muscle is compared across different treatment muscles from other test animals. As one of the more homogenous muscles the *M. longissimus dorsi* (LD) is often chosen for standardised sensory experiments. Additionally, it is imperative that these standardised samples are then cooked in an identical fashion and then served to assessors at a standardised temperature. Such studies could be designed to determine the effects of a feed, production method or process on the subsequent cooked sensory quality (Zakrys et al., 2008; Zakrys et al., 2009; Zakrys-Waliwander et al., 2010; Zakrys-Waliwander et al., 2011).

Most importantly the sample presentation order must be unbiased with samples presented in a randomised order to prevent first-order and carry-over effects (MacFie et al., 1989). This is to ensure that the after-effects of evaluating a sample do not carry over and thus influence the subsequent sample being tested. By presenting a different sampling order for each panellist, this error is minimised.

PANEL SCREENING

Panellists included in descriptive panels must be screened prior to inclusion in panel rosters so that it can be established that they meet the selection criteria. In effect they must have a relatively normal sensory response and also be able to demonstrate that they are capable of discriminating between samples. Initially a large volume of individuals should be screened as not all will meet the selection criteria and those that do are not always available to participate in panels for various reasons. Panel selection can be internal or external, both of which have their inherent advantages and disadvantages. Internal panels could be recruited from a company's personnel. This has the advantages of not having to additionally pay panellists as well as them being always easily contactable. However, panel work means they have to take leave from their normal day-to-day duties which can be difficult, perhaps reducing productivity and also developing resentment and ultimately a loss in motivation to participate as they may have more work to do in order to catch up once released from their panel work. External panels generally do not suffer from these problems, but they need to be paid and getting their participation can be logistically more complicated. However, they are impartial and as long as recruitment and retraining protocols are kept up to date, they are the source of choice for sensory panels. It is estimated to screen two to three times more individuals than are required to account for those not making the grade and unplanned for attrition (Murray et al., 2001). Selection of panellists should be assessed using a questionnaire targeting certain criteria including: availability, health, motivation, suitable character and dislikes (foods), etc. Selected panellists in effect should be punctual, in good health, have an interest in participating and be able to communicate their responses without being too dominant or weak in character as these individuals may introduce bias into the panel. Also it is important to produce a map for each panellists of the products they may wish to avoid for health reasons which might manifest allergic responses or if they particularly dislike a food or beverage. It is not recommended to include females who may be pregnant as their dietary needs can exclude them from being able to assess some products for dietary and safety reasons and also their sensory response may also be skewed.

Panel candidates are required to go through a screening procedure to ensure they have a suitable sensory response. They should have normal colour vision and be screened for ageusia, which is the inability to taste and anosmia, the inability to detect odours (ISO, 1991). Screening tests could include difference testing, sensory threshold testing or ranking tests which are described in detail in ISO (1993). Obviously, those that do not fit the correct profile should be excluded from further panel participation, but a certain level of leeway should be considered with candidates who may have potential as their level of acuity will increase with practice. It is really up to the discretion of the panel screening leader to decide on this.

CASE STUDY – DEMONSTRATION OF QUANTITATIVE DESCRIPTIVE ANALYSIS. THE DESCRIPTIVE PROFILE TRAINING PROCESS

Flavour research is concerned with developing improved methods of characterising and measuring overall flavour quality and individual attributes of food and beverage products. This can be achieved by studying the influence on flavour of changes in food materials and procedures at all stages in the food chain to protect established standards of flavour quality (Land, 1977).

In summary, there are generally two types of sensory panels that can be used in the sensory evaluation of food products, i.e., 'difference' testing panels and 'descriptive attribute testing' panels. Descriptive analysis has been used to quantify the sensory attributes of food and beverage products within the industry. The method has a number of advantages over difference testing in that it is quantitative and can be used to describe differences between products and the main sensory drivers (be they positive or negative, identified within products or especially when combined with objective consumer testing and multivariate data analysis). However, the method can be expensive and time-consuming because of the necessity to train and profile individual panellists over extended periods of time; days or even weeks. It is also not a method that can be readily used for routine analysis. In the next chapter we will discuss 'flash profiling' as a compromise method of analysis (O'Sullivan et al., 2011).

The following example demonstrates a training and sensory term reduction process for a complex food product, in this case warmed-over flavour (WOF) in meat (O'Sullivan et al., 2002). In order to provide a comprehensive overview of the method, WOF will initially be described in detail followed by the QDA protocol for its complete elucidation. This example was chosen as it is relatively complex and provides a good demonstration of training and the sensory term reduction process.

Tims and Watts (1958) were amongst the first workers to recognize WOF as an organoleptic challenge to meat products. In the last few decades there has been a rapid development of fast food facilities and widespread use of precooked frozen meals (Dethmers and Rock, 1975). However, the storage of precooked meat for a short period results in the development of a characteristic 'old, stale, rancid and painty' flavour and odour, caused by the catalytic oxidation of unsaturated fatty acids. This objectionable flavour becomes most noticeable when refrigerated cooked meat is reheated. The meat is said to have WOF. It is a rapidly occurring phenomenon contrasted to the slowly developing oxidative rancidity which results from long-term storage of frozen raw meat or storage of vegetable oils at ambient temperature (Angelo and Bailey, 1987). The predominant oxidation catalyst is iron from myoglobin and haemoglobin, which becomes more available following heat denaturation of the protein moiety of these complexes. Iron, of the iron porphyrins, also can become more available as oxidation catalysts during physical disruption of cells and cell membranes. Traditionally, lipid oxidation has been attributed to haeme catalysts such as haemoglobin, myoglobin and cytochromes (Tappel, 1962). This was a logical hypothesis, as the high concentrations of myoglobin and other haeme compounds make them likely candidates as prooxidants in muscle tissues (Asghar et al., 1988). However, Sato and Hegarty (1971) concluded that nonhaeme iron rather than haeme iron was the active catalyst responsible for the rapid oxidation of cooked meat. Haeme compounds were reported to have little influence on the development of off-flavours or thiobarbituric acid reactive substances in meat.

Rapid oxidation changes can also occur in raw meat when processed by present-day technologies (Angelo and Bailey, 1987). Although raw meat is generally considered less susceptible to WOF than is heated meat, after grinding and exposure to air, it rapidly develops odours that are similar to those in oxidized cooked meats (Sato and Hegarty, 1971). This phenomenon is also referred to as WOF (Pearson et al., 1977).

A number of sensory profiles have been conducted to date on WOF in various meat products. Byrne et al. (2001a) investigated the effects of preslaughter stress on WOF development in porcine meat. They found that preslaughter stress appeared in general to manifest itself as a separate sensory dimension to WOF in meat samples, but there were indications that increasing preslaughter stress may have reduced perceived WOF development. Byrne et al. (2002a) evaluated the effects of cooking temperature on WOF development in chicken. Cooking at higher temperatures produces Maillard reaction products, which are known to have an antioxidant effect (Bailey, 1988). However, Byrne et al. (2002a) showed that temperature increased the formation of Maillard-derived compounds, but did not show strong effects on the prevention of WOF in the cooked chicken patties. Byrne et al. (2002b) investigated the effects of the RN gene on WOF development in pork. A higher glycogen content in muscle results in lower postmortem ultimate pH in RN⁻ carriers, the dominant gene (Enfält et al., 1997). Byrne et al. (2001a) concluded that WOF, cooking temperature and genotype in the meat samples profiled were independent phenomena and that lactic/fresh sour flavour was a significant descriptor for describing meat from RN⁻ carriers. O'Sullivan et al. (2003) determined the sensory effects of iron supplementation on WOF development in pork meat patties made from *M. longissimus dorsi* and *M. psoas major*, respectively. They concluded that *M. psoas major* was more susceptible to WOF development as determined by sensory profiling compared to *M. longissimus dorsi* for all treatments.

Descriptive analysis is a method where defined sensory terms are quantified by sensory panellists during a training regime. Once an initial list of terms is decided upon the next step is to reduce these terms through the training and term reduction process. In order for a term to be included during subsequent profiling, it must fit four criteria. (1) The sensory terms selected must be relevant to the samples, (2) discriminate between the samples, (3) have cognitive clarity and (4) be nonredundant (Byrne et al., 1999a,b, 2001b; O'Sullivan et al., 2002). Various means can be employed in this term reduction process and have included principal component analysis (PCA) in conjunction with assessor suggestions (Byrne et al., 1999a,b, 2001b; O'Sullivan et al., 2002). FCP can also be used and involves panellists coming up with their own descriptive terms (Delahunty et al., 1997). The problem with this method is the subjective correlation of terms derived by different assessors. Generalised Procrustes analysis (GPA) has also been used in term reduction (Byrne et al., 2001b) and a method known as descriptor leverage. Both these methods are similar in that they involve a level-and-range correction of assessors, but the descriptor leverage method does not include the rotational aspect of GPA. The descriptor leverage method is rapid and unambiguous and was investigated by O'Sullivan et al. (2002) to determine whether it provided any discriminative improvement in the term reduction process. These authors concluded that the use of descriptor leverage provided a greater amount of information regarding the elimination of sensory terms as opposed to PCA and assessor suggestions alone. Descriptor leverage displayed the uniqueness of sensory terms for all the relevant principal components in a model and provided a higher degree of confidence with respect to sensory term reduction. O'Sullivan et al. (2002) further suggest that a combination of descriptor leverage, graphical interpretation of bilinear models and assessor suggestions may be a useful strategy in future vocabulary development studies. A schematic overview of the vocabulary development methodology over five training sessions is presented in Fig. 2.1. After training, sensory profiling can be undertaken.

Sensory Terms	Session 1 36 terms	Session 2 36 terms	Session 3 29 terms	Session 4 27 terms	Session 5 22 terms	Final List 21 terms
Cardboard-F						
Cardboard-O						√
Linseed-F						
Linseed-O						√
Rubber-F						
Rubber-O						√
Green-O						√
Rancid-F						√
Vegetable Oil-F						√
Fish-F						√
Lactic-AT						
Lactic-O						
Lactic-F						√
Fatty-Mouthcoating-AT						
Fatty-O						√
Sweet-T						√
Sour-T						√
Salt-T						√
Bitter-T						√
MSG-T (Monosodium Glutamate)					*MSG-T/ Bouillon-O*	√
Bloody-F						
Metallic-F					*Metallic-F/ Bloody-F*	√
Metallic-AT						
Tinny-F						√
Livery-F						√
Nut-F						
Nut-O						√
Meat-F						
Fresh Cooked Pork-O						
Fresh Cooked Pork-F						√
Bouillon-AT						
Bouillon-O						
Roasted-F						
Piggy-O						
Piggy-F						√
Astringent-AT						√

FIGURE 2.1 A schematic overview of the vocabulary development methodology over five training sessions. Blanks, removed sensory terms; italics, merged sensory terms; √, sensory terms in final list; Italic, merged terms; -AT, Aftertaste; -F, Flavour; -O, Odour; -T, Taste. *From O'Sullivan, M.G., Byrne D.V., Martens, M., 2002. Data analytical methodologies in the development of a vocabulary for evaluation of meat quality, Journal of Sensory Studies 17, 539–558.*

INITIAL VOCABULARY DEVELOPMENT

In the presented case study a preliminary evaluation of experimental samples by the panel leader and experts with product knowledge was to develop the initial training list of sensory terms (O'Sullivan et al., 2002; Byrne et al., 1999a,b, 2001a,b). The samples used covered the sensory variation in the experimental treatments (Table 2.2). Again, as mentioned above, the sensory terms selected had to fulfil the criteria: (1) must to be relevant to the samples; (2) discriminate between the samples; (3) have cognitive clarity; (4) be nonredundant (Byrne et al., 1999a,b, 2001b; O'Sullivan et al., 2002). From this an initial list of 36 sensory terms consisting of odours, flavours, the basic tastes and aftertastes was derived (Table 2.4) (O'Sullivan et al., 2002).

An eight-member sensory panel (four males/four females, aged from 24 to 62 years) was recruited from the public and students of The Royal Veterinary and Agricultural University, Frederiksberg, Denmark. Selection criteria for panellists were availability and motivation to participate on all days of the experiment. Sensory training was carried out in the panel booths at the university sensory laboratory that conforms to ASTM (2008) and ISO (1988). All panellists were prescreened for the ability to discriminate odours and tastes (ISO, 1991, 1992, 1993) (O'Sullivan et al., 2002).

Prior to the first training session the sensory methodology was explained to the panellists and the subsequent training session was used as a familiarisation exercise. Throughout the experiment, analysis was performed with assessors not having any knowledge of sample history. Training took approximately 1 h per day for 5 days. On the first day the sensory evaluation procedure was described to the participants. During subsequent days, samples (Table 2.3) were coded and presented to the group where they were asked to evaluate the odour, flavour, taste and aftertaste of the samples using the sensory terms outlined in Table 2.4. Unstructured 15-cm line scales anchored on the left by the term 'none' and on the right by the term 'extreme' were used for all the sensory descriptors (Meilgaard et al., 1999). The responses of panellists were recorded by measuring the distance in mm (1−150) from the left side of the scale. Reference materials were provided on all days of sensory training and were selected as representative of each sensory term from literature sources, through consultation with experts with product knowledge (O'Sullivan et al., 2002; Byrne et al., 1999a,b, 2001a,b) and using lexicons of descriptive terms (Civille and Lyon, 1996). Table 2.1 displays the final 21 terms and their corresponding reference substances. Panellists were asked to familiarise themselves with each reference before assessing the training samples (Table 2.3) and were instructed to refer to each sensory reference prior to line scale scoring and again to refer to a particular reference if they encountered difficulty in assessing its corresponding sensory term. Term grouping sheets were provided to panellists on all days of training to allow them to propose

TABLE 2.1 The Final List of 21 Descriptive Terms Developed for Further Sensory Profiling

Sensory Term	Definition (With Appropriate Reference)
Odour	*Odour Reference*
1. Cardboard-O	Wet cardboard
2. Linseed oil-O	Warmed linseed oil/linseed oil-based paint
3. Rubber/Sulphur-O	Warmed rubber/the white of a boiled egg
4. Nut-O	Crushed fresh hazel nuts
5. Green-O	Fresh green French beans
6. Fatty-O	Pig back fat (fresh, nonoxidised)
Taste	*Taste Reference*
7. Sweet-T	Sucrose 1 g/L aqueous solution
8. Salt-T	Sodium Chloride 0.5 g/L aqueous solution
9. Sour-T	Citric acid monohydrate 0.3 g/L aqueous solution
10. Bitter-T	Quinine chloride 0.05 g/L aqueous solution
11. MSG/Umami-T	Monosodium glutamate 0.5 g/L aqueous solution
Flavour	*Flavour Reference*
12. Metallic-F/Bloody-F	Ferrous sulphate 0.1 g/L aqueous solution
13. Fresh cooked pork-F	Oven-cooked pork without browning
14. Rancid-F	Oxidised vegetable oil
15. Lactic acid/fresh sour-F	Natural yoghurt
16. Vegetable oil-F	Fresh vegetable oil
17. Piggy/Animal-F	Skatole 0.06 µg/mL refined vegetable oil
18. Fish-F	Fish stock in boiling water
19. Tinny-F	Stainless steel strip
20. Livery-F	Cooked beef liver
Aftertaste	*Aftertaste Reference*
21. Metallic/Bloody-AT	Ferrous sulphate 0.1 g/L aqueous solution

Suffix to sensory terms indicates method of assessment by panellists; -AT, Aftertaste; -F, Flavour; -O, Odour; -T, Taste.
From O'Sullivan, M.G., Byrne D.V., Martens, M., 2002. Data analytical methodologies in the development of a vocabulary for evaluation of meat quality, Journal of Sensory Studies 17, 539–558.

TABLE 2.2 Supplemental Diets Fed to Experimental Pig Groups

Code	Dietary Treatment
E	Supplemental vitamin E (200 mg dl-α-tocopheryl acetate/kg of feed) + vitamin C.
Co	Control diet + vitamin C (9 g/kg feed).
I/E	Supplemental vitamin E (200 mg dl-α-tocopheryl acetate/kg of feed) and supplemental iron (3000 mg iron (II) sulphate/kg feed for 1 month prior to slaughter) + Vitamin C.
I	Supplemental iron (3000 mg iron (II) sulphate/kg feed for 1 month prior to slaughter) + Vitamin C.

From O'Sullivan, M.G., Byrne D.V., Martens, M., 2002. Data analytical methodologies in the development of a vocabulary for evaluation of meat quality, Journal of Sensory Studies 17, 539–558.

TABLE 2.3 Samples Selected for Training

	M. longissimus dorsi (L)					*M. psoas major* (P)		
Treatment Code	Day 0	Day 1	Day 3	Day 5	Day 0	Day 1	Day 3	Day 5
E (vitamin E)	LE0[T]	LE1	LE3	LE5	PE0	PE1	PE3	PE5[T]
Co (control)	LCo0	LCo1	LCo3	LCo5	PCo0[T]	PCo1	PCo3	PCo5
IE (iron/ vitamin E)	LIE0	LIE1	LIE3	LIE5[T]	PIE0	PIE1	PIE3	PIE5
I (iron)	LI0[T]	LI1	LI3	LI5	PI0	PI1	PI3	PI5[T]

First Letter: Muscle; Second Letter: Treatment (Co, Control group); Number: Days of WOF; [T] = Samples Selected For Training.
From O'Sullivan, M.G., Byrne D.V., Martens, M., 2002. Data analytical methodologies in the development of a vocabulary for evaluation of meat quality, Journal of Sensory Studies 17, 539–558.

redundant terms and also to suggest new sensory terms not presented in the list of terms provided (O'Sullivan et al., 2002; Byrne et al., 1999a,b, 2001a,b).

DATA ANALYSIS DURING VOCABULARY DEVELOPMENT

At the end of each session (day) in the training period, redundant or unreliable sensory terms were eliminated or merged in order to select a reduced list of terms for use in the next session. The goal was to obtain, at the end of the training period, a reduced descriptor list where each remaining term was relevant to the product, nonredundant, sample-discriminating and cognitively

TABLE 2.4 The Initial List of Terms Presented to the Panel

1. Cardboard-O	13. Sour-T	25. Nut-F
2. Linseed oil-O	14. Bitter-T	26. Piggy/Animal-F
3. Egg/Sulphur-O	15. MSG/Umami-T	27. Meat-F
4. Bouillon-O	16. Metallic-F	28. Fish-F
5. Fresh cooked pork-O	17. Bloody-F	29. Roasted-F
6. Nut-O	18. Fresh cooked pork-F	30. Tinny-F
7. Piggy/Animal-O	19. Rancid-F	31. Livery-F
8. Green-O	20. Cardboard-F	32. Lactic acid/fresh sour-AT
9. Fatty-O	21. Linseed oil-F	33. Metallic/Bloody-AT
10. Lactic acid/fresh sour-O	22. Egg/Sulphur-F	34. Fatty-mouthcoating-AT
11. Sweet-T	23. Lactic acid/fresh sour-F	35. Astringent-AT
12. Salt-T	24. Vegetable oil-F	36. Bouillon-AT

Suffix to sensory terms indicates method of assessment by panellists; -AT, Aftertaste; -F, Flavour; MSG, Monosodium Glutamate; -O, Odour; -T, Taste.
From O'Sullivan, M.G., Byrne D.V., Martens, M., 2002. Data analytical methodologies in the development of a vocabulary for evaluation of meat quality, Journal of Sensory Studies 17, 539–558.

clear to the assessors (O'Sullivan et al., 2002; Byrne et al., 1999a,b, 2001a,b). The profile reduction on each day was carried out on the basis of panel discussions, term grouping sheet recommendations and interpretation of PCA performed on the data from that day of training (based on mean-centred, unweighted variables, assessed by full cross-validation). In order to ensure description of variation types possibly absent in the six training samples, terms deemed to be casually important were also retained in the profile. After 5 days of training and profile reduction following the above procedures, the initial list of 36 sensory terms (Table 2.4) for odour, flavour, taste and aftertaste was reduced to 21 sensory terms (Table 2.1). After the project concluded, the data from the training period were reassessed more thoroughly in order to determine the potential usefulness of more advanced bilinear modelling methods.

PROTOCOL FOR THE EXAMINATION OF THE EFFECTIVENESS OF VOCABULARY DEVELOPMENT

The effect of panel training and profile reduction was assessed by (analysis of variance) ANOVA-partial least squares regression (APLSR) on panel mean data from individual training days. For each of training day 2 and 5 the panel profile was computed over the 8 panellists, resulting in two tables with 6 samples × 36 or 22 terms, respectively. Each table was in turn defined as Y-variables and regressed on six indicator X-variables (values 0 or 1) representing the six samples (i.e., the identity matrix for treatments). This way of using APLSR is very similar to a PCA of the sensory variables, but has the advantage of letting the samples also be represented as variables (Martens and Martens, 2001). With so few and so diverse samples, it is difficult to determine the optimal number of PCs to be interpreted. The first two principal PLS components (PCs), displayed here, described most of the sensory variation. The explained sensory variance in the PLSR for the panel mean data on training day 2 (Fig. 2.2A) was PC $1 = 43.7\%$ and PC $2 = 25.0\%$ and for training day 5 (Fig. 2.2B) was PC $1 = 49.0\%$ and PC $2 = 19.0\%$.

METHOD FOR QUANTIFICATION OF DISCRIMINATIVE ABILITY

The reliability of the sensory data was assessed for each of the training days 1, 2, 3, 4 and 5 in terms of the mean signal/noise ratio for the sensory terms used (36, 36, 29, 27, 22, 21 terms, respectively). For each day, this was performed as follows:

In order to have enough samples to avoid overoptimistic assessment, the data tables from the different assessors were merged so that the assessors could be regarded as replicates of each other. Between-assessor level effects were removed in a preprocessing step, because general differences in how the assessors used the sensory scale were considered irrelevant: For each of the eight assessors, the mean level for each word over the six samples was computed and subtracted. Then the 8 mean-centred matrices were merged into a level-corrected data table with 48 samples × the number of terms for that day.

The tables of level-corrected terms for training days were defined as Y-variables and regressed on a table of six X-variables (0/1 design indicators for the six samples) by discriminant partial least squares regression (DPLSR), with no prior scaling of the X- or Y-variables. Full leave-one-out cross-validation (i.e., 48 cross-validation segments) was used for model assessment. For each of the five-days models the cross-validation indicated 4 PCs to be optimal or near-optimal model rank.

For each day (#d), the signal/noise ratio (S/N) was estimated for each sensory assessor (#m), from the cross-validated mean square errors of prediction of the Y-variables MSEP(Y) (Martens and Martens, 2001), averaged over all 48 samples and all of the sensory terms for that day and assessor. The total initial variance before the modelling (i.e., at zero factors, MSEP(Y)$_{0\ \mathrm{PCs}}$) was defined as signal and the residual cross-validated variance after the modelling (MSEP(Y)$_{4\ \mathrm{PCs}}$) was defined as the estimated noise level:

$$S/N_{d,m} = \mathrm{MSEP}(Y)_{0\ \mathrm{PCs},d,m} / \mathrm{MSEP}(Y)_{4\ \mathrm{PCs},d,m} \qquad (2.1a)$$

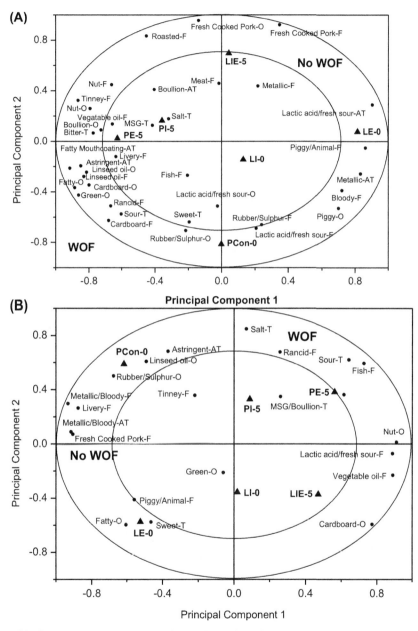

FIGURE 2.2 (A, B). An overview of the variation found in the mean data from the ANOVA-partial least squares regression (APLSR) correlation loadings plot for each of the six training samples (Table 2.3) for training session 2 (A) and 5 (B). Shown are the loadings of the x- and y-variables for the first 2 PCs for level-corrected data. • = Sensory term and ▲ = Sample. The concentric circles represent 100% and 50% explained variance, respectively. -AT, Aftertaste; -F, Flavour; -O, Odour; -T, Taste. *From O'Sullivan, M.G., Byrne D.V., Martens, M., 2002. Data analytical methodologies in the development of a vocabulary for evaluation of meat quality, Journal of Sensory Studies 17, 539–558.*

The average signal/noise ratio for each day d was defined as the mean over the eight assessors:

$$S/N_d = \sum_{m=1}^{8} \frac{S/N_{d,n}}{8} \qquad (2.1b)$$

The standard error of the mean S/N_d was defined as

$$s(S/N_d) = \sqrt{\frac{\sum_{m=1}^{8} \left(S/N_{d,m} - S/N_d\right)^2}{(8-1)\cdot 8}} \qquad (2.1c)$$

Comparison of means was performed by ANOVA using the Tukey honestly significant differences-test at the 5% level (SPSS, Chicago, IL), which was selected from a table of percentage points of the standardised range where $\alpha = 0.05$ and $k =$ number of steps between ordered means, was used to differentiate the S/N_d between days $d = 1, 2, ..., 5$.

THE DESCRIPTOR LEVERAGE METHOD

The uniqueness of the various sensory terms was assessed for each of the five days of training. The purpose of this was to see if the sensory descriptor leverage might have been used with greater success than the original PCA for term reduction at the end of each day.

Within each day, the data were first preprocessed to correct for scale differences between assessors: For each assessor, each term was standardized to zero mean (to correct for level effects in scale use) and a standard deviation of one (to correct for range effects in scale use) over the six samples. These corrected data were merged to yield 48 samples for the given number of sensory terms that day.

These level-and-range corrected sensory data were now defined as X-variables $k = 1, 2, ..., K$, and the six indicator variables for the products were used as six Y-variables and submitted to PLSR, again without any further scaling. The full leave-one-out cross-validation again showed 4 PCs to yield optimal or near-optimal predictive ability. Hence the leverage of the X-variables was estimated for up to 4 PCs.

Descriptor leverages value $h_{k,A}$ in a model with $A = 1, 2, 3, 4$ PCs are defined (Martens and Martens, 2001) for each X-variable k (sensory term m) as:

$$h_{k,A} = \sum_{a=1}^{A} w_{k,a}^2 \qquad (2.2)$$

The descriptor leverage values $h_{k,A}$ display the uniqueness of sensory terms used as X-variable in an A-dimensional bilinear model, and lie between 0 (a variable not affecting the model) and 1 (a dominant X-variable) (Martens and Martens, 2001).

All bilinear modelling was performed using the Unscrambler Software, version 7.6 (www.CAMO.com).

DEVELOPMENT OF PRELIMINARY LIST OF TERMS

A preliminary evaluation of experimental samples by the panel leader and experts with product knowledge was undertaken one week prior to sensory training to develop the initial list of sensory terms (Byrne et al., 1999a,b, 2001a,b). The samples used covered a range of days of WOF for each of the experimental treatments (Table 2.2). The sensory terms selected had to fulfil all selection criteria (see Section Initial Vocabulary Development). From this an initial list of 36 sensory terms consisting of odours, flavours, the basic tastes and aftertastes was derived (Table 2.4) (O'Sullivan et al., 2002).

EXAMINATION OF THE EFFECTIVENESS OF VOCABULARY DEVELOPMENT

Fig. 2.1 shows a schematic diagram of the vocabulary development over the five-days training period. PCA of sample scores, term grouping information and panel suggestions were used to reduce the sensory terms from 36 to 21 terms by day 5 (Fig. 2.1). The PCA models for panel mean results were assessed for PC 1 versus 2, PC 2 versus 3 and PC 1 versus 3 after each training session. After the first day of training, none of the sensory terms were eliminated as this session was used as a familiarisation period for the assessors. For session 2, seven terms were removed, Fresh Cooked Pork-O (odour), Piggy-O, Cardboard-F (flavour), Linseed-F, Nut-F, Lactic-AT (aftertaste) and Bouillon-AT. For training session 3, two terms were removed, *Lactic-O* and *Meat-F*. Again, using PCA covariance and assessor suggestions. In session 4, three terms were removed, *Roasted−F* and *Fatty-Mouthcoating-AT* and *Rubber-F* and four terms were merged. *Bouillon-O* and *MSG-T* became *MSG-T/Bouillon-O* and *Metallic-F* and *Bloody-F* were merged as *Metallic-F/Bloody-F*. For session 5, one term was removed, *Metallic−AT*.

This method of sensory term reduction is well documented (Byrne et al., 1999a,b, 2001a,b) and will not be discussed in detail here. The next part of this study was to examine the effectiveness of this protocol for term reduction and to examine if there was an improvement in assessor discriminative ability. The assessor discriminative ability was quantified and descriptor leverage, was used to measure the overall effectiveness of sensory term reduction (O'Sullivan et al., 2002).

The explained sensory variance in the PLSR for the panel mean data on training day 2 (Fig. 2.2A) was PC 1 = 43.7% and PC 2 = 25.0%. The corresponding percentages for training day 5 (Fig. 2.2b) were PC 1 = 49.0% and

PC 2 = 19.0%. As training progressed, PC 1 better described the main sources of variation than PC 2. In Fig. 2.2A the sensory terms *Fresh Cooked Pork-O* and *Fresh Cooked Pork-F* correlate well in this APLSR plot, but they also covary with the sample LIE-5, a sample that has been stored in a refrigerator at 4°C for 5 days. Two of the basic tastes (T), *Sour-T* and *Sweet-T* also covary indicating that the panellists are not fully clear with respect to scoring these descriptors. Fig. 2.2B (training day 5), however, displays a dramatic improvement in the discriminative abilities of assessors.

Compared to Fig. 2.2A, larger variation in WOF development was explained in Fig. 2.2B and the training samples fit the model space to a greater degree. PCon-0 and LE-0 correlate well with the fresh sensory term, *Fresh Cooked Pork-F* and PI-5 and PE-5 covary with *Rancid-F* and *Fish-F*. There also seems to be a separation between *M. longissimus dorsi* and *M. psoas major* on the basis of WOF development. *M. psoas major* covaries with the oxidative terms with PCon-0 covarying with *Linseed-O* and *Rubber-O* and PI-5 and PE-5 covarying with *Rancid-F* and *Fish-F*. However, the differentiation of *M. longissimus dorsi* appears to be from the fresh sensory terms *Sweet-T* and *Fatty-O* (nonoxidised fat) for LE-0 to the more oxidative term *Green-O* for LI-0 to *Cardboard-O* for LIE-5. Perhaps these sensory descriptors describe a less intense degree of WOF development compared to the oxidative descriptors associated with *M. psoas major*. *M. longissimus dorsi* is a white muscle and contains a lesser concentration of phospholipids, particularly the more unsaturated fatty acids compared to the red muscle *M. psoas major* and as such is much less susceptible to oxidative attack (Lawrie, 1991). This may explain this separation due to varying degrees of WOF development.

Overall, the discriminative abilities of the sensory panel improved over the course of sensory training (O'Sullivan et al., 2002).

QUANTIFICATION OF DISCRIMINATIVE ABILITY

Fig. 2.3 shows the average assessor signal-to-noise ratios S/N_d for the sensory variables for training days $d = 1, 2, 3, 4$ and 5. The S/N ratio is essentially the signal (total initial variance) divided by the noise (residual cross-validated variance). After 5 days of training, the assessors signal was significantly ($P < .05$) greater than the noisy elements. This can be used as an index of assessor sample discriminative ability. There is a large significant ($P < .05$) initial increase in S/N ratio in session 2 of training, followed by nonsignificant increase in session 3. Session 4 produced a slightly greater increase in assessor S/N, followed by the large significant ($P < .05$) increase in S/N in session 5. In general, as training progressed the S/N ratio increased, indicating an increase in assessor discriminative ability over time, culminating in the greatest level of discriminative ability at training day 5 (O'Sullivan et al., 2002).

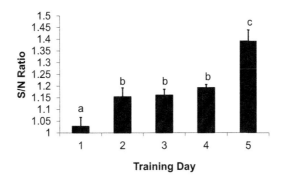

FIGURE 2.3   Shown are the mean assessor *S/N* ratios for training sessions 1, 2, 3, 4 and 5. along the *Y*-axis lies the total initial variance (signal)/residual cross-validated variance (noise). Means with different letters (a, b, c) were significantly different ($P < .05$). *From O'Sullivan, M.G., Byrne D.V., Martens, M., 2002. Data analytical methodologies in the development of a vocabulary for evaluation of meat quality, Journal of Sensory Studies 17, 539–558.*

DESCRIPTOR LEVERAGE TO DETERMINE THE EFFECTIVENESS OF THE SENSORY TERM REDUCTION PROTOCOL

After the project had finished, the descriptor leverage ($h_{k,A}$) for various sets of sensory terms $k = 1, 2, ..., K$ in PLSR models using PCs $a = 1, 2, ..., A$ was used to assess the effectiveness of the vocabulary reduction procedure used during the actual training period to develop the final sensory vocabulary (involving PCA and assessor suggestions) (O'Sullivan et al., 2002). In Fig. 2.4A the descriptor leverage for each sensory term on training day 2 is presented as consecutive bars corresponding to PCs 1, 2, 3 and 4, respectively. During sensory training session 2, eight sensory terms (*Fresh Cooked Pork-O, Piggy/Animal-Like-O, Cardboard-F, Linseed oil-F, Nut-F, Lactic acid/fresh sour-AT* and *Bouillon-AT*) were removed by interpretation of information gathered through PCA of sample scores, term grouping information and panel suggestions. The principal criteria for removal of a term at this stage of training were whether there was redundancy with respect to terms, e.g., *Cardboard-O* and *Cardboard-F*. If a term covaried to a greater degree in the PCA plot, or if the panel preferred one term over another, then the latter term was deemed redundant and thus removed from the descriptor list for the next training day (O'Sullivan et al., 2002).

Examination of Fig. 2.4A confirmed that the terms removed from the training list correlate with the actual method of removal, PCA covariance and assessor suggestions. For example, the descriptor leverage for the terms *Piggy/Animal-O, Cardboard-F, Linseed oil-F, Nut-F, Lactic acid/fresh sour-AT* and *Bouillon-AT* was lower than the descriptor leverage for their corresponding terms, *Piggy/Animal-F, Cardboard-O, Linseed oil-O, Nut-O, Lactic*

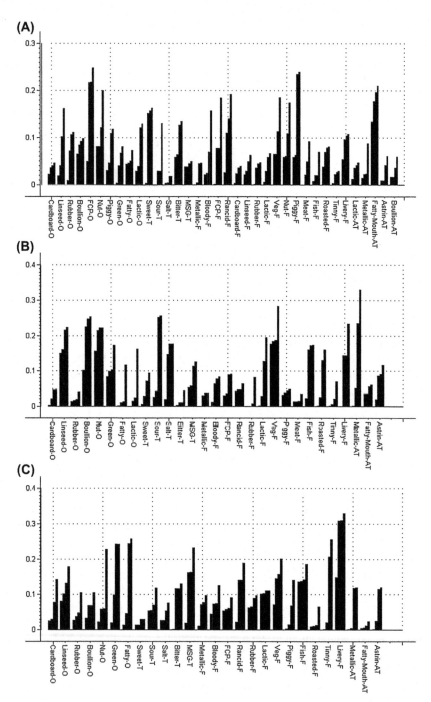

FIGURE 2.4 (A, B and C). Shown are the leverage (hx) values (bars) for each sensory term across four principal components (left to right, bars indicate pc1 to pc4, respectively) from discriminant partial least squares regression (DPLSR) for vocabulary development session 2 (A), 3 (B) and 4 (C). *From O'Sullivan, M.G., Byrne D.V., Martens, M., 2002. Data analytical methodologies in the development of a vocabulary for evaluation of meat quality, Journal of Sensory Studies 17, 539–558.*

acid/fresh sour-F and *Bouillon-O*, respectively. The former in each case was thus deemed redundant and correlated with the selection process using PCA and assessor suggestions (Fig. 2.1). However, *Fresh Cooked Pork-O* was removed in preference for *Fresh Cooked Pork-F* using this protocol (Fig. 2.4A) even though *Fresh Cooked Pork-O* had a much greater descriptor leverage. The leverage is still quite high for *Fresh Cooked Pork-F*, but *Fresh Cooked Pork-O* perhaps would have proven a more appropriate descriptor (O'Sullivan et al., 2002).

For training day 3, *Lactic acid/fresh sour-F* was explained better in the PCA plot than *Lactic acid/fresh sour-O* and was thus removed. *Meat-F* was also removed as *Fresh Cooked Pork-F* was also better explained in the PCA plot (Fig. 2.1). These results agree with Fig. 2.4B with *Lactic acid/fresh sour-F* and *Fresh Cooked Pork-F* producing greater descriptor leverages than *Lactic acid/fresh sour-O* and *Meat-F*, respectively.

For training day 4 the sensory terms, *Bouillon-O* and *MSG-T*, *Metallic-F* and *Bloody-F* were merged to form the descriptors *MSG-T/Bouillon-O* and *Metallic-F/Bloody-F*, respectively (Fig. 2.1). This decision was again based on a combination of panellist suggestion and PCA covariance. The term *Roasted-F* was dropped from the training list because it did not fulfil two of the criteria for term inclusion, relevance to the samples (i.e., samples were not roasted) and discrimination between the samples. The term *Fatty-Mouthcoating-AT* was dropped in favour of *Fatty-O*, again due to covariance in the PCA plot. The term redundancy process is in general in agreement with Fig. 2.1 except in the cases of the sensory terms *Cooked Pork-O* and *Rubber-F*, which were dropped in preference to *Cooked Pork-F* and *Rubber-O*, respectively, even though both produced a lower descriptor leverage across the four principal components (Fig. 2.4C). For the final day of training the term *Metallic−AT* was dropped as it covaried with the merged descriptor *Metallic-F/Bloody-F*.

In conclusion, the descriptor leverage appeared to indicate that vocabulary development was largely correct based on PCA and assessor suggestions, with few exceptions. Therefore, the use of leverage plus PLSR or PCA and assessor suggestions may in the future be used in combination to instil a greater degree of confidence in sensory vocabulary development (O'Sullivan et al., 2002).

SUMMARY

The sensory and data analytical methodology employed during the 5 days of training was effective in reducing the initial list of sensory terms from 36 to 21 words. The sensory descriptors used in the final list reflected the variation (Muscle type, degree of WOF) present in the meat samples (O'Sullivan et al., 2002).

After 5 days of training assessors could separate both *M. longissimus dorsi* and *M. psoas major* and in turn separate these different muscle samples with respect to their degree of WOF development. The six training samples also spanned the model space better for training day 5 compared to training day 2. As training progressed the assessor mean S/N ratio increased, indicating an increase in assessor discriminative ability over time culminating in the greatest level of discriminative ability between the training samples on the last training day 5. Thus the discriminative abilities of the sensory panel improved over the course of sensory training (O'Sullivan et al., 2002).

The use of descriptor leverage provided a greater amount of information regarding the elimination of sensory terms as opposed to PCA and assessor suggestions alone. Descriptor leverage displayed the uniqueness of sensory terms for all the relevant principal components in a model and provided a higher degree of confidence with respect to sensory term reduction. Thus, it is suggested that a combination of descriptor leverage, graphical interpretation of bilinear models and assessor suggestions may be a useful strategy in future vocabulary development studies (O'Sullivan et al., 2002).

REFERENCES

Angelo, A.J.St., Bailey, M.E., 1987. Warmed-over Flavor of Meat. Academic Press, Florida, pp. vii—viii.

Asghar, A., Gray, J.I., Buckley, D.J., Pearson, A.M., Booren, A.M., 1988. Perspectives on warmed-over flavor. Food Technology 42 (8), 102—108.

ASTM, 2008. Guidelines for the Sensory Evaluation Professionals Who Undertake the Development of a New Facility or Remodelling of an Existing Sensory Laboratory. ISBN:13: 978-0-8031-5893-1.

Bailey, M.E., 1988. Inhibition of warmed-over flavor with emphasis on Maillard reaction products. Food Technology 42, 123—126.

Byrne, D.V., Bak, L.S., Bredie, W.L.P., Bertelsen, G., Martens, M., 1999a. Development of a sensory vocabulary for warmed-over flavour 1: in porcine meat. Journal of Sensory Studies 14, 47—65.

Byrne, D.V., Bredie, W.L.P., Martens, M., 1999b. Development of a sensory vocabulary for warmed-over flavour: Part II. In chicken meat. Journal of Sensory Studies 14, 67—78.

Byrne, D.V., Bredie, W.L.P., Bak, L.S., Bertelsen, G., Martens, H., Martens, M., 2001a. Sensory and chemical analysis of cooked porcine meat patties in relation to warmed-over flavour and pre-slaughter stress. Meat Science 59, 229—249.

Byrne, D.V., O'Sullivan, M.G., Dijksterhuis, G.B., Bredie, W.L.P., Martens, M., 2001b. Sensory panel consistency during development of a vocabulary for warmed-over flavour. Food Quality Preference 12, 171—187.

Byrne, D.V., Bredie, W.L.P., Mottram, D.S., Martens, M., 2002a. Sensory and chemical investigations on the effect of oven cooking on warmed-over flavour development in chicken meat. Meat Science 61, 127—139.

Byrne, D.V., O'Sullivan, M.G., Bredie, W.L.P., Martens, M., 2002b. Descriptive sensory profiling and physical/chemical analyses of warmed-over flavour in meat patties from carriers and non-carriers of the RN⁻ allele. Meat Science 63, 211—224.

Cairncross, S.E., Sjostrom, L.B., 1950. Flavour profiles: a new approach to flavour problems. Food Technology 4, 308—311.

Caul, J.F., 1957. The profile method of flavor analysis. In: Mrak, E.M., Stewart, G.F. (Eds.), Advances in Food Research, vol. 7. Academic Press, New York, NY, pp. 1–40.

Civille, G.V., Lyon, B.G., 1996. In: examples, Civille, G.V., Lyons, B.G. (Eds.), Aroma and Flavour Lexicon for Sensory Evaluation: Terms, Definitions, References. American Society for Testing and Materials, Philadelphia, USA, pp. 46–49.

Civille, G.V., Szczesniak, A.S., 1973. Guidelines to training a texture profile panel. Journal of Texture Studies 4, 204–223.

Delahunty, C.M., McCord, A., O'Neill, E.E., Morrissey, P.A., 1997. Sensory characterisation of cooked hams by untrained consumers using free-choice profiling. Food Quality and Preference 8, 381–388.

Desai, N.T., Shepard, L., Drake, M.A., 2013. Sensory properties and drivers of liking for Greek yogurts. Journal of Dairy Science 96 (12), 7454–7466. http://dx.doi.org/10.3168/jds.2013-6973.

Dethmers, A.E., Rock, H., 1975. Effect of added sodium nitrite on sensory quality and nitrosamine formation in Thuringer sausage. Journal of Food Science 40, 491–495.

Drake, M.A., Yates, M.D., Gerard, P.D., Delahunty, C.M., Sheehan, E.M., Turnbull, R.P., Dodds, T.M., 2005. Comparison of differences between lexicons for descriptive analysis of Cheddar cheese flavour in Ireland, New Zealand, and the United States of America. International Dairy Journal 15 (5), 473–483. http://dx.doi.org/10.1016/j.idairyj.2004.09.006.

Enfält, A.C., Lundström, K., Hansson, I., Johansen, S., Nyström, P.E., 1997. Composition of non-carriers and heterozygous carriers of the RN⁻allele for carcass composition, muscle distribution and technological quality. Meat Science 45, 1–45.

Galán-Soldevilla, H., Ruiz Pérez-Cacho, P., Jiménez, S., Villarejo, M., Manzanares, A.B., 2005. Deveopment of a preliminary sensory lexicon for floral honey. Food Quality and Preference 16, 71–77.

ISO, 1988. International Standard 8589. Sensory Analysis. General Guidance for the Design of Test Rooms Ref. No. ISO 8589:1988 (E). International Organization for Standardization, Genève, Switzerland.

ISO, 1991. International Standard 3972. Sensory Analysis Methodology. Method of Investigating Sensitivity of Taste. Ref. No. ISO 3972:1991 (E). International Organization for Standardization, Genève, Switzerland.

ISO, 1992. International Standard 5496. Sensory Analysis Methodology. Initiation and Training of Assessors in the Detection and Recognition of Odours. Ref. No. ISO 5496:1992 (E). International Organization for Standardization, Genève, Switzerland.

ISO, 1993. International Standard 8586-1. Sensory Analysis Methodology. General Guidance for the Selection, Training and Monitoring of Assessors. Ref. No. ISO 8586-1:1993 (E). International Organization for Standardization, Genève.

ISO, 1994. International Standard. 11035. Sensory Analysis-identification and Selection of Descriptors Establishing a Sensory Profile by a Multidimensional Approach. Ref. No. ISO 11035:1994 (E). International Organization for Standardization, Genève.

ISO, 2007. International Standard 8589. Sensory Analysis — General Guidance for the Design of Test Rooms. International Organization for Standardization.

Johnsen, P.B., Civille, G.V., Vercellotti, J.R., Sanders, T.H., Dus, C.A., 1998. Development of a lexicon for the description of peanut flavour. Journal of Sensory Studies 3, 9–17.

Johnson, P.B., Civille, G.V., 1986. A standardized lexicon of meat WOF descriptors. Journal of Sensory Studies 1, 99–104.

Land, D.G., 1977. Flavour research in the ARC. ARC Research Review 3, 58.

Lawless, H.T., Heymann, H., 1998a. Sensory evaluation in quality control. In: Lawless, H.T., Heymann, H. (Eds.), Sensory Evaluation of Food, Principles and Practices. Chapman and Hall, New York, pp. 548–584.

Lawless, H.T., Heymann, H., 1998b. Descriptive analysis. In: Lawless, H.T., Heymann, H. (Eds.), Sensory Evaluation of Food, Principles and Practices. Chapman and Hall, New York (pp 117–138, pp. 341–378).

Lawrie, R.A., 1991. Meat Science, fifth ed. Pergamon Press, Oxford, United Kingdom, pp. 71–72.

Lyon, B.G., 1987. Development of chicken flavour descriptive attribute terms aided by multivariate statistical procedures. Journal of Sensory Studies 2, 55–67.

MacFie, H.J., Bratchell, N., Greenhoff, K., Vallis, L.V., 1989. Designs to balance the effect of order of presentation and first-order carry-over effects in hall tests. Journal of Sensory Studies 4, 129–148.

Martens, H., Martens, M., 2001. Multivariate Analysis of Quality. An Introduction, first ed. John Wiley and Sons, Chichester, United Kingdom.

Meilgaard, M.C., Civille, G.V., Carr, B.T., 1999. In Sensory Evaluation Techniques (Chapter 5), third ed. Academic Press, Florida, pp. 54–55.

Michon, C., O'Sullivan, M.G., Delahunty, C.M., Kerry, J.P., 2010a. Study on the influence of age, gender and familiarity with the product on the acceptance of vegetable soups. Food Quality and Preference 21, 478–488.

Michon, C., O'Sullivan, M.G., Delahunty, C.M., Kerry, J.P., 2010b. The investigation of gender related sensitivity differences in food perception. Journal of Sensory Studies 24, 922–937.

Michon, C., O'Sullivan, M.G., Sheehan, E., Delahunty, C.M., Kerry, J.P., 2010c. Investigation of the influence of age, gender and consumption habits on the liking for jam-filled cakes. Food Quality and Preference 21, 553–561.

Muñoz, A.M., Civille, G.V., 1992. The spectrum descriptive analysis method. In: Hootman, R.C. (Ed.), ASTM Manual on Descriptive Analysis. American Society for Testing and Materials, Pennsylvania.

Muñoz, A.M., Civille, G.V., Carr, B.T., 1992. Comprehensive descriptive method. In: Sensory Evaluation in Quality Control. Van Nostrand Reinhold, New York, pp. 55–82.

Muñoz, A.M., 2002. Sensory evaluation in quality control: an overview, new developments and future opportunities. Food Quality and Preference 13, 329–339.

Murray, J., Delahunty, C., Baxter, I., 2001. Descriptive sensory analysis: past, present and future. Food Research International 34 (6), 461–471.

Nelson, J.A., Trout, G.M., 1964. Judging Dairy Products. The Olson publishing Co., Milwaukee, Wiss., U.S.A.

O'Sullivan, M.G., Byrne, D.V., Martens, M., 2002. Data analytical methodologies in the development of a vocabulary for evaluation of meat quality. Journal of Sensory Studies 17, 539–558.

O'Sullivan, M.G., Byrne, D.V., Nielsen, J.H., Andersen, H.J., Martens, M., 2003. Sensory and chemical assessment of pork supplemented with iron and vitamin E. Meat Science 64, 175–189.

O'Sullivan, M.G., Kerry, J.P., Byrne, D.V., 2011. Use of sensory science as a practical commercial tool in the development of consumer-led processed meat products. In: Kerry, J.P., Kerry, J.F. (Eds.), Processed Meats. Woodhead Publishing Ltd., United Kingdom.

Pearson, A.M., Love, J.D., Shorland, F.B., 1977. 'Warmed-over' flavour in meat, poultry and fish. Advances in Food Research 23, 1–74.

Peryam, D.R., Pilgrim, F.J., 1957. Hedonic scale method of measuring food preferences. Food Technology 11 (9), 9–14.

Pérez-Cacho, P.R., Galán-Soldevilla, H., Molina Recio, G., León Crespo, F., 2005. Determination of the sensory attributes of a Spanish dry-cured sausage. Meat Science 71, 620−633.

Rétiveau, A., Chambers, D.H., Esteve, E., 2005. Developing a lexicon for the flavor description of French cheeses. Food Quality and Preference 16 (6), 517−527. http://dx.doi.org/10.1016/j.foodqual.2004.11.001.

Sato, K., Hegarty, G.R., 1971. Warmed-over flavour in cooked meat. Journal of Food Science 36, 1098−1102.

Stone, H., Sidel, J., Oliver, S., Woolsey, A., Singleton, R.C., 1974. Sensory evaluation by quantitative descriptive analysis. Food Technology 28, 24−34.

Tappel, A.L., 1962. Heme compounds and lipoxidase as biocatalysts. In: Symposium on Foods: Lipids and Their Oxidation. AVI Pub. Co., Inc. Westport Conn. 122.

Tims, M.J., Watts, B.M., 1958. Protection of cooked meats with phosphates. Food Technology 12, 240−243.

Van Hekken, D.L., Drake, M.A., Corral, F.J.M., Prieto, V.M.G., Gardea, A.A., 2006. Mexican chihuahua cheese: sensory profiles of young cheese. Journal of Dairy Science 89 (10), 3729−3738. http://dx.doi.org/10.3168/jds.S0022-0302(06)72414-6.

Williams, A.A., Arnold, G.M., 1985. A comparison of the aroma of six coffees characterised by conventional profiling, free-choice profiling and similarity scaling methods. Journal of the Science of Food and Agriculture 36, 204−214.

Yusop, S.M., O'Sullivan, M.G., Kerry, J.F., Kerry, J.P., 2009a. Sensory evaluation of Indian-style marinated chicken by Malaysian and European naïve assessors. Journal of Sensory Studies 24, 269−289.

Yusop, S.M., O'Sullivan, M.G., Kerry, J.F., Kerry, J.P., 2009b. Sensory evaluation of Chinese-style marinated chicken by Chinese and European naïve assessors. Journal of Sensory Studies 24, 512−533.

Zakrys, P.I., Hogan, S.A., O'Sullivan, M.G., Allen, P., Kerry, J.P., 2008. Effects of oxygen concentration on sensory evaluation and quality indicators of beef muscle packed under modified atmosphere. Meat Science 79, 648−655.

Zakrys, P.I., O'Sullivan, M.G., Allen, P., Kerry, J.P., 2009. Consumer acceptability and physiochemical characteristics of modified atmosphere packed beef steaks. Meat Science 81, 720−725.

Zakrys-Waliwander, P.I., O'Sullivan, M.G., Allen, P., O'Neill, E.E., Kerry, J.P., 2010. Investigation of the effects of commercial carcass suspension (24 and 48 hours) on meat quality in modified atmosphere packed beef steaks during chill storage. Food Research international 43, 277−284.

Zakrys-Waliwander, P.I., O'Sullivan, M.G., Walshe, H., Allen, P., Kerry, J.P., 2011. Sensory comparison of commercial low and high oxygen modified atmosphere packed sirloin beef steaks. Meat Science 88, 198−202.

Chapter 3

Sensory Affective (Hedonic) Testing

INTRODUCTION

In order for any given food product to be commercially successful, consumer desires and demands must be addressed and met with respect to the sensory properties of such products, before other quality dimensions become relevant (Chambers and Bowers, 1993). Consistent product quality is a key focus in the food industry. Ensuring superior quality, however, it is defined, is clearly required in the production and distribution of food products. Additionally, product quality directly relates to customer satisfaction and ultimately to repeat sales (Pecore and Kellen, 2002).

Affective sensory testing methods use hedonics (liking) to capture the emotive sensory response of naïve assessors which can be either consumers or assessors analogous to the consumer. They are being scored on their preference or on their 'Liking' of things like appearance, flavour or texture and ultimately their overall impression of a product or 'Overall Acceptability'.

Ideally the numbers of assessors required for affective testing, with the exception of focus groups, are much greater than with descriptive tests as the sample size must be representative of a larger consumer population. It is sensory science convention that hedonic tests and more analytical tests should not be undertaken with the same respondents. This is certainly true of trained sensory panels for descriptive profiling where hedonic elements should never be included. The reason for this being that as the panels are trained to respond to the defined sensory attributes any hedonic response they may have can be biased and thus unreliable.

Affective sensory tests can be qualitative (focus groups) or quantitative (preference, sensory acceptance tests, consumer tests). Qualitative and quantitative affective types of tests usually sit on opposite ends of the research and development spectrum. Focus groups generally use relatively small numbers of suitable screened individuals and are good for testing product and packaging concepts, ideas and are not that expensive. Sensory acceptance testing can be employed during the development and optimisation processes as a means of assessing variant suitability in a hedonic fashion and involves

A Handbook for Sensory and Consumer-Driven New Product Development.
http://dx.doi.org/10.1016/B978-0-08-100352-7.00003-8

anywhere from 25 to 75 individuals. Whereas consumer testing is generally involved after the various stages of product development is complete as a means of final product validation before product launch and involves large numbers of consumers (>100).

Optimisation of food or beverage products is undertaken to maintain or improve the liking of that product with respect to consumer sensory quality. Situations where it is required to maintain quality rather than improve quality could be in an ingredient substitution strategy or perhaps as part of a least cost formulation process. It brings to mind an old adage a former General Manager used say to me in my role, many years ago, as a Process and Product Development technologist for a large ingredients company, 'Give me the same not better'. This adage is also very much applied to the little spoken about reverse engineering strategies that many companies undertake whereby the objective is to replicate the sensory profile of competitor products. Maintaining sensory quality is also employed for reasons of nutritional optimisation by replacing an ingredient with a healthier ingredient. Thus acceptance and preference tests can be used to qualitatively and quantitatively determine if consumers like the modified product as well as account for differences due to age and gender as well as cultural influences (Yusop et al., 2009a,b; Michon et al., 2010a,b,c). Thus the product and individual attributes can be scored using hedonic and diagnostic scales to determine if formulation modifications to the product are necessary and cost-effective. Forced preference can be used to determine if one product is significantly preferred over others. For two or more samples, a ranking test can be used to determine preference.

QUALITATIVE AFFECTIVE TESTING-FOCUS GROUPS

Focus group interviews are a rapid method for collecting data through group interaction on topics determined by the researcher. Free-flowing conversation is moderated by a focus group leader, in an informal manner, to promote group discussion and participation in order to obtain insights into consumer thoughts and behaviour regarding a particular product concept. The setting is usually very informal and should put the group at ease. Typically a conference room is used with participants sitting around the conference table. The proceedings are sometimes viewed and filmed by way of a two-way mirror in an adjoining room so that the discussion can be observed by other individuals who may take notes or record the proceedings for later discussion. The opinions, body language and facial expressions of the participants can be also observed to assist in assessing their reaction to the product concept in the discussion. The researcher's interest provides the focus, whereas group interaction produces the data. This can be achieved through direct questioning, role-playing or other projective techniques. Typically, sessions can last up to 2 h. Focus groups can be used to build an understanding of a product category from the consumer viewpoint, inspiring idea generation, but can also be used to evaluate new product concepts and even samples (Edmunds, 1999). The test is subjective in that members of the group are asked about their feelings towards a product. The aim is to determine eating habits, preferences and opinions of the panel for the particular product in question (Bryhni et al., 2002).

The leader of the focus group or moderator should be a trained individual. This is necessary in order for consumer insights and behaviour to be determined rather than just superficial responses by the group. The group should be carefully selected and reflect the ultimate target market of the product (package) to be assessed. As the group of individuals, usually requires 6 to 10 people, which are paid, it is the responsibility of the moderator to clearly determine the suitability and motives of candidates for participation. Perhaps if their main priority is payment, they may be less than truthful about their true demographic situation. Such individuals can be classed as 'cheaters', 're-peaters' or 'cheater/repeaters'. Cheaters lie about their situation in order to qualify for the focus group. Whereas repeaters lie about the length of time since they last participated in a group, which should be not less than 6 months. Obviously the 'cheater/repeater' practices both of these untruthful tactics and can even do this on a professional or semiprofessional basis. The 'cheater/repeater' causes a particular problem for the undertaking of focus groups and can be extremely difficult to prevent. For this reason, where possible, candidates should be asked to provide suitably reliable identification or be paid directly to a bank account. In this manner, false identities can be eliminated and a record can be kept to monitor future participation in other groups if necessary. Also, the screening questionnaire can help in weeding out unsuitable individuals. A carefully constructed recruitment 'screening' questionnaire should be produced as well as an informal conversation in order for the moderator to determine their suitability. This screening questionnaire should be thorough, but concise, and can be written or filled in verbally or over the telephone. A lack of knowledge of the product to be tested or an ambiguous demographic situation is ground for noninclusion. Contradictory responses or ambiguity could identify such candidates. However, this does not stop the 'cheater/repeater' participating in other external focus groups in a particular area. How can this be countered? One way is to explain to candidates that a criterion for participation is to allow the moderator or other such individual to contact other known focus group practitioners in an area, such as other sensory service providers. In the United States and Canada, focus group tracking service can be used out to help identify 'cheater/repeaters'. Their details and contact number are searched through a database of focus group participants to identify and then exclude these individuals.

Once an appropriate group is selected, focus groups can prove very useful. According to Powell and Single (1996), focus groups are critical in order to generate concepts for an area when previous knowledge about the area to be investigated quantitatively is limited. However, a poorly selected focus group can do exactly the opposite and be incredibly damaging with misinformed responses perhaps being used to formulate a new product thus initiating a costly product development process which is ultimately doomed to failure. One way of preventing this is to use at least two or multiple focus groups as a means of validation. The obtaining of similar or trending responses will be indicative of consensus and the moderator can have confidence that the data obtained are correct. Focus participants should not be too meek or too bold. Individuals who do not give their opinion are of little use and equally un-helpful are the very extroverted who can take over and perhaps bias a free-flowing discussion.

SENSORY ACCEPTANCE TESTING

The sensory acceptance test is a small-panel guidance test that usually involves 25–75 participants who are regular consumers of the product to be tested. The selection of the panellists is crucial in insuring useful information is collected. Panellists should be carefully screened and should represent the target audi-ence of the product or concept. The assessors can come from internal sources such as employees or they can be externally recruited. There are advantages and disadvantages to both schemes. Employees are always on site, but may not always be available due to work pressure, or appropriately motivated, but are certainly cost-effective. External panels generally require payment and must go through the process of recruitment and the various logistical challenges of getting them to attend a session at a particular time and location. However, as external panels are paid, they are usually available for longer and the panel leader is not under pressure to return them to their work. The test should only be used to measure acceptance or preference. Product characteristics are sometimes included in this test, but this is not recommended as the results can be confounded (Stone and Sidel, 2004; Stone et al., 2012a). They can assess only hedonic (Liking) descriptors such as liking of: appearance, aroma, flavour, texture and overall acceptability or they can assess only preference or the attributes that determine preference. In this manner, large numbers of experimental prototypes can be screened. Ultimately, selected formulations which have achieved the necessary sensory acceptance target will be submitted for further affective sensory testing using conventional consumer preference studies using consumers ($n > 100$).

Sensory acceptance tests can be undertaken in a laboratory, a central location or in the home. In the laboratory the presentation situation is stand-ardised for each assessor with respect to product presentation, lighting, tem-perature, surroundings (standard sensory booth with serving hatch) and for this reason it is possible to use between 25 and 50 assessors up to 75, but 40 is recommended (Stone et al., 2012a). Stone et al. (2012a) also state that a laboratory acceptance test could be conducted with as little as 24 assessors if all the assessors evaluate all the products (balanced block design). However, the data obtained become statistically stronger the greater the number of assessors used. The greater the number of subjects used in a test, the less the likelihood of type I and type II errors occurring. As previously discussed in Chapter 1, type I error occurs when differences are found between samples when really there are not any. The opposite can also occur, by not rejecting the null hypothesis and is called beta risk, or type II error. Here no differences are found between samples where differences really exist. Alpha and beta risk can be reduced by increasing the number of observations or the amount of data needed to make a decision.

Also, no more than six products should be presented and the test should be restricted to acceptance or preference. Greater numbers of products can be evaluated if a break or interval is included and only acceptance is measured in which case up to 12 products could be evaluated (Kamen et al., 1969).

Central location testing should use 100+ members of the general public, not employees, who are users of the product to be tested. Again no more than five or six products should be tested and acceptance or preference should only be measured.

Finally for home testing, 50–100 member of the public can be used to assess only one or two products. However, in addition to acceptance or preference attribute, they may also be used to assess intensity and marketing information. In effect the product can be tested under actual use as well as information on the entire households opinion, pricing, frequency of use, etc. However, such tests can be expensive (Stone et al., 2012a).

The numbers of assessors used in the aforementioned tests are recommendations and as such should not be considered a go or no-go scenario if target numbers cannot be met.

The classical sensory acceptance test involves the assessment of product liking or preference using the nine-point hedonic scale (Fig. 3.1). This scale has been ubiquitously used for such sensory evaluation since its development (Peryam and Pilgrim, 1957). It is a tried and tested scale which is reliable and easy to use by naïve or untrained assessors. However, continuous line scaling may also be employed (Fig. 3.1), where the assessors mark their scores on a

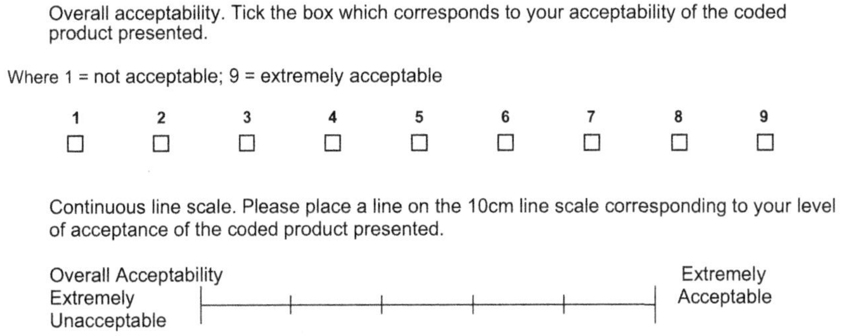

Overall acceptability. Tick the box which corresponds to your acceptability of the coded product presented.

Where 1 = not acceptable; 9 = extremely acceptable

1	2	3	4	5	6	7	8	9
☐	☐	☐	☐	☐	☐	☐	☐	☐

Continuous line scale. Please place a line on the 10 cm line scale corresponding to your level of acceptance of the coded product presented.

Overall Acceptability
Extremely   Extremely
Unacceptable ├──────┼──────┼──────┼──────┤   Acceptable

FIGURE 3.1 Nine-point hedonic scale and continuous line scale.

10- or 15-cm scale anchored on the left by none and the right by the term extreme. This is not a problem when the data are collated electronically using software such as Fizz, Compusense or RedJade, but when paper ballots are used, the use of this scale is more challenging. The line marking on the line must be then measured with a ruler and then inputted manually into the computer software such as Excel. Magnitude scaling is also discussed in the literature, but is not widely used.

Consumer acceptance tests are much cheaper to conduct compared to larger scale consumer tests, which will be discussed in the next section. However, the value of these tests is in assisting the product developer in filtering their experimental prototypes in a standardised and scientific manner using hedonics. Sensory acceptance tests are not a substitute for the more costly large-scale consumer tests, so ultimately the final variants produced through product development and optimisation should be validated by appropriately conducted larger scale consumer testing. Here again the sensory aspects of the product are measured. Additionally, market research tests can be conducted which are broader in their design also capturing information perhaps on pricing, purchase intent or use of the product. In this fashion the product developer can then examine the data and decide if the product is good enough to go to market or not.

CASE STUDIES — SENSORY ACCEPTANCE TESTS

The following case studies will demonstrate the use of sensory acceptance testing in product development and optimisation and its integration with other tests. These tests included descriptive sensory testing (ranking descriptive analysis (RDA)) as well as physicochemical analysis. The data were interpreted using chemometrics which will be discussed in detail in Chapter 5. Sensory acceptance testing combined with RDA with the resulting data analysed by multivariate data analysis is an approach that has been successfully demonstrated for a number of products including: chocolate pudding (Richter et al., 2010), white pudding (Fellendorf et al., 2015, 2016b), black Pudding (Fellendorf et al., 2016a,c), corned beef (Fellendorf et al., 2016d), coffee (Stokes et al., 2017), butter (O'Callaghan et al., 2017) and mozzarella cheese (Henneberry et al., 2016).

Fellendorf et al. (2016a,c) used consumer acceptance studies to screen a large numbers of experimental prototypes in two experiments designed to reduce the salt and fat levels in black pudding sausage.

CASE STUDY 1

In this first study (Fellendorf et al., 2016a), 25 black pudding formulations with varying fat contents of 2.5%, 5%, 10%, 15%, 20% (w/w) and sodium contents of 0.2%, 0.4%, 0.6%, 0.8%, 1.0% (w/w) were manufactured on commercial meat process equipment on a pilot scale and then cooked. The study was undertaken as part of a salt and fat reduction strategy for processed meats in Ireland as a means of improving the nutritional quality of the products but by also maintaining the sensory quality, safety, functionality and commercial viability of products. This case study presents a typical product development scenario for many companies who wish to perhaps collaborate with national and consumer initiatives for healthier food products and by doing so in a sensory optimised fashion maintain maximum sensory quality.

Processed meats have moved more into public focus over the past 20 years for many reasons, but particularly with respect to health concerns and present a good example of sensory optimisation for nutritional reasons. Irrespective of public opinion, the processing of underutilised meat is necessary on ethical grounds alone as it is responsible for converting inedible material to a more palatable form, thereby reducing food waste and generating more protein-based food products which also present product diversity in the marketplace. Additionally, and more specifically, processing extends shelf life, improves texture and enhances overall flavour (Fellendorf et al., 2016a). However, meat processing inevitably leads to products having higher amounts of salt, saturated fatty acids and preservatives such as nitrates which have health implications (Barcellos et al., 2011), particularly where overconsumption of such products occurs. A higher risk of coronary heart disease, stroke, cancer and obesity has been tenuously linked with their consumption (Demeyer et al., 2008; Gilbert and Heiser, 2005; Verbeke et al., 2010; Tobin et al., 2012a,b, 2013; WHO, 2003, 2009).

Black pudding, also known as blood sausages or blood pudding, is a kind of meat product made by blood from pigs, ducks, chickens, cattle, sheep or lamb and is popular in Asia, Europe and America. Recipes and servings differ dramatically from country to country. In Estonia and Italy, for example, blood sausages are mostly eaten in winter. Black pudding-style products can contain fillings such as breadcrumbs, pine nuts, rice, mashed potatoes, apples, oatmeal, barley, buckwheat, onions, rice, milk and salt, but also chocolate, raisins, sugar or butter. Blood sausages are spread on bread, served on sticks as a snack or in slices eaten together with mashed potatoes, fried bacon, beans, eggs, loganberry jam, butter or sour cream (Fellendorf et al., 2016a,c; Adesiyun and Balbirsingh, 1996; Marianski and Marianski, 2011; Predika, 1983; Santos et al., 2003; Stiebing, 1990). However, all black pudding-style products are unified in composition owing to the presence of blood or blood by-product as an ingredient, thereby providing a unique source of proteins and iron. The blood products used are composed of blood cells, containing erythrocytes, leucocytes, platelets, as well as haemoglobin, suspended in blood plasma. The iron present in blood is highly bioavailable as it is bound to haemoglobin which also acts as colour agent. In general, blood proteins possess emulsifying, foaming, gelling and/or water-binding properties. Moreover, no allergenic potential has been associated with blood-based ingredients (Pares et al., 2011; Reizenstein, 1980; Young et al., 1973).

Compositional analysis was performed on all the experimental prototypes and the analysis involved the measurement of protein, ash, moisture, fat and salt content for each pudding formulation, as well as physicochemical analysis, including colour, texture analysis and cooking loss.

Sensory acceptance testing and a rapid sensory method known as RDA were conducted on all samples. The sensory acceptance test was conducted using untrained assessors ($n = 25-28$) (Stone et al., 2012a; Stone and Sidel, 2004) in the age range of 21−60. They were chosen on the basis that they consumed black pudding products regularly. The experiment was conducted in panel booths which conform to the International Standards (ISO, 1988). The sensory test was split into five sessions, whereby five reheated samples (coded and presented in a randomised order) were served to the assessors. The assessors were asked to assess, on a continuous line scale from 1 to 10 cm, the following attributes: liking of appearance, liking of flavour, liking of texture, liking of colour and overall acceptability (hedonic). Black pudding samples were presented in duplicate (Stone et al., 2012b). The assessors then participated in a RDA (Richter et al., 2010) using the consensus list of sensory descriptors including grain quantity, fatness, spiciness, saltiness, juiciness, toughness and off-flavour (intensity), which was also measured on a 10-cm line scale. All samples were again presented in duplicate (Stone et al., 2012b). The latter sensory technique will be discussed in detail in the next chapter of this book on Rapid Sensory Methods.

Fellendorf et al. (2016a) found that samples high in sodium (0.6−1.0%) were scored higher in juiciness, toughness, saltiness, fatness and spiciness. These samples were the most accepted, whereas samples containing 0.2% sodium were the least accepted. Black pudding samples containing 0.6% sodium and 10% fat displayed a positive ($P < 0.05$) correlation to liking of flavour and overall acceptability. This in fact meets the sodium target level set by the Food Safety Authority of Ireland (FSAI, 2011, 2014) and shows additionally, that a fat reduction in black pudding products is more than achievable.

CASE STUDY 2

In a second more elaborate study, Fellendorf et al. (2016c) used the findings from the first experiment as a basis for further optimisation using commercially available salt and fat replacers. Thus, 22 black puddings possessing different fat (10%, 5%) and sodium (0.6%, 0.4%) levels, based on the first study's findings, were produced with 11 different salt and fat replacers (Table 3.1, from Fellendorf et al., 2016c). Again products were manufactured on commercial meat processing equipment on a pilot scale. Compositional analysis was performed on all the experimental prototypes and included the measurement of protein, ash, moisture, fat and salt content for each pudding formulation, as well as physicochemical analysis, including colour, texture analysis and cooking loss.

TABLE 3.1 Black Pudding Formulations With Different Replacers

	Formulation (%)										
Sample [a]	Meat	Fat	Salt	Water	Blood Powder	Seasoning	Oatmeal	Onion	Boiled Barley	Rusk	Replacer
Control 1	35.60	15.38	1.02	27.00	3.00	1.10	6.55	2.50	3.00	4.85	—
Control 2	42.79	7.69	1.52	27.00	3.00	1.10	6.55	2.50	3.00	4.85	—
Wheat bran 1	34.60	15.38	1.02	27.00	3.00	1.10	6.55	2.50	3.00	4.85	1.00
Wheat bran 2	41.79	7.69	1.52	27.00	3.00	1.10	6.55	2.50	3.00	4.85	1.00
Sodium citrate 1	35.15	15.38	0.97	27.00	3.00	1.10	6.55	2.50	3.00	4.85	0.50
Sodium citrate 2	42.34	7.69	1.47	27.00	3.00	1.10	6.55	2.50	3.00	4.85	0.50
Carrageen 1	35.10	15.38	1.02	27.00	3.00	1.10	6.55	2.50	3.00	4.85	0.50
Carrageen 2	42.29	7.69	1.52	27.00	3.00	1.10	6.55	2.50	3.00	4.85	0.50
Pectin 1	35.20	15.38	1.02	27.00	3.00	1.10	6.55	2.50	3.00	4.85	0.40
Pectin 2	42.39	7.69	1.52	27.00	3.00	1.10	6.55	2.50	3.00	4.85	0.40
KCl 1	34.58	15.38	1.02	27.00	3.00	1.10	6.55	2.50	3.00	4.85	1.02
KCl 2	42.11	7.69	1.52	27.00	3.00	1.10	6.55	2.50	3.00	4.85	0.68
KClG 1	34.08	15.38	1.02	27.00	3.00	1.10	6.55	2.50	3.00	4.85	0.91/0.61
KClG 2	41.77	7.69	1.52	27.00	3.00	1.10	6.55	2.50	3.00	4.85	0.61/0.41

Continued

TABLE 3.1 Black Pudding Formulations With Different Replacers—cont'd

Sample [a]	Meat	Fat	Salt	Water	Blood Powder	Seasoning	Oatmeal	Onion	Boiled Barley	Rusk	Replacer
						Formulation (%)					
CMC 1	35.10	15.38	1.02	27.00	3.00	1.10	6.55	2.50	3.00	4.85	0.50
CMC 2	42.29	7.69	1.52	27.00	3.00	1.10	6.55	2.50	3.00	4.85	0.50
Seaweed 1	32.30	15.38	1.02	27.00	3.00	1.10	6.55	2.50	3.00	4.85	3.30
Seaweed 2	39.49	7.69	1.52	27.00	3.00	1.10	6.55	2.50	3.00	4.85	3.30
PuraQ 1	32.60	15.38	1.02	27.00	3.00	1.10	6.55	2.50	3.00	4.85	3.00
PuraQ 2	39.79	7.69	1.52	27.00	3.00	1.10	6.55	2.50	3.00	4.85	3.00
KCPCl 1	34.10	15.38	1.02	27.00	3.00	1.10	6.55	2.50	3.00	4.85	0.5/0.5/0.5
KCPCl 2	41.29	7.69	1.52	27.00	3.00	1.10	6.55	2.50	3.00	4.85	0.5/0.5/0.5
WMS 1	32.60	15.38	1.02	27.00	3.00	1.10	6.55	2.50	3.00	4.85	3.00
WMS 2	39.79	7.69	1.52	27.00	3.00	1.10	6.55	2.50	3.00	4.85	3.00

CMC, carboxymethylcellulose; KCl, potassium chloride; KClG, mixture of potassium chloride and glycine; seaweed, seaweed wakame; PuraQ, PuraQ Arome NA4; KCPCl, combination of potassium citrate, potassium phosphate and potassium chloride; WMS, waxy maize starch.
[a]Sample code: 1—10% fat and 0.4% sodium, 2—5% fat and 0.6% sodium.

The sensory acceptance test was conducted using untrained assessors ($n = 25-30$) (Stone et al., 2012a; Stone and Sidel, 2004) in the age range of $21-60$. They were again chosen on the basis that they consumed black pudding products regularly. The experiment was conducted in asimilar fashion to the first case study in panel booths which conform to the International Standards (ISO, 1988). The sensory test was again split into five sessions, whereby five reheated samples (coded and presented in a randomised order) were served to the assessors. The assessors were again asked to assess, on a continuous line scale from 1 to 10 cm, the following attributes: liking of appearance, liking of flavour, liking of texture, liking of colour and overall acceptability (hedonic) with samples were presented in duplicate (Stone et al., 2012b). The assessors also participated in an RDA (Richter et al., 2010) using the consensus list of sensory descriptors including grain quantity, fatness, spiciness, saltiness, juiciness, toughness and off-flavour (intensity), which was also measured on a 10-cm line scale. All samples were again presented in duplicate (Stone et al., 2012b). This RDA method will be discussed in the next chapter.

Black pudding samples with 5% fat and 0.6% sodium containing potassium chloride (KCl), potassium chloride and glycine mixture (KClG), and seaweed, respectively, and 10% fat and 0.4% sodium containing carrageen were rated higher ($P < 0.05$) for spiciness and saltiness. Samples with 10% fat and 0.4% sodium containing KClG were rated positive ($P < 0.05$) to fatness. Samples with 5% fat and 0.6% sodium containing pectin and a combination of potassium citrate, potassium phosphate and potassium chloride (KCPCl), as well as samples containing 10% fat and 0.4% sodium with waxy maize starch (WMS) were liked ($P < 0.05$) for flavour and overall acceptance by assessors. The Food Safety of Ireland (FSAI, 2011, 2014) recommends a sodium target level of 0.6% and an even lower sodium level (0.4%) was achieved to produce a highly sensory accepted product (Fellendorf et al., 2016b).

PREFERENCE TESTS

The paired comparison test, as discussed in Chapter 1, of this book, is also categorised as an affective test as it can be used to measure preference. Whereas acceptance tests measure the degree of liking of the products presented to the panel preference tests simply measure which products assessors prefer over another or others. Directions are simply given to the respondent 'Which product of those presented do you prefer?'. They may also be asked to rank products in order of preference or use a rating scale. The latter method involves asking the assessor to mark on a category hedonic scale (e.g., seven or nine point) how they rate a presented product from dislike extremely to like extremely (See Fig. 3.1). As with hedonic tests, it is sensory science convention that hedonic tests and descriptive tests should not be undertaken with the same respondents. This is certainly true of trained sensory panels for descriptive profiling where hedonic elements should never be included. The reason for this being that as the panels are trained to respond to the defined sensory attributes any hedonic response they may have can be biased and thus unreliable.

RANKING PREFERENCE TESTS

In the Ranking preference test, assessors are asked to order a selection of coded samples (4–6) with respect to their preference similar to the ranking attribute tests described in Chapter 1. During a session, consumers or naïve assessors receive coded randomised samples and are asked to rank or order them in either descending or ascending preference. Again, where fatigue is less likely more samples may be used.

CONSUMER ACCEPTANCE TESTS

Consumer acceptance testing can be considered a larger scale sensory acceptance test. The logistics of organising these are much more complicated but the number of samples presented is usually reduced. The larger the number of respondents ($n \geq 100$), the greater the reliability of the data obtained. Essentially a particular demographic is recruited based on the company's own knowledge of their products consumers. This information can come from market research sources and the methods they employ which will be discussed in the next section. The demographic can be organised on a very large number of potential factors, but gender, age and employment situation are usually important. Additionally, where the consumers live e.g., rural or urban could also be considered. Acceptance testing data can be combined with more descriptive data and analysed using chemometrics which has been discussed above in the presented case studies.

MARKET RESEARCH

Market research involves the gathering of information regarding products from individuals who represent the buyers or future buyers of that product. A large number of consumers can be surveyed using questionnaires either conducted by telephone, on paper (by mail, online) or by using smaller groups of individuals in focus groups as discussed earlier in this chapter. However, market research is usually undertaken on large statistically valid groups which fit the projected demographic. Telephone surveys or online tests should be short, no longer than 15 min as any longer will not hold the consumers attention and could potentially annoy them. For online surveys a link can be embedded in an email or on a website along with a tempting invitation for the consumer to participate, perhaps an incentive such as a prize or draw is sometimes offered to reward those who take the questionnaire.

Before the commencement of any research, it is important that defined goals are set so that the right questions can be asked to establish and what information is required. The questionnaire is then designed which should be tested in advance to ensure its suitability and to identify any errors or clumsiness. The questionnaire is designed to capture information regarding their eating habits or behaviour for the products presented.

QUESTIONNAIRE DESIGN

The questionnaire should be concise, easy to read with short questions which applies to both consumer acceptance testing and market research. Initially, general questions are asked which then become more specific. The consumer can be led through the questionnaire using carefully worded layered questions. Those that do not fit the demographic can thus be weeded out. For example, if the survey was on a 'blue cheese' product perhaps more general questions regarding dairy products could become more specific like 'which of these cheese do you regularly consumer?', and those that do not tick the blue cheese box in a list of other cheeses can thus be excluded from the survey.

The survey should not present ambiguous questions or ones that can prompt the consumer. The scales used should also be clear and easily understood and be consistent throughout the survey by having similar ranges. The Likert scale is often used. This scale asks whether the consumer strongly disagrees or strongly agrees with a series of questions, i.e., This product is good value for money!, I enjoy eating this product as part of my daily diet!, I am more likely to buy a product if it is healthy!, etc.

Likert scales are very useful as they can measure a very broad range of topics and are easy to understand (Fig. 3.2). They are also more precise than just a yes or no answer. However, the right question must be asked to be relevant and Likert scales do not offer any reasons for agreeing or disagreeing with the question asked.

Semantic differential scaling can also be used which measure, on bipolar scales, concepts or meaning anchored at each end by opposing adjectives and perhaps also a neutral adjective in the centre, e.g., low quality to high quality, cheap to expensive, smooth to rough, etc (Fig. 3.3). Bipolar adjective scales

Where 1 = 1 strongly disagree and 7 = strongly agree

1	2	3	4	5	6	7
☐	☐	☐	☐	☐	☐	☐

FIGURE 3.2   Likert scale.

	1	2	3	4	5	6	7	
Good	☐	☐	☐	☐	☐	☐	☐	**Bad**

FIGURE 3.3   Semantic differential scale.

are a simple, economical means for obtaining data on people's reactions. With adaptations, such scales can be used with adults or children, persons from all walks of life and persons from any culture (Heise, 1970). This method is intuitive for the consumer and it is easy to measure the results, but it is also subjective and like the Likert scale it does not offer any information on why the consumer answers in this fashion.

Initial survey questions should be used to verify that the correct consumer is being targeted. For example, if the survey is on dairy products, ensure that the consumer is a consumer of dairy products. If they answer no to a validation question, their data can then be excluded. The questions used in the survey should potentially help develop consumer profiles, segments within these profiles and establish consumer characteristics and behaviour. In this manner, products and services can be optimised to improve consumer satisfaction as well as highlight changes in their opinion or even promote awareness of the product being tested.

Multiple choice questions ask the consumer to select an answer from a number, at least two, from a given list of alternatives (Fig. 3.4). Again these are easy to compile and input and are easy for the consumer to understand, but the list provided may not be broad enough and no explanation is offered as to why a particular box is ticked.

Open-ended questions allow the respondent to write what they wish without scale or limitation. They are most beneficial when the researcher is conducting exploratory research, especially when the range of responses is unknown (Zikmund and Babin, 2006). The respondent answers with anything that comes into their mind. This is useful as finally we may have a scale which captures the 'why factor' and the individual can really focus on what is important to them. However, the consumer can find this approach hard and confusing (if too many questions are asked) and the results can be difficult to process and interpret. It can also take longer than other scales and thus may increase the cost of the survey. The question given may also prompt the respondent and introduce bias.

Rank order questions require the respondent to choose and rank from among a list of alternatives. They are usually forced in that the consumer must respond and are similar to the rank preference test described above. They allow for forced discrimination so the best and worst items are clearly identified, but again do not offer why an item was chosen. They also can be tiring

How often do you eat beef? (In any form such as Steaks, Roasts, Stews, Casseroles, Kebabs, BBQ etc.). Please tick the appropriate box.

Daily [ ] 4-5 times a week [ ] 2-3 times a week [ ] Weekly [ ]

Fortnightly [ ] Monthly [ ] Never eat [ ]

FIGURE 3.4 Multiple choice questions.

for the respondent and be limited to the number of alternatives provided and thus may not have a broad enough range of alternatives.

OPEN-ENDED QUESTIONS AND CLOSED-OPTION QUESTIONS

Open-ended questions are easy to write, allow consumers to respond in their own words and communicate their unbiased opinions and suggestions regarding the product they have been testing. They are very useful in obtaining consumer suggestion of product use, improvement or optimisation. Open-ended questions are well suited to areas where the respondent has the information readily in mind but the interviewer is unable to anticipate all answers or provide a checklist (Lawless and Heymann, 1999). They also can capture anything that could have been omitted in other forms of questions. However, open-ended questions can sometimes have a lower response compared to other types of questions. They can also be difficult to interpret as the subject's individual subjective responses may be ambiguous or nonaligned and it is ultimately up to the presenter to determine interpretation and data assignments, input and analysis. In such cases of misinterpretation of data by the presenter can introduce bias into the subsequent collation and statistical analysis. The hand writing of respondents can also be difficult to sometimes read and extroverted respondents may dominate the data also introducing a bias (Lawless and Heymann, 1999).

Closed-option questions limit responses in a controlled manner thus minimising bias and allowing easier data entry, interpretation and statistical analysis. These questions do not require the same level of thought as open-ended questions and are easier to answer for the respondent. However, they are more rigid and may not capture what the subject really thinks and the assumption is made that the individual is competent or suitable as a subject when there is a possibility they do not understand what they are answering (Lawless and Heymann, 1999).

CONJOINT ANALYSIS

The essence of conjoint analysis is that it as a statistical technique used to identify the value placed by individuals on different product attributes. The process measures a mapping from more detailed descriptors of a product or service onto an overall measure of the customer's evaluation of that product (Hauser and Rao, 2004). This method determines what combination of attributes influences consumer choice and decision-making and is used frequently in market research to study the effects of controlled stimuli or information on a particular consumer response. Statistical experimental design, analysis of variance and cluster analysis enable the response of each consumer to be analysed for the relative importance of each factor, and, similarly performing subjects can be clustered. It is a method used for understanding how consumers trade off product features (Green and Rao, 1971).

Specific products or concepts are presented to consumers and the manner in which they make preferences can be analysed by modelling to determine information such as profitability and market share.

SAMPLE PRESENTATION (RANDOMISATION, PRESENTATION ORDER, BLIND TASTING AND BRANDING EFFECTS)

For sensory acceptance testing and consumer testing, generally, samples should be presented randomly coded (three digit) and in a randomised presentation order. This is to minimise first order and carryover effects (MacFie et al., 1989). In sequences of human sensory assessments, the response to a stimulus may be influenced by previous stimuli. When investigating this phenomenon experimentally with several types or levels of stimulus, it is useful to have treatment sequences which are balanced for first-order carryover effects (Nonyane and Theobald, 2007). First-order effects occur when the first sample in a sensory sequence of samples is scored differently to the other samples. This can be rectified by randomising the presentation order so that, where possible, each sample to be tested has an equal chance of being assessed first. Additionally, a warm-up sample could be inserted for the first slot. As carryover effects may also occur, it is a good idea to randomise presentation order anyway.

Carryover effects occur when a flavour, taste, aftertaste or other effects (trigeminal heating, chilli, paprika, etc.) are not completely cleansed from the palate between samples. This results in carryover which can interfere with the next sensory samples.

Sensory testing is generally undertaken 'blind' with samples presented with random three-digit codes. In this manner, only the sensory attributes of that product are the only assessments being made, free from any personal views or biases on brand or product configuration the assessor may have. The assessors should be given minimal information prior to a sensory test because this too can influence their behaviour. If, for example, assessors were told they were assessing samples in order to determine if off-flavours exist in the product or not, then a bias is introduced with the assessor then focusing on this attribute which may not have an influential effect. When branding is included in sensory tests, again bias may be introduced. Preferential brands will have a 'halo effect'.

According to Hutchings (1977), the importance of different product attributes varies with the situation and time. For example, a product seen on a supermarket shelf may have different attributes affecting perception when compared to the same product seen on a plate. The product on a plate would be affected by anticipatory and participatory attributes (Hutchings, 1977). Moskowitz (1981) studied the 'relative importance of perceptual factors to consumer acceptance' found that branding can encourage a product's acceptability. Similarly, Martin (1990) found that a beer tasted blind was rated higher than when the test was repeated but this time it was identified. Therefore, branding will influence the outcome either positively or negatively on a sensory assessment. To determine the sensory quality, assessments must be performed using blind coded samples.

REFERENCES

Adesiyun, A.A., Balbirsingh, V., 1996. Microbiological analysis of "black pudding", a Trinidadian delicacy and health risk to consumers. International Journal of Food Microbiology 31 (1−3), 283−299. http://dx.doi.org/10.1016/0168-1605(96)00944-0.

de Barcellos, M.D., Grunert, K.G., Scholderer, J., 2011. Processed meat products: consumer trends and emerging markets. In: Kerry, J.P., Kerry, J.F. (Eds.), Processed Meats. Improving Safety, Nutrition and Quality. Woodhead Publishing Ltd, Oxford, Cambridge, Philadelphia, New Dehli, pp. 30−53.

Bryhni, E.A., Byrne, D.V., Claudi-Magnussen, C., Agerhem, H., Johansson, M., Lea, P., Rødbotten, M., Martens, M., 2002. Consumer perceptions of pork in Denmark, Norway and Sweden. Food Quality and Preference 13, 257−266.

Chambers, E., Bowers, J., 1993. Consumer perception of sensory quality in muscle foods: sensory characteristics of meat influence consumer decisions. Food Technology 47, 116−120.

Demeyer, D., Honikel, K., De Smet, S., 2008. The World Cancer Research Fund report 2007: a challenge for the meat processing industry. Meat Science 80, 953−959.

Edmunds, H., 1999. The Focus Group Handbook. NTC Contemporary Publishing group, Chicago, USA.

Fellendorf, S., O'Sullivan, M.G., Kerry, J.P., 2015. Impact of varying salt and fat levels on the physiochemical properties and sensory quality of white pudding sausages. Meat Science 103, 75−82.

Fellendorf, S., O'Sullivan, M.G., Kerry, J.P., 2016a. The reduction of salt and fat levels in black pudding and the effects on physiochemical and sensory properties. International Journal of Food Science and Technology (accepted).

Fellendorf, S., O'Sullivan, M.G., Kerry, J.P., 2016b. Effect of using ingredient replacers on the physicochemical properties and sensory quality of low salt and low fat white puddings. European Food Research and Technology (Accepted).

Fellendorf, S., O'Sullivan, M.G., Kerry, J.P., 2016c. Impact of using replacers on the physico-chemical properties and sensory quality of reduced salt and fat black pudding. Meat Science 113, 17−25.

Fellendorf, S., O'Sullivan, M.G., Kerry, J.P., 2016d. Impact on the physicochemical and sensory properties of salt reduced corned beef formulated with and without the use of salt replacers. Meat Science (Submitted).

FSAI, 2011. Salt Reduction Programme (SRP), p. 85, 2011 to 2012.

FSAI, 2014. Monitoring of Sodium and Potassium in Processed Foods Period: September 2003 to July 2014, p. 44.

Gilbert, P.A., Heiser, G., 2005. Salt and Health: The CASH and BPA Perspective, 30. British Nutrition Foundation, pp. 62−69.

Green, P.E., Rao, V.R., August, 1971. Conjoint measurement for quantifying judgmental data. Journal of Marketing Research 8, 355−363.

Hauser, J.R., Rao, V., 2004. Conjoint analysis, related modeling, and applications. In: Green, P.E., Wind, Y. (Eds.), Advances in Marketing Research: Progress and Prospects, 2004. Springer Science & Business Media, pp. 141—168.

Henneberry, S., O'Sullivan, M.G., Kilcawley, K.N., Kelly, P.M., Wilkinson, M.G., Guinee, T.P., 2016. Sensory quality of unheated and heated Mozzarella-style cheeses with different fat, salt and calcium levels. International Journal of Dairy Science 69, 38—50.

Heise, D.R., 1970. The semantic differential and attitude research (Chapter 14). In: Summers, G.F. (Ed.), Attitude Measurement, 1970. Rand McNally, Chicago, pp. 235—253.

Hutchings, J.B., 1977. The importance of visual appearance of foods to the food processor and the consumer. Journal of Food Quality 1, 267—278.

ISO, 1988. Sensory Analysis. General Guidance for the Design of Test Rooms. Ref. No, 8589. International Organization for Standardization, Genève, Switzerland, p. 1988.

Kamen, J.M., Peryam, D.R., Peryam, D.B., Kroll, B.J., 1969. Hedonic differences as a function of number of samples evaluated. Journal of Food Science 34, 475—479.

Lawless, H.T., Heymann, H., 1999. Sensory Evaluation of Food: Principles and Practices. Kluwer Academic/Plenum Publishers, New York.

MacFie, H.J., Bratchell, N., Greenhoff, K., Vallis, L.V., 1989. Designs to balance the effect of order of presentation and first-order carry-over effects in hall tests. Journal of Sensory Studies 4 (2), 129—148.

Marianski, S., Marianski, A., 2011. Making Healthy Sausages, pp. 251—262 (Bookmagic).

Martin, D., 1990. The impact of branding and marketing on perception of sensory qualities. Food Science and Technology 4, 44—49.

Michon, C., O'Sullivan, M.G., Delahunty, C.M., Kerry, J.P., 2010a. Study on the influence of age, gender and familiarity with the product on the acceptance of vegetable soups. Food Quality and Preference 21, 478—488.

Michon, C., O'Sullivan, M.G., Delahunty, C.M., Kerry, J.P., 2010b. The investigation of gender related sensitivity differences in food perception. Journal of Sensory Studies 24, 922—937.

Michon, C., O'Sullivan, M.G., Sheehan, E., Delahunty, C.M., Kerry, J.P., 2010c. Investigation of the influence of age, gender and consumption habits on the liking for jam-filled cakes. Food Quality and Preference 21, 553—561.

Moskowitz, H.R., 1981. Relative importance of perceptual factors to consumer acceptance: linear versus quadratic analysis. Journal of Food Science 46, 244—248.

Nonyane, B.A.S., Thoebald, C.M., 2007. Design sequences for sensory studies: achieving balance for carry-over and position effects. British Journal of Mathematical and Statistical Psychology 60, 339—349.

O'Callaghan, T., O'Sullivan, M.G., Kerry, J.P., Kilcawley, K.N., Stanton, C., 2017. Effects of feeding grass, clover or concentrate on the sensory and physicochemical quality of butter. International Journal of Dairy Science (submitted).

Pares, D., Saguer, E., Carretero, C., 2011. Blood by-products as ingredients in processed meat. In: Kerry, J.P., Kerry, J.F, (Eds.), Processed Meats. Improving Safety, Nutrition and Quality. Woodhead Publishing Ltd, Oxford, Cambridge, Philadelphia, New Dehli, pp. 218—242.

Pecore, S., Kellen, L., 2002. A consumer-focused QC/sensory program in the food industry. Food Quality and Preference 13, 369—374.

Powell, R.A., Single, H.M., 1996. Focus groups. International Journal for Quality in Health Care 8, 499—504.

Peryam, D.R., Pilgrim, F.J., 1957. Hedonic scale method of measuring food preferences. Food Technology 11 (9), 9—14.

Predika, J., 1983. The Sausage-Making Cookbook. Stackpole Books, p. 192.

Reizenstein, P., 1980. Hemeoglobin fortification of food and prevention of iron deficiency with heme iron. Acta Medica Scandinavian Supplement 629, 1–47.

Richter, V., Almeida, T., Prudencio, S., Benassi, M., 2010. Proposing a ranking descriptive sensory method. Food Quality and Preference 21 (6), 611–620. http://dx.doi.org/10.1016/j.foodqual. 2010.03.011.

Santos, E.M., González-Fernández, C., Jaime, I., Rovira, J., 2003. Physicochemical and sensory characterisation of Morcilla de Burgos, a traditional Spanish blood sausage. Meat Science 65 (2), 893–898. http://dx.doi.org/10.1016/S0309-1740(02)00296-6.

Stiebing, A., 1990. Blood sausage technology. Fleischwirtschaft 70 (4), 424–428.

Stokes, C., O'Sullivan, M.G., Kerry, J.P., 2017. Ranking acceptance analysis and the investigation of simultaneous or monadic sample presentation with affective and descriptive sensory evaluation methods. Journal of Sensory Studies (Submitted).

Stone, H., Bleibaum, R.N., Thomas, H.A., 2012b. Test strategy and design of experiments. In: Stone, H., Bleibaum, R.N., Thomas, H.A. (Eds.), Sensory Evaluation Practices, fourth ed. Elsevier Academic Press, USA, p. 135.

Stone, H., Bleibaum, R.N., Thomas, H.A., 2012a. Affective testing. In: Stone, H., Bleibaum, R.N., Thomas, H.A. (Eds.), Sensory Evaluation Practices, fourth ed. Elsevier Academic Press., USA, pp. 306–309.

Stone, H., Sidel, J.L., 2004. Affective testing. In: Stone, H., Sidel, J.L. (Eds.), Sensory Evaluation Practices. Food Science and Technology, International Series, third ed. Academic Press/ Elsevier., USA, pp. 247–277.

Tobin, B.D., O'Sullivan, M.G., Hamill, R.M., Kerry, J.P., 2012a. Effect of varying salt and fat levels on the sensory quality of beef patties. Meat Science 4, 460–465.

Tobin, B.D., O'Sullivan, M.G., Hamill, R.M., Kerry, J.P., 2012b. Effect of varying salt and fat levels on the sensory and physiochemical quality of frankfurters. Meat Science 92, 659–666.

Tobin, B.D., O'Sullivan, M.G., Hamill, R.M., Kerry, J.P., 2013. The impact of salt and fat level variation on the physiochemical properties and sensory quality of pork breakfast sausages. Meat Science 93, 145–152.

Verbeke, W., Pérez-Cueto, F.J.A., De Barcellos, M.D., Krystallis, A., Grunert, K.G., 2010. European citizen and consumer attitudes and preferences regarding beef and pork. Meat Science 84, 284–292.

WHO, 2003. Diet, Nutrition and the Prevention of Chronic Diseases, p. 148.

WHO, 2009. Recommendations for Preventing Cardiovascular Diseases. WHO Technical Report Series. No. 916 (TRS 916).

Young, C.R., Lewis, R.W., Landmann, W.A., Dill, C.W., 1973. Nutritive value of globin and plasma protein fractions from bovine blood. Nutrition Reports International 8, 211–217.

Yusop, S.M., O'Sullivan, M.G., Kerry, J.F., Kerry, J.P., 2009a. Sensory evaluation of Indian-style marinated chicken by Malaysian and European naïve assessors. Journal of Sensory Studies 24, 269–289.

Yusop, S.M., O'Sullivan, M.G., Kerry, J.F., Kerry, J.P., 2009b. Sensory evaluation of Chinese-style marinated chicken by Chinese and European naïve assessors. Journal of Sensory Studies 24, 512–533.

Zikmund, W., Babin, B., 2006. Exploring Marketing Research, tenth ed. Cengage learning.

Chapter 4

Rapid Sensory Profiling Methods

INTRODUCTION – THE NEED FOR NEW METHODS

Several changes have occurred in the sensory arena over the last few years with new tools developed for the sensory scientist and product developer. In the previous chapters we have already discussed classical descriptive profiling methods as well as hedonic sensory techniques with their inherent advantages and disadvantages.

Descriptive methods involve the quantitative determination of the appearance, aroma, flavour, texture, taste and aftertaste in a set of samples. Firstly an appropriately screened panel is selected based on their performance including a normal sensory response with a high level of acuity. The procedure involves the use of sensory terms that are produced in collaboration with the panellists and the panel leader, which is called the quantitative descriptive analysis (QDA) method or the spectrum method which uses a strict technical sensory vocabulary using reference materials. The method has a number of advantages over difference testing in that it is quantitative and can be used to describe differences between products and the main sensory drivers, be they positive or negative, identified within products or especially when combined with objective consumer testing and objective multivariate data analysis (O'Sullivan et al., 2011). With QDA the panellists are trained, using a subset of the samples to be profiled which reflect main sensory variation in the whole set of samples and their performance monitored using statistical methods. In the case of QDA, a reference is provided for each sensory attribute to assist the panellist in quantifying that particular attribute (see Chapter 2). When training is complete, these references are removed and profiling takes place on the full sample set. The language is descriptive and nonhedonic, in that assessors are not asked, for example, how much they rate or like the product being tested. The evaluation process aims at providing a subject-independent description, free of hedonic judgements. Quantification is performed on descriptive attributes that are clearly identified (Delarue and Sieffermann, 2004).

The different methods for descriptive profiling as described in Chapter 2 are the most powerful, sophisticated and most extensively used tools in

A Handbook for Sensory and Consumer-Driven New Product Development.
http://dx.doi.org/10.1016/B978-0-08-100352-7.00004-X

sensory science, which provides a complete description of the sensory characteristics of food products (Varela and Ares, 2012). However, the method can be expensive and time-consuming because of the necessity to train and profile individual panellists over extended periods of time; days or even weeks. It is also not a method that can be readily used for routine analysis (O'Sullivan et al., 2011). Descriptive profiling methods can also be labour-intensive and costly to undertake. In a dynamic product development environment where it can be necessary to get the product to market with all due speed, especially considering the fickleness of the consumer, descriptive profiling does not always fit well. Additionally, new line extension variants of a product with perhaps a new flavour profile make it necessary to undertake completely new descriptive profiles. Also, a single stimulus can be perceived quantitatively and qualitatively different from one subject to the other, especially when chemical senses are involved (Lawless, 1999). A weakness inherent in regular descriptive analysis is related to the use of attributes that are common to all the judges. Indeed, the ability of subjects to communicate their sensory perceptions on the same vocabulary bases is sometimes doubtful, even after thorough training (Delarue and Sieffermann, 2004).

On the opposite end of the sensory spectrum the affective methods described in Chapter 3 are restricted to only measure hedonics, which includes acceptability or preference, but do not describe the product. Traditionally, this issue was solved by combining descriptive data and hedonic data using data analytical tools involving predominantly chemometrics in a method called preference mapping. Internal and external preference mapping will be discussed in detail in the next chapter of this book. From these, statistically valid correlations could be determined from the data set so that sensory and product development scientists could understand how their modifications were affecting the sensory profile of the product with respect to the modalities, appearance, aroma, flavour, texture, taste and aftertaste as well as consumer opinion.

The 'intensive' use of consumers for sensory tests is not accepted by everybody in the sensory community (Worch et al., 2014). Consumers can only tell you what they 'like or dislike' (Lawless and Heymann, 1999). Some sensory scientists consider that using consumers for sensory descriptive tasks is not appropriate as consumers lack consensus and repeatability as well as comprehension of the meaning of the sensory attributes (Stone and Sidel, 1993; Lawless and Heymann, 1999). In contrast, other sensory scientists have shown through different studies that consumers can describe the sensory characteristics of products with a precision comparable to those obtained by experts (Worch et al., 2014). However, only 'simple' sensory attributes can be used (cannot use technical or chemical terms) (Worch et al., 2010, 2014) and larger numbers of consumers are required to make up for a lack of appropriate training.

In this chapter, new sensory methods will be presented which deliver rapid and more cost-effective solutions to these problems and to a certain extent narrow the divide between the rigid rules of classic descriptive profiling and the emotional responses involved with affective sensory methods.

Rapid sensory evaluation methods can provide the industry with quick results with respect to the end user and a reduction in resources needed to complete them. These methodologies are also more flexible and can be used with semitrained assessors and even naïve assessors, providing sensory maps very close to a classic descriptive analysis with highly trained panels (Varela and Ares, 2012). A number of different techniques have been published over the years on such rapid methods including: projective mapping (Risvik et al., 1994), polarised sensory positioning (PSP) (Teillet et al., 2010), ranking test (Rodrigue et al., 2000), flash profiling (Loescher et al., 2001; Dairou and Sieffermann, 2002, 2004), ultraflash profiling (Pagès, 2003), napping (Pagès, 2003, 2005), free sorting (Rosenberg et al., 1968; Lawless et al., 1995), optimised descriptive profiling (ODP) (Silva et al., 2012, 2013, 2014), check-all-that-apply (CATA) methods (Adams et al., 2007; Ares et al., 2010a,b), temporal dominance of sensations (TDS) (Pineau et al., 2003) and the ideal profile method (IPM) (Worch et al., 2014; Worch and Ennis, 2013). Some of these methods are simple ranking tests which are easier to perform compared to very time-consuming traditional descriptive methods as only slightly fam-iliarised subjects are required (Meilgaard et al., 1991). These tests are rapidly executed, and they can be performed on specific product attributes and will be discussed in detail in the presented chapter.

RANKING TEST

Rodrigue et al. (2000) were some of the first to suggest that a ranking test could be used in the description of a product instead of traditional profiling. They used a panel of eight assessors to evaluate 10 attributes of sweet corn as well as untrained ($n = 20$) assessors evaluating the same products and attri-butes using a ranking procedure. The first panel received suitable training to perform a descriptive profile, whereas the second, with limited training, were asked to rank the products according to the same attributes as the first panel. Ten attributes were thus common to both groups and included: yellow and grey colours, free substances, grain integrity, global intensity, fresh corn on the cob flavour, sweetness, firmness, crunchiness and residues. Samples were pre-sented monadically to the trained subjects who rated intensity of each attribute on an unstructured 10-point scale. Whereas, the untrained subject presented with all the products simultaneously and were asked to assess them for each attribute. Ties among products were not allowed for ranking (Rodrigue et al., 2000). The great advantage here was that only one familiarisation or training session was required for the ranking test compared to the much more elaborate procedure followed for descriptive profiling. The results from both methods were similar in terms of overall product discrimination, with generally very comparable results between the trained and the untrained assessors. Rodrigue et al. (2000) concluded that whenever one has not enough time to train a descriptive panel, one could consider using an untrained panel and conduct a ranking test. In their case the cost of the ranking test was less than a third of the cost involved for the traditional profile. This recommendation may still hold even if the number of products to compare is large, as a ranking test using a balanced incomplete block design could be used (Rodrigue et al., 2000).

FLASH PROFILING

In 2000, one of the first new generation of rapid descriptive sensory methods was presented by Sieffermann (2000). In this paper, he suggested combining free choice profiling (FCP) with a comparative evaluation of the product set in a technique named flash profiling. FCP involved panellists developing their own descriptive terms (Williams and Langron, 1984; Delahunty et al., 1997). Assessors are allowed to develop their own individual vocabularies to describe sensory perceptions and to use these to score sets of samples. As a consequence of removing the need to agree vocabularies, FCP requires little training, only instruction in the use of the chosen scale (Kilcast, 2000). Assessors develop their own attributes to describe the products, with their own vocabulary and in any number, limited only by their sensory skills, they then quantify their personal attributes using line scales. The method is based on the assumption that panellists do not differ in their perceptions but solely in the manner they describe them (Murray et al., 2001).

With flash profiling, assessors receive an introduction to the samples and after a short instructional presentation are told to generate their own vocabulary free-of-choice, based on their own sensory perception and to cover the sensory variations in the samples. After generating relevant attributes, they are allowed to see other assessors' vocabularies and to add or substitute attributes in their own list as they wish. For each attribute, samples are ranked according to their intensity on an ordinal scale anchored from 'lower' to 'higher'. Unlike the ranking test proposed by Rodrigue et al. (2000) where ties are not allowed, flash profiling allows the assessors to apply the same rank to two or more samples if no difference was perceived (Dehlholm et al., 2012). This method offers a compromise over conventional descriptive methods. In flash profiling, because assessors choose their own words to evaluate the whole product set comparatively (Dairou and Sieffermann, 2002), the profiling is thus a very rapid sensory profiling technique. Flash profiling has been compared in a number of research papers with classical descriptive methods (Loescher et al., 2001; Dairou and Sieffermann, 2002, 2004). Additionally, Delarue and Sieffermann (2004) compared flash profiling to conventional profiling for strawberry-blended yoghurts and apricot 'fromages frais', and found that flash profiling was slightly more discriminating than the conventional profile. The method was less time-consuming than the conventional profile and provided interesting alternative method to evaluate quickly an array of products

(Dairou and Sieffermann, 2002). Also, Rason et al. (2003) performed a flash profile on 12 traditional French dry sausages where the test subjects generated their own list of sensory terms (appearance, texture, aroma) which they used to evaluate the test products simultaneously. This comparative evaluation enabled a quick positioning of the traditional dry sausages on a sensory map (Sieffermann, 2000). Flash profiling can be carried out using semi-trained or trained assessors (Dairou and Sieffermann, 2002, 2004) or with consumers (Lassoued et al., 2008; Moussaoui and Varela, 2010; Veinand et al., 2011).

There are certain advantages and disadvantages with flash profiling compared to conventional descriptive profiling methods. The principal advantage of flash profiling is that the method is less time-consuming compared to traditional profiling because extensive training is not undertaken. This is because absolute consensus is not required between subjects for the definition of descriptive sensory term development as attributes are generated individually by free choice. Flash profiling takes into account the diversity of the point of view of the subjects about the products. Subjects choose their vocabulary according to their own sensitivity and perception. The combination of each subject description reflects different points of view depending on the importance given by subject to each sensory modality. This combination enriches the description (Dairou and Sieffermann, 2002). Additionally, all the samples are prepared before the sessions and presented to the assessors at the same time. The simultaneous presentation of products is very time-saving because subjects do not need any familiarisation phase with samples. From the start, they can discriminate any relevant attributes using very intuitive ordinal scales. Flash profiling is essentially a two-step process where assessors firstly evaluate samples comparatively to generate the individual sensory lexicon of descriptors followed by a second step where the sample set is ranked on the attribute scales developed, from low to high and they can designate the same rank to two or more samples if no difference is perceived (Dehlholm et al., 2012). Another benefit of the flash profile is the ease of session organisation. It needs four times less sessions than the conventional profile. There are no panellists schedules matching problem as each session is individual. Then the presence of the experimenter is not required (Dairou and Sieffermann, 2002).

Disadvantages of flash profiling include the fact that the number of samples that can be profiled can be limited. Large sample sets could possibly cause confusion with assessors. Tarea et al. (2007) demonstrated that up to 49 samples could be assessed in a flash profile of pear puree in one session (2−5 h). A significant contributing factor to this was the fact that the assessors were highly motivated, experienced and also trained and could take breaks during the profiling session. Further issues with flash profiling could include one of the foundations of its flexibility, i.e., the inclusion of free choice in individual sensory lexicon selection. These individual lexicons can be quite varied and are thus open to semantic interpretation (Dairou and

Sieffermann, 2002), but the core attributes should have some level of consensus.

This method of sensory profiling may have useful applications in the sensory evaluation of foods and beverages. This is even more relevant considering the convenient methodology employed compared to the more conservative methods and the necessity to minimise costs within the very competitive processed food sector (O'Sullivan et al., 2011). However, its use must be tempered with caution as it does not offer the discriminating power of the classical sensory descriptive profile such as the QDA. Also, the profile obtained for flash profiling reflects the relative variation amongst the samples in that particular time point. Traditional descriptive profiling is more appropriate when there is a need to compare samples in different moments in time or when comparing different sample sets with a few samples in common, or when a very detailed sensory description is required (Varela and Ares, 2012). The semantic consensus obtained in the conventional profile allows for a more accurate description of the products (Dairou and Sieffermann, 2002). The flash profiling technique is also not recommended for studies of stability and quality control, since it does not indicate the magnitude of the difference between the products. On the other hand, due to semantic terminology, this methodology is presented as a communication tool between research, development and marketing (Silva et al., 2012).

Rapid sensory methods such as flash profiling have been demonstrated for a diverse selection of products including: pear/apple puree and fresh cheese (Loescher et al., 2001) yoghurt (Delarue and Sieffermann, 2004), jams (Dairou, and Sieffermann, 2002), beer (Hempel et al., 2013a), bakery products (Lassoued et al., 2008; Hempel et al., 2013d), Gouda cheese (Cavanagh et al., 2014; Yarlagadda et al., 2014b), Cheddar cheese (Yarlagadda et al., 2014a), French dry sausages (Rason et al., 2003), beef patties (Tobin et al., 2012a), Frankfurters (Tobin et al., 2012b), breakfast sausages (Tobin et al., 2013) and ready-to-eat mixed salad products (Hempel et al., 2013b,c).

RANKING DESCRIPTIVE ANALYSIS

Ranking descriptive analysis (RDA) is a modification of flash profiling developed by Richter et al. (2010). In this method, chocolate puddings with different sugar and sweetener contents were compared using RDA and two descriptive methods, QDA and FCP. Unlike flash profiling, the sensory lexicon was not developed in a free choice-type manner with the attributes the same for both RDA and the traditional method so there was no issue with differences in semantic consensus as described above for flash profiling. Twenty one assessors were utilised for the RDA compared to 12 for QDA and 14 for FCP. These untrained assessors came to a consensus on the quantity of the sample to be served as well as the procedure and then ranked each of the chocolate pudding samples for the RDA, using ordinal scales for each of the defined (qualitative) attributes, whereas interval scales were used in the traditional descriptive techniques. Samples were presented monadically to the trained panel but for the RDA all the products were presented simultaneously to the assessors and ranked for appearance and aroma attributes followed by another session for texture and flavour attributes. This method allowed the discrimination of the samples with efficiency similar to that displayed by the descriptive methods of the QDA and FCP. Richter et al. (2010) also suggest that when time is insufficient to train a panel, the use of an untrained panel and a ranking test should be considered. Although larger numbers of assessors (21 for Richter et al., 2010) were required for RDA, it could be conducted with minor costs compared to QDA and with smaller amounts of product and sessions. However, Richter et al. (2010) observed it was important to train a panel in order to obtain good descriptor conceptualisation and greater panel consensus. Perhaps a more intense qualitative training for RDA would allow more consistent results, primarily for the complex attributes of texture (Richter et al., 2010).

RDA has been successfully demonstrated for a number of other products including: chocolate pudding (Richter et al., 2010), white pudding (Fellendorf et al., 2015, 2016b), black pudding (Fellendorf et al., 2016a,c), corned beef (Fellendorf et al., 2016d), butter (O'Callaghan et al., 2016) and mozzarella cheese (Henneberry et al., 2016).

Fellendorf et al. (2015) used RDA as well as sensory acceptance testing to investigate 25 white pudding formulations produced with varying fat contents (20%, 15%, 10%, 5%, 2.5%w/w) and varying sodium contents (1.0%, 0.8%, 0.6%, 0.4%, 0.2%w/w). The sensory acceptance test was conducted using untrained assessors ($n = 25-30$) (Stone et al., 2012a; Stone and Sidel, 2004) split into five sessions, whereby five reheated samples (coded and presented in a randomised order) were served to the assessors. The assessors were asked to assess, on a continuous line scale from 1 to 10 cm, the following attributes: liking of appearance, liking of flavour, liking of texture, liking of colour and overall acceptability (hedonic). Samples were presented in duplicate (Stone et al., 2012b). The assessors then participated in an RDA (Richter et al., 2010) using the consensus list of sensory descriptors including grain quantity, fatness, spiciness, saltiness, juiciness, toughness and off-flavour, which was also measured on a 10-cm line scale. All samples were again presented in duplicate in a separate session (Stone et al., 2012b; Fellendorf, O'Sullivan and Kerry, 2015). These authors found that puddings containing higher sodium levels (1.0%, 0.8%) were the most accepted, with the exception of those with the lowest fat content. Lower fat and salt containing puddings were tougher, less juicy, less spicy and lighter in colour. However, the pudding sample containing 15% fat and 0.6% sodium was highly accepted ($P < .05$), thereby satisfying the sodium target (0.6%) set by the Food Safety Authority of Ireland (FSAI, 2011, 2014). In this case, Fellendorf et al. (2015) assessed a large number of samples ($n = 25$) using RDA but had a relatively short attribute list

($n = 7$). Samples were also presented in a duplicate separate session with good reproducibility.

FREE SORTING

Another rapid sensory method that can be used to differentiate samples is 'sorting' where subjects examine the set of items and group them according to similarity (Lawless et al., 1995). It allows either trained or untrained assessors to group samples, all presented in a single session, usually in to two or more groups, according to their familiarities or differences using their own personal criteria. Groups cannot contain only one sample. Similar samples are grouped together, whereas very different samples are placed in other groups. Once complete, the assessors describe the groups using their own descriptive terms (Lawless et al., 1995; Popper and Heymann, 1996).

Trained assessors will find this component of sorting easier than untrained panellists and for this reason it is recommended that naïve assessors receive a list of predefined sensory characteristics from which they choose to describe the sample groups (Lelièvre et al., 2008). Larger numbers of assessor are generally used for untrained assessors ($n = 20-50$) compared to trained panels ($n = 9-15$) (Cartier et al., 2006).

The data collected from sorting can be analysed by either multidimensional scaling (MDS) (Lawless et al., 1995; Varela and Ares, 2012) or a method called DISTATIS. With MDS, a similarity matrix is created by counting the number of times that each pair of samples is sorted within the same group (Varela and Ares, 2012). It is calculated for the entire group of assessors and not individually. On the other hand, DISTATIS uses three-way distance tables and does takes into account the sample grouping provided by each assessor Abdi et al. (2007).

Sorting procedures have been applied to a large selection of different food and beverage products, including: cheese (Lawless et al., 1995), yogurts (Saint Eve et al., 2004), jellies (Tang and Heymann, 1999), breakfast cereals (Cartier et al., 2006), olive oil (Santosa et al., 2010), red wine (Gawel et al., 2001), drinking waters (Falahee and MacRae, 1997), beers (Chollet and Valentin, 2001), coffee (Moussaoui and Varela, 2010) and orange-flavoured powdered drinks (Ares et al., 2011).

Heymann (1994) was one of the first sensory scientists to investigate sorting as a technique and compared FCP and MDS of vanilla samples. The method has its origins in the late 1960s in the social sciences. Rosenberg et al. (1968) employed a multidimensional approach to the structure of personality impressions.

Lawless et al. (1995) employed MDS of sorting data for a range of cheeses (Roquefort, Cheddar, Muenster, Gouda, Edam, Monterey Jack, Havarti, Jarlsberg, Emmenthaler, Provolone, Boursin, Brie, Camembert and Feta). The 16 different cheese samples were presented together in a blind (coded) randomised fashion to two different groups of assessors, an experienced group very familiar with most of the cheeses to be tested and a consumer group who were much less familiar. Both groups were asked to taste the samples and sort them into at least two groups, but they could make more groups if they so wished. Lawless et al. (1995) concluded that the consumer group individuals were generally similar in their perceptual maps compared to the experienced group.

Blancher et al. (2007) in a cross-cultural study of jellies between France and Viet Nam compared a conventional profiling with flash profiling and free sorting and found that the flash profile was configurationally closer to the conventional profile than free sorting.

Free multiple sorting (FMS) as described by Dehlholm et al. (2012) appears to be an elaboration of the sorting technique. They state that the difference between 'sorting' and 'FMS' is that with sorting, assessors group the samples only once, whereas with FMS, multiple groupings can be made. FMS appears to be a development of the free sorting method described by Steinberg (1967). Dehlholm et al. (2012) presented nine liver pâté samples to two panels ($n = 13$ and $n = 10$) who were asked to sort the same set of samples multiple times until they had exhausted their own individual permutations. In each sorting, they could generate between two and eight groups containing one to eight samples. The average of the sortings per assessor were 8 for Panel A and 10 for Panel B, ranging between 3 and 12 per assessor for Panel A and 6 and 14 per assessor for Panel B. These authors conclude that the results produced by FMS were quite variable as the RV coefficients between panels FMS configurations showed only a 70% similarity (0.71) (Dehlholm et al., 2012).

The method has advantages over traditional profiling in that it is easy to perform using minimum training, it does not require any quantitative rating system, there is no forced agreement among panellists and the methods are less fatiguing and can handle large numbers of samples. It is also described as being an easy and enjoyable task for participants (Coxon, 1999). However, for more complex products, smaller numbers of samples should be assessed because of the potential for assessor fatigue and it is an easier technique for trained panellists compared to untrained (Bijmolt and Wedel, 1995; Cartier et al., 2006; Varela and Ares, 2012). Although sorting tasks can be applied to a large sample set, it is important to take into account that all samples should be presented simultaneously in a single session (Varela and Ares, 2012). Sorting also does not really stand alone as a technique as in order to interpret the data effectively, it must be combined with other sensory or instrumental data (Pagès, 2005).

PROJECTIVE MAPPING: NAPPING, PARTIAL NAPPING, SORTED NAPPING AND ULTRAFLASH PROFILING

Projective mapping or napping is a technique with its origins in market research (Risvik et al., 1994; Pagès, 2005) where presented samples are grouped by subjects on typically an A3, A4 or 60-cm^2 sheets (King et al., 1998; Kennedy and Heymann, 2009; Nestrud and Lawless, 2010).

The term napping originates from the French word nappe which means tablecloth, i.e., grouping samples on a bidimensional space like a tablecloth (Varela and Ares, 2012). In a single session the assessors taste and smell the products and then orientate them into groups on the paper as they wish. A control sample may also be inserted into the sample set to validate assessor groupings. Samples which are close are similar or correlated and those that are far apart are the opposite. Assessors can be trained ($n = 9-15$) (Risvik et al., 1997; Perrin et al., 2008; Varela and Ares, 2012) or untrained ($n = 15-50$) (Ares et al., 2010a,b; Albert et al., 2011; Kennedy and Heymann, 2009; Nestrud and Lawless, 2010; Varela and Ares, 2012). The coordinates on the individual assessor sheet groupings (A3, A4), where the samples are placed or marked, can then be inputted into a spreadsheet and can be combined for later multivariate data analysis to determine the specific sensory profile. Data are entered as position coordinates (x and y), with an origin that can be placed anywhere (Perrin et al., 2008).

A further development of napping is 'ultraflash profiling'. This is an addition to napping whereby after the specific grouping of samples by assessors a second task may be employed with napping whereby the subjects are asked to describe their samples or groupings of samples. This procedure is called 'ultraflash profiling' and adds a descriptive component to the mapping process (Dehlholm et al., 2012; Varela and Ares, 2012). Similar to this method is 'sorted napping' which involved the assessors basically categorising their product maps by circling groupings on their mapping sheets. Pagès et al. (2010) were the first to suggest this modification and presented 8 smoothies to 24 panellists and asked them to taste and smell the products and perform the classical napping procedure and group on the napping sheet. Then in a second step the assessors were asked to circle the groups (more than two) and to associate words with the groups.

Partial napping is another simplified derivative of napping whereby assessors perform the napping procedure on one or specific sensory modalities instead of the intuitive association of conventional projective mapping (Pfeiffer, and Gilbert, 2008; Grygorczyk et al., 2013). It can be very useful in determining information on discrimination between products in terms of attributes that are important but not easily explained or quantified or for market exploration and product ideation studies and for developing new vocabularies, especially in studies that require a certain degree of focus, e.g., on product texture (Dehlholm et al., 2012).

Napping has been applied to various food products and beverages such as brandy (Louw et al., 2015), beer (Reinbach et al., 2014), wines (Perrin and Pagès, 2009), coffee (Moussaoui and Varela, 2010), chocolate (Risvik et al., 1994), chocolate dairy desserts (Ares et al., 2010b), soups (Risvik et al., 1997), snack bars (King et al., 1998), citrus juices (Nestrud and Lawless, 2008), fish nuggets (Albert et al., 2011) and powdered drinks (Ares et al., 2011).

Napping is a very rapid user-friendly technique (Risvik et al., 1994), is easy to perform using minimum training, and like sorting does not require any quantitative rating systems and there is no forced agreement among panellists.

However, disadvantages of napping can include a limitation in the maximum number of samples that can be mapped because of adaptation or fatigue, 10−20 (Schifferstein, 1996), whereas Pagès (2005) states a maximum of 12 samples that can be presented in a single session. As in the case with sorting, napping does not really stand alone as a technique as in order to interpret the data effectively, it must be combined with other sensory or instrumental data (Pagès, 2005) or with a verbalisation task to better understand the perceptual dimensions (Moussaoui and Varela, 2010).

RAPID PROFILING USING REFERENCES: POLARISED SENSORY POSITIONING, THE OPTIMISED SENSORY PROFILE METHOD, OFF-FLAVOUR QUANTIFICATION, POLARISED PROJECTIVE MAPPING AND RANKED-SCALING

In PSP, samples are assessed by either trained or untrained assessors through the comparison of a sample (or set) to provided reference products. The method is convenient and rapid and because references are used, samples analysis does not have to be completed in a single session. First proposed by Teillet et al. (2010) to explore the sensory aspects of mineral water, assessors ($n = 15$) were asked to compare the similarity of samples to reference water samples (in this case represented by the commercial brands Evian, Volvic and Vittel) on unstructured line scales anchored on one end with the term 'Same Taste' and the other with 'Totally Different Taste'. Thus, how similar the sample was to each of the three references was determined with a line marked on these three different similarity scales. Multivariate data analysis can then be used to analyse the raw data collected and the samples can be correlated directly to their similarity to the reference products. A description phase should also be performed in order to gather information about the sensory characteristics responsible for the similarities and differences between products (Varela and Ares, 2012). Teillet et al. (2010) also state that references can also be considered as 'global' descriptors where the data are analysed by calculating average scores, and sample representation is obtained by principal component analysis (Teillet et al., 2010; Varela and Ares, 2012).

O'Sullivan et al. (2010, 2011) used a similar method to PSP for off-flavour quantification of gas-initiated flavour taints in modified atmosphere packed (MAP) beef steaks. In MAP, O_2 is used to maintain meat colour, CO_2 inhibits microbiological growth and N_2 to maintain desired pack shape (O'Sullivan et al., 2010, 2011). However, O_2 in packs can produce rancid off-flavours and

CO_2 can accumulate in MAP packed meat and manifest as CO_2 taint or off-flavour. In this study, product sensory references were solely provided to assist in identifying only the off-flavours developed in products. The semi-trained panel had experience of the other descriptors. The premise of this experiment was to investigate if a semi-trained panel could quantify off-flavour (O_2 and CO_2) in four MAP (0% O_2/100% CO_2, 50% O_2/20% CO_2/30% N_2, 70% O_2/30% CO_2 and 80% O_2/20% CO_2) packed beef steak treatments using appropriate products as sensory references and also to determine if taint could be dissipated from the meat if allowed to rest before cooking after immediately opening packs. Steaks were stored at 4°C for 15 days and tested for pH, colour, drip loss and cooking loss. Panellists were asked to evaluate the meat for tenderness, juiciness, CO_2 off-flavour and oxidative flavour in two separate sets of samples. One set cooked immediately on pack opening and the other set cooked after 30 min to facilitate the dissipation of the CO_2 flavour from the meat (O'Sullivan et al., 2010, 2011). The CO_2 flavour sensory descriptor reference sample was presented as a steak packed in 100 mL CO_2/100 mL pack gas (stored for the same time as the samples presented on the test day) and cooked immediately on pack opening and served on each test day to the panellists. It had a distinctively sour and acidic taste. Similarly, the sensory descriptor for oxidised flavour, described as rancid, cardboard or linseed oil like flavour was presented as a 100 mL O_2/100 mL gas packed steak sample (stored for the same time as the samples presented on the test day), cooked and served on each test day to the panellists (O'Sullivan et al., 2011). In effect the references provided on the given test day represented the maximum loading of the off-flavour in the same sample test samples. In this fashion, off-flavour development could be tracked over the course of the shelf life to an end point determined by the maximum safe microbiological threshold. Assessors were not given prior information about the coded test products. ANOVA-partial least squares regression was used to process the raw data accumulated from the 10 test subjects during the sensory evaluation and shelf life/retail display using instrumental methods (Fig. 4.1). The X-matrix was designated as 0/1 design variables for treatment and days of storage. The Y-matrix was designated as sensory and instrumental variables (O'Sullivan et al., 2010, 2011). The sensory protocol effectively quantified the taints as 'CO_2 flavour' increased over time for all treatments and also the most CO_2 off-flavours were found in the samples cooked immediately after opening MA packs and also in the meat packed under 100% CO_2 (O'Sullivan et al., 2010, 2011). These authors concluded that opening retail MAP packs up to 30 min prior to cooking has a benefit towards the perceived sensory quality of the cooked product. This is a potentially important finding for producers as the provision of package guidance to optimise flavour quality to the consumer could improve the overall sensory quality of the product.

Navarro da Silva et al. (2012) proposed a method called the 'ODP' (optimised sensory profile). Like the method described by O'Sullivan

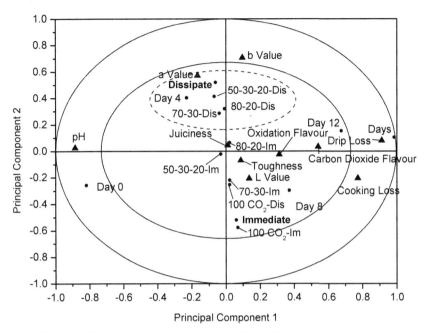

FIGURE 4.1 ANOVA-partial least squares regression (APLSR) correlation loadings plot for each of the four MAP treatment groups (0% O_2/100% CO_2, 50% O_2/20% CO_2/30% N_2, 70% O_2/30% CO_2 and 80% O_2/20% CO_2). Shown are the loadings of the X- and Y-variables for the first 2 PCs for ▲ = days and the individual MAP treatments, • = sensory descriptor and instrumental variables. Im (Immediate) = meat samples cooked immediately after opening of the MA packaging and presented to panellists. Dis (Dissipate) = meat samples left 30 min in ambient air to let CO_2 dissipate, then cooked and presented to panellists. The validated explained variance for the model constructed was 41% and the calibrated variance was 42.7%. *From O'Sullivan, M.G., Cruz, M. and Kerry, J.P., 2011. Evaluation of carbon dioxide flavour taint in modified atmosphere packed beef steaks. LWT-Food Science and Technology 44, 2193−2198.*

et al. (2010, 2011), reference materials were presented during evaluation of the products, but in this case for all the attributes and anchored with 'weak and strong' references and not just the maximum intensity of off-flavours as described in the method of O'Sullivan et al. (2011). This allows the semi-trained judges to compare the samples with these references and facilitates the allocation of attribute intensity on an unstructured scale (Navarro da Silva et al., 2012).

Polarised projective mapping is another technique recently described in the plethora of rapid descriptive methods. It is a hybrid method combining projective mapping and PSP where assessors are asked to locate a set of samples on a sheet of paper in which three reference samples have been previously located (Ares et al., 2013). Assessors taste the references and samples and then place each of the samples on the sheet according to the perceived similarities and differences between the sample and each of the references or poles. Again, like projective mapping, the samples which are placed close to each other are similar and those that are far apart are not similar. Then the assessors are asked to describe their groupings which they have made in a fashion very similar to ultraflash profiling. Ares et al. (2013) say that this method enables the results from different sessions to be compared as well as obtaining verbal descriptive information about the sensory characteristics of the products tested and not a just descriptive comparison between the samples and the reference products.

Pecore et al. (2015) utilised a method called ranked-scaling which involves the initial tasting by a previously trained panel of a set of specific spectrum intensity references, then the side-by-side presentation of all samples. The samples are tasted for the presence of the first attribute, and ranked in order of intensity for that attribute. The panellists then assign an intensity rating for that attribute for each sample and then progress through the rest of the attributes. The method is useful when studying subtle product differences and when striving for accuracy in correlating to instrumental measurements. However, it is potentially fatiguing thus Pecore et al. (2015) suggest the method be limited to six or fewer attributes, and for five or fewer samples.

CHECK-ALL-THAT-APPLY

CATA is a very easy rapid method that is used to gather consumer information by means of assessors answering multiple choice questions (MCQs) on a survey type questionnaire. Quite simply it is a method where a product is presented to the assessor, usually the consumer, and they are asked to either answer questions relating to the product (packaging, usage, hedonics, situation, etc.) or on the sensory attributes of the product. They answer the questions provided by ticking the box (or boxes) from a list of multiple terms provided which they decide are pertinent or fit the product. The method has applications in sensory science, particularly with consumer studies and has its origins in the market research realm (Rasinski et al., 1994). Generally, between 50 and 100 consumers are used in CATA surveys with samples presented monadically in the standard randomised fashion. Consumers then answer the provided questions by checking or ticking as many of the boxes provided as they deem fit the question or statement (Adams et al., 2007; Ares et al., 2010a,b; Dooley et al., 2010; Lado et al., 2010; Varela and Ares, 2012). The method is quick, but only provides qualitative data, large numbers of consumers are required and the technique is only as good as the questions provided. If very similar products are tested, then differentiation may not be achieved as the CATA questions may not provide suitable levels of separation (Dooley et al., 2010; Varela and Ares, 2012). Also, rotation of terms within a CATA question is required as consumers sometimes select terms based solely on their position on the list, biased to those at the top, therefore an unchecked term could still actually apply to the product (Krosnick, 1999; Castura,2009; Sudman and Bradburn, 1982).

TEMPORAL DOMINANCE OF SENSATIONS

TDS describes a method that measures the specific dynamic sensory sensation of food or beverage products while they are being perceived in the mouth during a specific time period. Food and beverage products are not unidimensional in that they do not only have one sensory dimension but are composed of many which change over the course of product consumption. TDS judges do not require lengthy training and several attributes can be evaluated over the evaluation period. TDS studies the sequence of dominant sensations of a product during a certain time period (Pineau et al., 2009). TDS consists of test subjects identifying and sometimes rating the intensity of sensations perceived as dominant until the perception ends and then selection of a new dominant attribute whenever they perceive a change in the dominant sensations (Di Monaco et al., 2014). Judges start the evaluation while putting samples into their mouths. Once the chronometer starts, they identify the sensations perceived as dominant and sometimes rate their intensities if required while performing the tasting protocol. The evaluation ends when judges can no longer perceive sensations, and stop the chronometer (Meillon et al., 2009). The statistical analysis of TDS data can be difficult and is restricted to the description of TDS curves (Lawless and Heymann, 2010). There were no valid statistical inference tests until 2010. Meyners and Pineau (2010) proposed a randomisation test based on distances between matrices. The TDS sequences are unfolded to data matrices with a single nonzero entry per time point (column). The sum of the Euclidean distances between these matrices is determined and serves as a test statistic for the global test (Meyners and Pineau, 2010). Growing usage of TDS by sensory analysts highlights certain limitations in TDS data acquisition and interpretation. To overcome these limitations, TimeSens, a web-based software dedicated to both TDS data acquisition and analysis has been developed. TimeSens runs as a plug-in web browser, requiring only an Internet access, and allowing data acquisition everywhere, even allowing interfacing with personal mobile phones (Visalli et al., 2011).

The study of temporal dominance is important in defining the specific flavour profile of many products of which cream liqueurs present an important example. The flavour of cream liqueurs involves a complex interaction of volatile and nonvolatile flavour compounds. These flavours originate from the liqueur components such as cream, cocoa, vanilla and whiskey. The polarity of these flavour compounds determines how they partition in the continuous phase of the cream liqueur emulsion, within the fat globule itself or across the globule aqueous phase interface (O'Sullivan, 2011). Fat has a significant effect on the partition of volatile compounds between the food and the air phases with lipophilic aroma compounds being the most affected (Bayarri et al., 2006). This flavour partitioning also affects flavour release and ultimately the flavour profile of the liqueur with more polar flavours perceived first and the more nonpolar perceived last perhaps even as aftertaste sensations (O'Sullivan, 2011). Thus, the TDS can be applied to specific branded products and can be useful in defining product characteristics and brand identity. One of the drawbacks of TDS is how to modify products once areas of optimisation have been identified. By attempting to subtly change a sensory dimension in the TDS spectrum, the flavour profile of the product might be knocked out of kilter. Additionally, an identified sensory dimension might have a synergistic or antagonistic effect on other sensory dimensions making modification more difficult. Thomas et al. (2015) presented a method for the assessment of the 'temporal drivers of liking' (TDL) to identify sensory attributes inducing a positive or negative trend of liking when they become dominant. The inclusion of hedonics is novel but it is yet to be seen how useful this test really is, as again by modifying a particular sensory dimension from identified trends of liking might result in unknown synergistic or antagonistic effects on other liking dimensions. For example, the flavour of cream liqueurs involves a complex interaction of volatile and nonvolatile flavour compounds, the polarity of which determines how they partition in the continuous phase of the cream liqueur emulsion, within the fat globule itself or across the globule aqueous phase interface (O'Sullivan, 2011). This flavour partitioning also affects flavour release and ultimately the flavour profile of the liqueur. With TDS analysis the more polar flavours are perceived first and the more nonpolar perceived last perhaps even as aftertaste sensations (O'Sullivan, 2011). One part of the TDS sequence can be a harsh alcohol flavour that in TDL could be described as a negative dimension. If a product developer was to use this as a basis for optimisation through reduction, this would fundamentally alter the balance of flavours across the entire flavour profile. This is because the flavour partitioning coefficients would change for all the polar and nonpolar flavour compounds resulting in a product with a flavour profile dramatically changed from the original. Therefore, it is important that users of the method recognise that a priori knowledge is essential in order to avoid artefacts.

JUST ABOUT RIGHT SCALES

Just about right (JAR) scales are used by sensory scientists and market researchers to measure the optimum level of a specific attribute (or attributes), by allowing the assessor to determine if an attribute is 'JAR'. They are often used in conjunction with liking (hedonic) and sensory intensity scales (Gacula et al., 2007). Usually the centre of these bipolar five or seven point category scales is labelled 'Just about Right' with the opposing anchor points labelled 'Much Too Little' and 'Much Too much' (Lawless and Hemann, 2010). Alternatively, unstructured line scaling can also be employed where the scales are anchored at one point with 'not nearly enough' at one end, 'much too much at the other' and 'JAR' in the centre (Anon, 2015). Only attributes that have an optimum can be used (attributes where more or less is always better are not suitable) and the attribute must not have a negative association (Lawless and Heymann, 2010). The acquired data can be used to assist in product optimisation and are, among others, often used in food research and development to determine the optimal level of a specific ingredient (López Osorino and Hough, 2010). Corporate marketing and R&D management appear to hold satisfactory opinions about the merits of the JAR scale, including its apparent directness and ease of use in the field at the time of data collection (Gacula et al., 2007).

A criticism of JAR scales is that there is a risk that the subject might misinterpret an attribute, or that the test is a too analytical in evaluation of the food or beverage than the consumer can handle. Also, it can be ambiguous whether the JAR level refers to the acceptance of the product or to a preference (Gacula et al., 2007). The inclusion of JAR scales can influence hedonic scores, thus introducing bias (Earthy et al., 1997). The Stuart—Maxwell frequency test and the McNemar test for JAR data are suitable statistical methods which can be used to test difference between products. Also, data penalty analysis can be used for data interpretation (Anon, 2015; Lawless and Heymann, 2010).

THE IDEAL PROFILE METHOD

As discussed previously, JAR scales can be used to determine the optimum level of an attribute intensity of a food or beverage product. A step forward from this is the IPM. IPM is a descriptive analysis technique in which consumers are asked to rate products on both their 'perceived' and 'ideal' intensities from a list of attributes. In addition, overall liking is asked. At the end of the test, a lot of information is obtained about the products and the consumer. Each consumer provides a sensory profile of the products, their hedonic ratings and their ideal profile (Worch et al., 2014; Worch and Ennis, 2013; Worch et al., 2013). In theory the method could be used to create ideal products for consumers, but the reliability of the data can be fragile. With this method, each consumer evaluates several products following standard randomisation and monadic presentation and rates products on both perceived and ideal intensity for a list of attributes. So, if the first question is: 'Please rate the sweetness of this product', the second question will be: 'Please rate your ideal sweetness for this product'. This methodology has been adopted with the aim to mimic the JAR scale, but using the perceived and ideal intensities instead of the difference with an imagined ideal. Additional hedonic questions are also asked for each tested product. The ideal profiles are directly actionable to guide for products' improvement. However, this particular information should be carefully managed since it is obtained from consumers and it describes virtual products (Worch and Ennis, 2013).

REFERENCES

Abdi, H., Valentin, D., Chollet, S., Chrea, C., Abdi, H., Valentin, D., et al., 2007. Analyzing assessors and products in sorting tasks: DISTATIS, theory and applications. Food Quality and Preference 18, 627–640.

Adams, J., Williams, A., Lancaster, B., Foley, M., 2007. Advantages and uses of check-all-that-apply response compared to traditional scaling of attributes for salty snacks. In: 7th Pangborn Sensory Science Symposium. Minneapolis, USA, 12– 16 August, 2007.

Albert, A., Varela, P., Salvador, A., Hough, G., Fiszman, S., 2011. Overcoming the issues in the sensory description of hot served food with a complex texture. Application of QDA®, flash profiling and projective mapping using panels with different degrees of training. Food Quality and Preference 22, 463–473.

Anon, 2015. Just about Right Scales. Society of Sensory Professionals. http://www.sensorysociety. org/knowledge/sspwiki/Pages/Just%20About%20Right%20Scales.aspx.

Ares, G., Barreiro, C., Deliza, R., Giménez, A., Gámbaro, A., 2010a. Application of a check-all-that-apply question to the development of chocolate milk desserts. Journal of Sensory Studies 25, 67–86.

Ares, G., Giménez, A., Barreiro, C., Gámbaro, A., 2010b. Use of an open-ended question to identify drivers of liking of milk desserts. Comparison with preference mapping techniques. Food Quality and Preference 21, 286–294.

Ares, G., Varela, P., Rado, G., Gimenez, A., 2011. Are consumer profiling techniques equivalent for some product categories? The case of orange-flavoured powdered drinks. International Journal of Food Science and Technology 46, 1600–1608.

Ares, G., de Saldamando, L., Vidal, L., Antúnez, L., Giménez, A., Varela, P., 2013. Polarized projective mapping: comparison with polarized sensory positioning approaches. Food Quality and Preference 28, 510–518

Bayarri, S., Taylor, A.J., Hort, J., 2006. The role of fat in flavor perception: effect of partition and viscosity in model emulsions. Journal of Agriculture and Food Chemistry 54, 8862–8868.

Bijmolt, T., Wedel, M., 1995. The effects of alternative methods of collecting similarity data for multidimensional scaling. International Journal of Research in Marketing 12, 363–371.

Blancher, G., Chollet, S., Kesteloot, R., Hoang, D.N., Cuvelier, G., Sieffermann, J.M., 2007. French and Vietnamese: how do they describe texture characteristics of the same food? A case study with jellies. Food Quality and Preference 18, 560–575.

Cartier, R., Rytz, A., Lecomte, A., Poblete, F., Krystlik, J., Belin, E., Martin, N., 2006. Sorting procedure as an alternative to quantitative descriptive analysis to obtain a product sensory map. Food Quality and Preference 17, 562–571.

Castura, J.C., 2009. Do panellists donkey vote in sensory choose-all-that-apply questions?. In: 8th Pangborn Sensory Science Symposium, July 26–30, Florence.

Cavanagh, D., Kilcawley, K.N., O'Sullivan, M.G., Fitzgerald, G.F., McAuliffe, O., 2014. Assessment of wild non-dairy lactococcal strains for flavour diversification in a mini Gouda type cheese model. Food Research International 62, 432–440.

Chollet, S., Valentin, D., 2001. Impact of training on beer flavour perception and description: are trained and untrained panelists really different? Journal of Sensory Studies 16, 601–618.

Coxon, A.P.M., 1999. Sorting Data: Collection and Analysis. Series: Quantitative Applications in the Social Sciences. Sage publications Inc., New York.

Dairou, V., Sieffermann, J.M., 2002. A comparison of 14 jams characterized by conventional profile and a quick original method, the Flash profile. Journal of Food Science 67 (2), 826–834.

Dehlholm, C., Brockhoff, P.B., Meinert, L., Aaslyng, M.D., Bredie, W.L.P., 2012. Rapid descriptive sensory methods — comparison of free multiple sorting, partial napping, napping, flash profiling and conventional profiling. Food Quality and Preference 26, 267—277.

Delahunty, C.M., McCord, A., O'Neill, E.E., Morrissey, P.A., 1997. Sensory characterisation of cooked hams by untrained consumers using free-choice profiling. Food Quality and Preference 8, 381—388.

Delarue, J., Sieffermann, J.M., 2004. Sensory mapping using Flash profile. Comparison with a conventional descriptive method for the evaluation of the flavour of fruit dairy products. Food Quality and Preference 15, 383—392.

Di Monaco, R., Su, C., Masi, P., Cavella, S., 2014. Temporal dominance of sensations: a review. Trends in Food Science & Technology 38, 104—112.

Dooley, L., Lee, Y.-S., Meullenet, J.-F., 2010. The application of check-all-that-apply (CATA) consumer profiling to preference mapping of vanilla ice cream and its comparison to classical external preference mapping. Food Quality and Preference 21, 394—401.

Earthy, P.J., Macfie, H.J.H., hedderley, D., 1997. Effect of question order on sensory perception and preference in central location trials. Journal of Sensory Studies 12, 215—237.

Falahee, M., MacRae, A., 1997. Perceptual variation among drinking waters: the reliability of sorting and ranking data for multidimensional scaling. Food Quality and Preference 8, 389—394.

Fellendorf, S., O'Sullivan, M.G., Kerry, J.P., 2015. Impact of varying salt and fat levels on the physiochemical properties and sensory quality of white pudding sausages. Meat Science 103, 75—82.

Fellendorf, S., O'Sullivan, M.G., Kerry, J.P., 2016a. The reduction of salt and fat levels in black pudding and the effects on physiochemical and sensory properties. International Journal of Food Science and Technology (Online).

Fellendorf, S., O'Sullivan, M.G., Kerry, J.P., 2016b. Effect of using replacers on the physico-chemical properties and sensory quality of low salt and low fat white puddings. European Food Research and Technology (Online).

Fellendorf, S., O'Sullivan, M.G., Kerry, J.P., 2016c. Impact of using replacers on the physico-chemical properties and sensory quality of reduced salt and fat black pudding. Meat Science 113, 17—25.

Fellendorf, S., O'Sullivan, M.G., Kerry, J.P., 2016d. Impact on the physicochemical and sensory properties of salt reduced corned beef formulated with and without the use of salt replacers. Meat Science (Submitted).

FSAI, 2011. Salt Reduction Programme (SRP) — 2011 to 2012, 85.

FSAI, 2014. Monitoring of Sodium and Potassium in Processed Foods Period: September 2003 to July 2014, 44.

Gacula, M., Rutenbeck, S., Pollack, L., Resurreccion, A.V.A., Moskowitz, H.R., 2007. The just about right intensity scale: functional analyses and relation to hedonics. Journal of Sensory Studies 22, 194—211.

Gawel, R., Iland, P.G., Francis, I.L., 2001. Characterizing the astringency of red wine: a case study. Food Quality of Preference 12, 83—94.

Grygorczyk, A., Lesschaeve, I., Corredig, M., Duizer, L., 2013. Impact of structure modification on texture of a soymilk and cow's milk gel assessed using the napping procedure. Journal of Texture Studies 44, 236—246.

Hempel, A., O'Sullivan, M.G., Papkovsky, D., Kerry, J.P., 2013a. Use of optical oxygen sensors to monitor residual oxygen in pre- and post-pasteurised bottled beer and its effect on sensory attributes and product acceptability during simulated commercial storage. LWT-Food Science and Technology 50, 226–231.

Hempel, A., O'Sullivan, M.G., Papkovsky, D., Kerry, J.P., 2013b. Non-destructive and continuous monitoring of oxygen levels in modified atmosphere packaged ready-to-eat mixed Salad products using optical oxygen sensors. Journal of Food Science 78, S1057–S1062.

Hempel, A., O'Sullivan, M.G., Papkovsky, D., Kerry, J.P., 2013c. Assessment and use of optical oxygen sensors as tools to assist in optimal product component selection for the development of packs of ready-to-eat mixed salads and for the non-destructive monitoring of in-pack oxygen levels using chilled storage. Foods 2, 213–224.

Hempel, A., O'Sullivan, M.G., Papkovsky, D., Kerry, J.P., 2013d. Use of smart packaging technologies for monitoring and extending the shelf-life quality of modified atmosphere packaged (MAP) bread: application of intelligent oxygen sensors and active ethanol emitters. European Food Research and Technology 237, 117–124.

Henneberry, S., O'Sullivan, M.G., Kilcawley, K.N., Kelly, P.M., Wilkinson, M.G., Guinee, T.P., 2016. Sensory quality of unheated and heated Mozzarella-style cheeses with different fat, salt and calcium levels. International Journal of Dairy Science 69, 38–50.

Heymann, H., 1994. A comparison of free choice profiling and multidimensional scaling of vanilla samples. Journal of Sensory Studies 9, 445–453.

Kennedy, J., Heymann, H., 2009. Projective mapping and descriptive analysis of milk and dark chocolates. Journal of Sensory Studies 24, 220–233.

Kilcast, D., 2000. Sensory evaluation methods for shelf-life assessment. In: Kilcast, D., Subramaniam, P. (Eds.), The Stability and Shelf-Life of Food. Woodhead Publishing Limited, Cambridge, UK, pp. 79–103 (Chapter 4).

King, M.C., Cliff, M.A., Hall, J.W., 1998. Comparison of projective mapping and sorting data collection and multivariate methodologies for identification of similarity-of-use of snack bars. Journal of Sensory Studies 13, 347–358.

Krosnick, J.A., 1999. Survey research. Annual Review of Psychology 50, 537–567.

Lado, J., Vicente, E., Manzzioni, A., Ares, G., 2010. Application of a check-all-that-apply question for the evaluation of strawberry cultivars from a breeding program. Journal of the Science of Food and Agriculture 90, 2268–2275.

Lassoued, N., Delarue, J., Launay, B., Michon, C., 2008. Baked product texture: correlations between instrumental and sensory characterization using flash profile. Journal of Cereal Science 48, 133–143.

Lawless, H.T., Heymann, H., 1999. Sensory Evaluation of Food: Principles and Practices. Kluwer Academic/Plenum Publishers, New York.

Lawless, H.T., Heymann, H., 2010. Sensory Evaluation of Food – Principles and Practices, second ed. Springer New York Dordrecht Heidelberg, London.

Lawless, H.T., Sheng, N., Knoops, S.S.C.P., 1995. Multidimensional-scaling of sorting data applied to cheese perception. Food Quality and Preference 6, 91–98.

Lawless, H.T., 1999. Descriptive analysis of complex odors: reality, model or illusion? Food Quality and Preference 10, 325–332.

Lelièvre, M., Chollet, S., Abdi, H., Valentin, D., 2008. What is the validity of the sorting task for describing beers? A study using trained and untrained assessors. Food Quality and Preference 19, 697–703.

Loescher, E., Sieffermann, J.M., Pinguet, C., Kesteloot, R., Cuvlier, G., 2001. Development of a List of Textural Attributes on Pear/apple Puree and Fresh Cheese: Adaptation of the Quantitative Descriptive Analysis Method and Use of Flash Profiling. 4th Pangborn, Dijon, France.

López Osorino, M.M., Hough, G., 2010. Comparing 3-point versus 9-point just-about-right-scales for determining the optimum concentration of sweetness in a beverage. Journal of Science 25 (Suppl. s1), 1–17.

Louw, L., Oelofse, S., Naes, T., Lambrechts, M., van Rensburg, P., Nieuwoudt, H., 2015. Optimisation of the partial napping approach for the successful capturing of mouthfeel differentiation between brandy products. Food Quality and Preference 41, 245–253.

Meilgaard, M.C., Civille, G.V., Carr, B.T., 1991. Sensory Evaluation Techniques. CRC Press, Inc., Boca Raton, FL.

Meillon, S., Urbano, C., Schlich, P., 2009. Contribution of the temporal dominance of sensations (TDS) method to the sensory description of subtle differences in partially dealcoholized red wines. Food Quality and Preference 20, 490–499.

Meyners, M., Pineau, N., 2010. Statistical inference for temporal dominance of sensations data using randomization tests. Food Quality and Preference 21, 805–814.

Moussaoui, K.A., Varela, P., 2010. Exploring consumer product profiling techniques and their linkage to a quantitative descriptive analysis. Food Quality and Preference 21, 1088–1099.

Murray, J.M., Delahunty, C.M., Baxter, I.A., 2001. Descriptive sensory analysis: past, present and future. Food Research International 34, 461–471.

Navarro da Silva, R.C.S., Rodrigues Minim, V.O., Simiqueli, A.A., da Silva Moraes, L.E., Gomide, A.I., Minim, L.A., 2012. Optimized Descriptive Profile: a rapid methodology for sensory description. Food Quality and Preference 24, 190–200.

Nestrud, M.A., Lawless, H.T., 2008. Perceptual mapping of citrus juices using projective mapping and pro ling data from culinary professionals and consumers. Food Quality and Preference 19, 431–438.

Nestrud, M.A., Lawless, H.T., 2010. Perceptual mapping of apples and cheeses using projective mapping and sorting. Journal of Sensory Studies 25, 390–405.

O'Callaghan, T., O'Sullivan, M.G., Kerry, J.P., Kilcawley, K.N., Stanton, C., 2016. Effects of feeding grass, clover or concentrate on the sensory and physicochemical quality of butter. International Journal of Dairy Science (submitted).

O'Sullivan, M.G., Cruz-Romero, M., Kerry, J.P., 2010. Carbon dioxide flavour taint in modified atmosphere packed lean beef. In: Fourth European Conference on Sensory and Consumer Research: A Sense of Quality, 5–8 Sept, Palacio Europa, Vitoria-gasteiz, Spain.

O'Sullivan, M.G., Kerry, J.P., Byrne, D.V., 2011. Use of sensory science as a practical commercial tool in the development of consumer-led processed meat products. In: Kerry, J.P., Kerry, J.F. (Eds.), Processed Meats. Woodhead Publishing Ltd., United Kingdom.

O'Sullivan, M.G., Cruz, M., Kerry, J.P., 2011. Evaluation of carbon dioxide flavour taint in modified atmosphere packed beef steaks. LWT-Food Science and Technology 44, 2193–2198.

O'Sullivan, M.G., 2011. Chapter 4, Sensory shelf-life evaluation. In: Piggott, J.R. (Ed.), Alcoholic Beverages: Sensory Evaluation and Consumer Research. Woodhead Publishing Limited, Cambridge, UK.

Pagès, J., Cadoret, M., Lê, S., 2010. The sorted napping: a new holistic approach in sensory evaluation. Journal of Sensory Studies 25, 637–658.

Pagès, J., 2003. Direct collection of sensory distances: application to the evaluation of ten white wines of the Loire Valley. Sciences des Aliments 23, 679–688.

Pagès, J., 2005. Collection and analysis of perceived product inter-distances using multiple factor analysis: application to the study of 10 white wines from the Loire Valley. Food Quality and Preference 16, 642–649.

Pecore, S., Kamerud, J., Holschuh, N., 2015. Ranked-Scaling: a new descriptive panel approach for rating small differences when using anchored intensity scales. Food Quality and Preference 40 (2015), 376–380.

Perrin, L., Pagès, J., 2009. Construction of a product space from the ultra-flash profiling method: application to 10 red wines from the Loire Valley. Journal of sensory Studies 24, 372–395.

Perrin, L., Symoneaux, R., Maître, I., Asselin, C., Jourjon, F., Pagès, J., 2008. Comparison of three sensory methods for use with the Napping procedure: case of ten wines from Loire Valley. Food Quality and Preference 19, 1–11.

Pfeiffer, J.C., Gilbert, C.C., 2008. Napping by modality: a happy medium between analytical and holistic approaches. In: Proceedings of Sensometrics Conference 2008, Canada.

Pineau, N., Cordelle, S., Schlich, P., 2003. Temporal dominance of sensations: a new technique to record several sensory attributes simultaneously over time. In: Abstract Book of 5th Pangborn Sensory Science Symposium, Boston, MA, USA, July 20–24, 2003. p. 121.

Popper, P., Heymann, H., 1996. Analyzing differences among products and panelists by multidimensional scaling. In: Naes, T., Risvik, E. (Eds.), Multivariate Analysis of Data in Sensory Science. Elsevier, Amsterdam, pp. 159–184.

Rasinski, K.A., Mingay, D., Bradburn, N.M., 1994. Do respondents really mark all that apply on self-administered questions? Public Opinion Quarterly 58, 400–408.

Rason, J., Lebecque, A., leger, L., Dufour, E., 2003. Delineation of the sensory characteristics of traditional dry sausage. I – typology of the traditional workshops in Massif Central. In: The 5th Pangborn Sensory Science Symposium, July 21–24, Boston, USA.

Reinbach, H.C., Giacalone, D., Machado Ribeiro, L., Bredie, W.L.P., Frøst, M.B., 2014. Comparison of three sensory profiling methods based on consumer perception: CATA, CATA with intensity and Napping®. Food Quality and Preference 32, 160–166.

Richter, V., Almeida, T., Prudencio, S., Benassi, M., 2010. Proposing a ranking descriptive sensory method. Food Quality and Preference 21 (6), 611–620. http://dx.doi.org/10.1016/j.foodqual 2010.03.011.

Risvik, E., McEwan, J.A., Colwill, J.S., Rogers, R., Lyon, D.H., 1994. Projective mapping: a tool for sensory analysis and consumer research. Food Quality and Preference 5, 263–269.

Risvik, E., McEwan, J.A., Rodbotten, M., 1997. Evaluation of sensory profiling and projective mapping data. Food Quality and Preference 8, 63–71.

Rodrigue, N., Guillet, M., Fortin, J., Martin, J.F., 2000. Comparing information obtained from ranking and descriptive tests of four sweet corn products. Food Quality and Preference 11, 47–54.

Rosenberg, S., Nelson, C., Vivekana, P.S., 1968. A multidimensional approach to structure of personality impressions. Journal of Personality and Social Psychology 9, 283.

Saint Eve, A., Paci-Kora, E., Martin, N., 2004. Impact of the olfactory quality and chemical complexity of the flavouring agent on the texture of low fat stirred yogurts assessed by three different sensory methodologies. Food Quality and Preference 15, 655–668.

Santosa, M., Abdi, H., Guinard, J.X., 2010. A modified sorting task to investigate consumer perceptions of extra virgin olive oils. Food Quality and Preference 21, 881–892.

Schifferstein, H.N.J., 1996. Cognitive factors affecting taste intensity judgments. Food Quality and Preference, Second Rose-Marie Pangborn Memorial Symposium 7 (3–4), 167–175.

Sieffermann, J.M., 2000. Le profil flash: un outil rapide et innovant d'évaluation sensoriel descriptive, Agoral 2000, 12ème rencontres, L'innovation: de l'idée au succés, 335–340.

Silva, R.C.S.N., Minim, V.P.R., Simiqueli, A.A., Moraes, L.E.S., Gomide, A.I., Minim, L.A., 2012. Optimized descriptive profile: a rapid methodology for sensory description. Food Quality and Preference 24, 190–200.

Silva, R.C.S.N., Minim, V.P.R., Carneiro, J.D.S., Nascimento, M., Della Lucia, S.M., Minim, L.A., 2013. Quantitative sensory description using the optimized descriptive profile: comparison with conventional and alternative methods for evaluation of chocolate. Food Quality and Preference 30, 169—179.

Silva, R.C.S.N., Minim, V.P.R., Silva, A.N., Peternelli, L.A., Minim, L.A., 2014. Optimized descriptive profile: how many judges are necessary? Food Quality and Preference 36, 3—11.

Steinberg, D.D., 1967. The Word Sort: an instrument for semantic analysis. Psychonomic Science 8, 541—542.

Stone, H., Sidel, J., 1993. Sensory Evaluation Practices. Academic Press, California.

Stone, H., Sidel, J.L., 2004. Affective testing. In: Stone, H., Sidel, J.L. (Eds.), Sensory Evaluation Practices. Food Science and Technology, International Series, third ed. Academic Press/Elsevier, USA, pp. 247—277.

Stone, H., Bleibaum, R.N., Thomas, H.A., 2012a. Affective testing. In: Stone, H., Bleibaum, R.N., Thomas, H. (Eds.), Sensory Evaluation Practices, fourth ed. Elsevier Academic Press, USA, pp. 291—325.

Stone, H., Bleibaum, R., Thomas, H., 2012b. Test strategy and design of experiments. In: Stone, H., Bleibaum, R.N., Thomas, H.A. (Eds.), Sensory Evaluation Practices, fourth ed. Elsevier Academic Press, USA, pp. 117—157.

Sudman, S., Bradburn, M.B., 1982. Asking Questions. Jossey-Bass, San Francisco.

Tang, C., Heymann, H., 1999. Multidimensional sorting, similarity scaling and free-choice profiling of grape jellies. Journal of Sensory Studies 17, 493—509.

Tarea, S., Cuvelier, G., Siefffermann, J.-M., 2007. Sensory evaluation of the texture of 49 commercial apple and pear purees. Journal of Food Quality 30, 1121—1131.

Teillet, E., Schlich, P., Urbano, C., Cordelle, S., Guichard, E., 2010. Sensory methodologies and the taste of water. Food Quality and Preference 21, 967—976.

Thomas, A., Visalli, M., Cordelle, S., Schlich, P., 2015. Temporal drivers of liking. Food Quality and Preference 40, 365—375.

Tobin, B.D., O'Sullivan, M.G., Hamill, R.M., Kerry, J.P., 2012a. Effect of varying salt and fat levels on the sensory quality of beef patties. Meat Science 4, 460—465.

Tobin, B.D., O'Sullivan, M.G., Hamill, R.M., Kerry, J.P., 2012b. Effect of varying salt and fat levels on the sensory and physiochemical quality of frankfurters. Meat Science 92, 659—666.

Tobin, B.D., O'Sullivan, M.G., Hamill, R.M., Kerry, J.P., 2013. The impact of salt and fat level variation on the physiochemical properties and sensory quality of pork breakfast sausages. Meat Science 93, 145—152.

Varela, P., Ares, G., 2012. Sensory profiling, the blurred line between sensory and consumer science. A review of novel methods for product characterization. Food Research International 48, 893—908.

Veinand, B., Godefroy, C., Adam, C., Delarue, J., 2011. Highlight of important product characteristics for consumers. Comparison of three sensory descriptive methods performed by consumers. Food Quality and Preference 22, 474—485.

Visalli, M., Monterymard, C., Duployer, G., Schlich, P., 2011. TimeSens, a web-based sensory software for temporal dominance of sensations. In: Pangborn Symposium, 4—8 September 2011, Toronto, Canada.

Williams, A.A., Langron, S.P., 1984. The use of free-choice profiling for the evaluation of commercial ports. Journal of the Science of Food and Agriculture 35, 558—568.

Worch, T., Ennis, J.M., 2013. Investigating the single ideal assumption using ideal profile method. Food Quality and Preference 29, 40—47.

Worch, T., Dooley, L., Meullenet, J.F., Punter, P., 2010a. Comparison of PLS dummy variables and fishbone method to determine optimal product characteristics from ideal profiles. Food Quality and Preference 21, 1077–1087.

Worch, T., Lê, S., Punter, P., 2010b. How reliable are the consumers? Comparison of sensory profiles from consumers and experts. Food Quality and Preference 21, 309–318.

Worch, T., Crine, A., Gruel, A., Lê, S., 2014. Analysis and validation of the ideal profile method: application to a skin cream study. Food Quality and Preference 32, 132–144.

Yarlagadda, A., Wilkinson, M.G., Ryan, S., Doolan, A.I., O'Sullivan, M.G., Kilcawley, K.N., 2014a. Utilisation of a cell free extract of lactic acid bacteria entrapped in yeast to enhance flavour development in Cheddar cheese. International Journal of Dairy Science Technology 67, 21–30.

Yarlagadda, A., Wilkinson, O'Sullivan, M.G., Kilcawley, K.N., 2014b. Utilisation of micro-fluidisation to enhance enzymatic and metabolic potential of lactococcal strains as adjuncts in Gouda type cheese. International Dairy Journal 38, 124–132.

Chapter 5

Multivariate Data Analysis for Product Development and Optimisation

INTRODUCTION

Data analysis is the key to unlocking the valuable information we gather as sensory and consumer scientists and is the essence of product development and optimisation. Consumer preferences can be correlated to descriptive sensory data as well as instrumental or physicochemical data to allow us better understand the products we are developing in order to optimise further. However, this process presents some challenges as depending on your perspective we are quite often trying to compare apples and oranges. On one end of the sensory spectrum we are using affective methods (described in Chapter 3) to determine consumer, or naïve assessor, information which is restricted to only measure the hedonic (liking) qualities of foods, beverages and objects (packaging). These data are acceptability or preference observations that do not describe the product, but are only subjective emotive responses or hedonic (liking). Consumers can be clear about which products they like and dislike, they are not always able to describe specifically why they like or dislike a product (Van Kleef et al., 2006).

On the other end of the sensory spectrum we are using descriptive analysis (described in Chapter 4) to quantify and accurately determine the sensory profile of products using trained panellists and are certainly not measuring hedonics. This presents a problem, how do we mesh these diametrically opposed types of data, the subjective and emotive to the objective and descriptive.

Traditionally this issue was solved by combining descriptive data and hedonic data using data analytical tools involving predominantly chemometrics in a method called preference mapping. From these statistically valid correlations could be determined from the data set so that sensory and product development scientists could understand how their modifications were affecting the sensory profile of the product with respect to the modalities, appearance, aroma, flavour, texture, taste and aftertaste as well as consumer opinion. Preference analyses techniques are able to relate external information

A Handbook for Sensory and Consumer-Driven New Product Development.
http://dx.doi.org/10.1016/B978-0-08-100352-7.00005-1

about perceived product characteristics to consumer preference ratings to understand what attributes of a product are driving preferences (Van Kleef et al., 2006).

Lastly, the statistical analysis of experimental data is not just solely about preference mapping and the analysis and interpretation of affective and descriptive sensory data. Other measurements such as physicochemical (compositional, pH, nutrient, etc.) as well as instrumental, texture profile analysis (TPA), gas chromatography/mass spectrometry (GC/MS) (flavour volatiles), electronic nose (volatile fingerprinting), HPLC (nonvolatile compounds) or FTIR (Fourier transfer infrared, spectral) type data can be analysed in a multivariate fashion with our sensory data to develop a greater understanding of the sensory, physical and chemical properties of food and beverage products to present reliable results. This is demonstrated in this chapter by way of a case study which shows how these various data streams can be analysed, and in this case, used to develop not a product but an instrument used for online at line analysis of food quality.

Modern multivariate data analytical software has come a long way in the last 10 years. Once the tool of the qualified chemometrician or multivariate statistician, analysis can now be undertaken reliably and with confidence by almost anybody with a little training. This user friendliness can only but increase the utilisation of multivariate data analysis in product development and optimisation projects and must contribute to improving the success profile of new products from inception to launch.

MULTIVARIATE DATA ANALYSIS

General Introduction to Bilinear Modelling

Bilinear modelling (BLM) is based on one single tool for mathematical extraction of relevant information from the input data, combined with extensive graphics and cross-validation (Martens and Martens, 2001). This soft modelling approach provides informative model maps that can be interpreted in light of background knowledge, with an eye for the unexpected. The statistical validation guards against wishful thinking and automatic outlier warnings alert the researcher to anomalies that otherwise could have been overlooked (Martens and Martens, 2001). Computerised data analysis allows the analysis of large data tables (Martens and Martens, 2001), and sensory profiling evaluation and profiling methodologies can generate a large amount of data. However, human cognitive capacity is limited. Therefore, the data analysis must be able to bring out the reliable and relevant information for us, without loss of essential dimensionality, while also not overwhelming us (Martens and Martens, 2001). The BLM methods presented here include principal component analysis (PCA), partial least squares regression (PLSR) and jack-knifing.

PRINCIPAL COMPONENT ANALYSIS

PCA is one of the most basic multivariate methods and can be used in the interpretation of large data matrices such as those produced through sensory evaluation. It is often used during descriptive panel training to assist in determining panel consensus as well as in the sensory term reduction process. It involves decomposing a data matrix X into a 'structure part' and a 'noise part'. The data matrix or data is the X matrix and is made up of objects and variables. The objects can be observations, samples, experiments, etc., while the variables could be for measurements within each experiment, attributes or whatever is related to the objects (Esbensen et al., 1996). The variables characterise the objects. In the context of the sensory analysis the objects are denoted by the samples and the variables are denoted by sensory attributes (O'Sullivan et al., 2002a,b,c). In PCA analysis the complete set of variables can be reduced to a smaller number of principal components (PCs) retaining the maximum amount of information, expressed by percentage of explained variance. PCs are estimated linear combinations of the original variables, explaining most of the variance, derived in increasing order of importance. PCA also reveals relationships between variables (loadings) and objects (scores). PCA identifies the main directions of variability in the multivariate space with the most important source of systematic variance set as the 'first PC' and then the second source of systematic variation set as the 'second PC'. These first two PCs generally reflect the main sources of structural variation in the data set. Additional PCs are identified in decreasing order of importance until the maximum amount of variation is extracted from the data set; however, these data become noisier as the number of PCs increases (Næs et al., 1996). The examining of the biplots produced by the PCA analysis gives an overview of the main information in the input data and reveals relationships between the variables (loadings) and samples (scores) (Esbensen et al., 1996).

How many PCs do I need to be able to effectively interpret my data? The answer to this crucial question can be determined by a very important process called cross-validation which is a statistical method that allows the determination of the number of significant components. In this technique, new samples or a subset of samples are repeatedly left out of the data and test the estimated model on the omitted samples. In this fashion an estimated model can be tested on samples that have not been a component of the estimation. This procedure is repeated until all the samples have been validated once and the results summarised as explained variance (Martens and Næs, 1989). A process called leverage correction may also be used for validation in PCA (Esbensen et al., 1996).

PCA is most used to assess the effectiveness of training of panellists (O'Sullivan et al., 2002c). PCA has been used in many studies to facilitate the removal of redundant sensory terms to investigate sensory characteristics of food products (Byrne et al., 1999, 2001). O'Sullivan et al. (2002b) used PCA and assessor suggestions for reducing the number of sensory terms to 21 terms during a 5-day training period developed to explore the main sensory dimensions of warmed-over flavour (WOF) in iron and vitamin E supplemented pork samples. The discriminative ability of the sensory panel improved over the course of sensory training and was quantified by using the mean assessor signal to noise ratios (S/N) for the sensory terms for each training session. Further detailed multivariate analysis found that the bilinear descriptor leverage was a particularly efficient method for term reduction.

NAÏVE ASSESSOR RELIABILITY

Computerised graphical displays may for many people simplify the understanding of statistical results (Martens and Martens, 2001). Certain techniques can be used to determine the reliability of naïve assessors for sensory assessment so that they in turn can be used in a reliable way. The S/N ratio is a common method employed to do just this (O'Sullivan et al., 2002a,b,c). It is a graphical way of presenting tables of P values. In the S/N plots the total variance (signal) is shown along the abscissa and the residual variance (noise) along the ordinate. Points which appear to the far right of the plot have high initial total variance (reliable elements) and points which appear higher up on the plot have a high estimated residual, nonsystematic variance (unreliable elements). Points on the diagonal line ($S/N = 1$) have estimated residual variances equal to initial total variance and are thus only noise (O'Sullivan et al., 2002c). Therefore, when assessors or replicates lie below this $S/N = 1$ line they show that signal is greater than noise. The proximity of replicates in this plot will also give an indication of the reliability of sensory evaluation.

Sensory profiling is a quantitative method by which a sensory panel utilises a vocabulary to describe perceived sensory characteristics in a product qualitatively and quantitatively (Meilgaard, 1999). The use of trained assessor and consumer (untrained) panels are common methods for the sensory evaluation of foods. O'Sullivan et al. (2002a) used a trained and an untrained panel for the sensory visual assessment of pork colour and showed that the trained panel was more reliable, but the untrained panel was effective in the evaluation of a familiar product. Byrne et al. (1999) and O'Sullivan et al. (2002b) developed sensory vocabularies for WOF in chicken meat and porcine meat respectively, during training of panellists prior to subsequent sensory profiling. Risvik et al. (1997) in a sensory study on seven blueberry soups found that a trained panel using sensory profiling and consumers using projective mapping perceived the samples in different ways, though the major dimensions were similar. These authors further postulated that while the trained panel was more discriminating, the results reiterate questioning the complete relevance of the trained panel results in relation to consumer preference. O'Sullivan et al. (2002c) investigated the reliability of untrained assessors for objective sensory evaluation and the validity of combining sensory data obtained from different groups of assessors using S/N plots and jack-knife estimated stability plots. The jack-knife estimated stability plots showed the stability of the X- and Y-variables' configuration against peculiarities in the individual assessor. Where both the X- and Y-variables display relatively small perturbations, relative distance from the mean value, it shows reliability. Inversely large perturbations signify greater variability. Generalised Procrustes analysis (GPA) may also be used for assessor reliability assessment and works in a similar manner to jack-knife estimated stability testing. O'Sullivan et al. (2002c) showed that the S/N and jack-knife estimated stability plots gave credence to combining the sensory data from three different assessor groups thus allowing the interpretation of the data as a combined assessor group. These authors concluded that the untrained groups of test subjects produced useful and reliable data and confirm that three panels of 15 naïve assessors each could give repeatable intersubjective description of the most dominant sensory variation dimensions. One potential application for such reliability testing is in the realm of consumer research. As in the above example, different groups of consumers may be compared in a consumer preference test and their reliability determined directly and in an objective manner.

GENERALISED PROCRUSTES ANALYSIS

GPA is a modification of the Procrustes analysis developed by Hurley and Cattell (1962) which was produced to match two different solutions from factor analysis. GPA is widely used in the analysis of sensory and consumer science data (Oreskovich et al., 1991; Dijkterhuis, 1995) and matches more than two data sets (Gower, 1975) where the data for each panellist are analysed separately, correcting through transformations, termed shifting, rotating, reflecting and stretching for individual differences in the use of the sensory scale and in the interpretation of the sensory attributes (Dijkterhuis, 1996; Arnold and Williams, 1986). The development of GPA as statistical tool (Gower, 1975) allowed the possibility of analysing the data coming from data sets which differed in the number of attributes per consumer and also having differences in the use of the scale (Varela and Ares, 2012). GPA is a very useful technique as it provides important information on products, terms and panellists and also has potential for use in sensory-instrumental relation (Dijkterhuis, 1994).

GPA has applications especially in analysis of free-choice profiling data, where different tasters are allowed to use their own vocabulary for charac-terising and scoring the samples (Arnold and Williams, 1986). Also, Napping is a technique where presented samples are grouped by subjects on typically an A3, A4 or 60 cm^2 sheets of paper (King et al., 1998; Kennedy and Heymann, 2009; Nestrud and Lawless, 2010). By placing products on a sheet, in a single session, assessors taste and smell the products and then orientate them into groups on the paper as they wish. A control sample may also be inserted into the sample set to validate assessor groupings. Samples which are close are similar or correlated and those that are far apart are the opposite. Each individual generates a two dimensional data matrix representing the coordinates of all the placed products. GPA is one of the most preferred statistical techniques which can be used to extract information about the tested products, which can be utilised for further product development or product optimisation (Tomic et al., 2015; Gower, 1975). What is obtained is essentially a 'consensus' or 'mean' product configuration across all individuals which gives important insight into the overall perception of the products. There are three steps in the GPA process for Napping and include translation, where all individual configurations are moved to the middle of the mapping sheet; rotation and reflection of individual configurations until they are in best possible agreement with one another; and finally isotropic scaling, where the data undergo shrinking or stretching of individual configurations until they are as alike as possible, but without changing the relative distances between the products in each configuration (Tomic et al., 2015).

PREFERENCE MAPPING

Preference mapping techniques involve statistically relating external information about perceived product characteristics to consumer preference or acceptance ratings to determine a product's characteristics and obtain a better understanding of what attributes of a product are driving preferences (Van Kleef et al., 2006; Meilgaard et al., 2007). For PLS preference mapping the Y-matrix could be the different consumer segment's preferences for a product and the X-matrix the sensory descriptive profile for the same products (Martens and Martens, 2007). Modern multivariate data analytical software packages have become very user friendly and readily facilitate the analysis of these different modes of data, the subjective hedonic determinations of consumers to demographic (age, gender, habitation) and more objective measurements such as sensory profiling data (trained descriptive panel data) and physicochemical analysis (rheological, compositional, volatile flavour (GC/MS), etc.) or least cost formulation (LCF) information. The plots produced through software programs such as Senstools, XLSTAT or Unscrambler (Camo, Norway) can provide graphical correlations of these data (PCA plots) as well as the underlying statistics. Consumers though clear in why they like or dislike products can be vague about the 'why' component of this hedonic determination. As consumers are not generally used to provide objective descriptive information, preference mapping using multivariate data analysis allows for the correlation of these data to the objective measurements already described. Thus, preference mapping methods can be very useful for consumer-driven product development and optimisation. A typical scenario might be the nutritional optimisation of a formulation by perhaps salt, sugar or fat reduction. The subset of samples selected for analysis must realistically and representatively reflect the product consumer space and truly span the sensory variation. This could utilise up to 12 products with a minimum of 6 (MacFie, 2007). This subset is selected from a complete set of samples which could represent all the competitor samples in the market place as well as the experimental prototypes. Consumer preference data can be mapped on to sensory descriptive and physicochemical data to determine which products are the most accepted, or more often, which products offer the best possible compromise with respect to consumer sensory quality, functionality, shelf life and even cost.

There are two types of preference analysis, internal and external. Internal preference analysis gives precedence to consumer preferences and uses perceptual information as a complementary source of information (Van Kleef et al., 2006). In effect the consumers' preference for the products (hedonic) is used to create a liking space upon which sensory descriptive ratings or other measurements are subsequently mapped. The consumer data are central to the space and these data form the core on to which the other information is projected. For external analysis, the opposite is the case, the sensory perceptual space is set by descriptive sensory data or instrumental data, and consumer preference information is subsequently overlaid onto this sensory space. It gives priority to perceptual information by building the product map based on attribute ratings and only fits consumer preferences at a later stage (Van Kleef et al., 2006). O'Sullivan et al. (2011a,b) presented data from external preference mapping of samples from each a holistic sensory-based QC data plus expert sensory description (X-matrix) in relation to liking of consumers ($n = 205$) of the samples (Y-matrix). The cluster of points indicates the direction of consumers' preferences for products.

The third method, sometimes called hybrid preference mapping, is based on the integrating of consumer and sensory data. PLS is commonly used to correlate consumer and descriptive data to determine the drivers of consumer preference. Internal or external preference mapping techniques extract PCs from the consumer data or descriptive data individually, whereas with PLS the consumer and descriptive data are considered together. The hedonic and descriptive scores are correlated to determine the sensory characteristics that are most important for consumer preference or acceptance (Meilgaard et al., 2007).

PARTIAL LEAST SQUARES REGRESSION

PLSR is the BLM (Bi-Linear Modelling) of two data tables X and Y. The process allows the extraction of the main variation patterns from one data table X that have relevance also for another data table Y from the same samples. This allows the interpretation of the structures within X and between Y (Martens and Martens, 2001).

PLS provides versatile, detailed overview and insight into very complex data structures, simplifies cross-disciplinary communication and thus facilitates easier interpretation than other methods. The method can handle 'missing data' works well in cases of far more variables than samples, which is often the case with data acquired through product development and optimisation studies (Martens and Martens, 2007). Similar to PCA the associations of the X and Y matrices are interpreted by numbers of PCs which explain the highest percentage of the total covariation. Therefore high explained variance represents systematic information whereas low explained variance can indicate random measurement and noise or error in the model. In general, for ANOVA partial least squares regression (APLSR), the X-matrix, can be designated as 0/1 design variables, e.g., assessor, replicate and product, etc. The Y-matrix can be designated as sensory, instrumental, chemical or physical variables (O'Sullivan et al., 2002a,b,c; O'Sullivan et al., 2003a,b,c,d). Additionally to the X-matrix being products or also potentially time-points (days), variables can be combined such as product X days. These are then modelled on to the Y-matrix, our measured data, the sensory, consumer, instrumental, physicochemical, GC/MS volatile data, etc. Additionally, 'all' these different modalities can be modelled simultaneously using predictive PLSR (PPLSR), as long as we standardise the data to account for different magnitude scales. This form of PLSR projects the response variables onto the design variables to determine to which degree each of the design variables in X contribute to the variation in the response variable Y (Martens and Martens, 2001). 0/1 design variables may also be placed in the Y-matrix and this application may be referred to as discriminant PLSR (DPLSR). The summary graphics of the PLSR model, as presented by multivariate software (Unscrambler, Camo, Norway), gives the analyst much better overview than stacks of tables or printouts from lots of different and separate univariate analysis outputs (Martens and Martens, 2007).

PPLSR may also be employed where sensory variables are put in X and chemical/instrumental data in Y. The prediction between sensory terms and instrumental and physical data can be explored using PLSR. The X-matrix can be set as sensory terms and the Y-matrix as the instrumental/physical measurements (O'Sullivan et al., 2003b,d). The goal is to determine what specifically in the X-matrix was relevant to describing and therefore predicting the Y-matrix.

Again validation techniques like cross-validation, the same as with PCA, are used to determine the number of PCs that are reliably interpretable and to determine the stability and explained variance of the model obtained. To derive significance indications for the relationships determined in the quantitative APLSR, DPLSR or PPLSR, regression coefficients can be analysed by jack-knifing which is based on cross-validation and stability plots (Martens and Martens, 1999, 2000, 2001). This allows the determination of the regression coefficients $\left(\hat{b}\right)$ with uncertainty limits that correspond to ± 2

standard uncertainties estimated by leave-one-replicate-out jack-knifing, i.e.,

$\hat{b} \pm 2\hat{s}(\hat{b})$ (Martens and Martens, 2001). From these, the significances

($P < .05$) of the variable relationships in the X- and Y-matrices can be determined, i.e., $\alpha \approx 0.05$, defined as the Type I probability that the observed effects could have been caused by random measurement errors (Martens and Martens, 2001).

L-PLSR allows PLS regression to connect three different data blocks (Sæbøa et al., 2008). In which case the X-matrix is defined as product characteristics, the Y-matrix is consumer liking data and the third matrix Z is consumer background and presented as an L-PLSR correlation loadings plot. Essentially we wish to determine the relationship between consumer liking and both consumer background characteristics and product descriptors (Martens and Martens, 2007).

PLSR has been used by a number of authors to investigate the sensory implications of many food types including fresh meat (Byrne et al., 2002; Yusop et al., 2010, chicken; O'Sullivan et al., 2002a, pork; Guerrero et al., 2015; O'Sullivan et al., 2011a,b, 2015, beef), processed meats (Tobin et al., 2012a,b; burgers; frankfurters; Tobin et al., 2013, sausages; Fellendorf et al., 2015, 2016b, white pudding; Fellendorf et al., 2016a,c,d, corned beef), cheeses (Cavanagh et al., 2014, Yarlagadda et al., 2014a, Gouda; Yarlagadda et al., 2014b, Cheddar), beer (Hempel et al., 2013a), bread (Hempel et al., 2013d), salads (Hempel et al., 2013b,c), soups (Michon et al., 2010a) and confectionary products (Michon et al., 2010b).

CASE STUDY – MULTIVARIATE DATA ANALYSIS OF SENSORY PROFILING AND INSTRUMENTAL DATA

BLM was used to determine the correlation of GC/MS and electronic nose data to sensory data in the presented case study. The sensory data were generated from cooked retail display-stored pork meat samples (Table 5.1) which had developed varying degrees of lipid oxidation, principally WOF (warmed-over flavor) after cooking. The principle aims of this study were (1) to use multivariate data analysis to correlate sensory and GC/MS analysis and to identify compounds that could be used as indices of lipid oxidation; (2) to use multivariate data analysis to assess the suitability of the electronic nose for the measurement of WOF development in different types of cooked pork meats; (3) to utilise multivariate data analysis to determine the reproducibility of electronic nose measurements from two separate data sets. Level correction was used to normalise the data from these two different electronic nose data sets (O'Sullivan et al., 2003b). An eight-member sensory was trained prior to sensory profiling of samples cooked in an oven under standardised conditions (Byrne et al., 1999; O'Sullivan et al., 2002a,b,c). Unstructured 15-cm line

TABLE 5.1 Sample Codes, Muscles, Supplemental Diets and Days of WOF[d]

Sample Code	Muscle	Vitamin E[a]	Iron[b]	Vitamin C[c]	Days of WOF[d]
LD-E-0-1,-2	M. longissimus dorsi	✔	✔	✔	0
LD-I-0-1,-2	M. longissimus dorsi	×	✔	✔	0
LD-IE-5-1,-2	M. longissimus dorsi	✔	✔	✔	5
PS-E-5-1,-2	M. psoas major	✔	×	✔	5
PS-I-5-1,-2	M. psoas major	×	✔	✔	5
PS-C-0-1,-2	M. psoas major	×	×	✔	0

-1, Sample set 1, analysed at The Royal Veterinary and Agricultural University, Copenhagen, Denmark; *-2*, Sample set 2, analysed at The *YTI Research Centre, P.O Box 181, FIN-50101, Mikkeli, Finland*, after 11 months frozen storage at −20°C. ✔ = present, × = absent.
[a]*200 mg dl-α-tocopheryl acetate/kg of feed.*
[b]*3000 mg iron (II) sulphate/kg feed.*
[c]*9 g/kg feed.*
[d]*Warmed-over flavour.*
From O'Sullivan, M.G., Byrne, D.V., Jensen, M.T., Andersen, H.J. Vestergaard, J., 2003b. A comparison of warmed-over flavour in pork by sensory analysis, GC/MS and the electronic nose. Meat Science 65, 1125–1138.

scales anchored on the left by the term 'none' and on the right by the term 'extreme' were used for all the sensory descriptors (Table 5.2) (Meilgaard et al., 1999). The responses of the panellists were recorded by measuring the distance in mm (1−150) from the left side of the scale for the odour, flavour, taste and aftertaste sensory terms as described in Table 5.2. For GC/MS analysis, solid-phase microextraction was used to measure volatiles from the same samples as used for sensory analysis. Electronic nose analysis (MGD-1, Environics Ltd., Finland) was conducted initially using the samples outlined in Table 5.1 (sample set 1). Eleven months later the experiment was repeated using similar samples to those presented in Table 5.1 (sample set 2) (O'Sullivan et al., 2003b). APLSR was used to analyse all sensory and instrumental data. Table 5.3 corresponds to compounds identified by GC/MS analysis of meat samples, Fig. 5.1 corresponds to data for sensory and GC data for *M. longissimus dorsi* and Fig. 5.2 for *M. psoas major*. Many of the compounds associated with oxidation of lipids were found to correlate with the oxidative sensory descriptors and the samples with the greater levels of WOF development, i.e., pentanal, 2-pentylfuran, octanal, nonanal, 1-octen-3-ol and hexanal. Additionally, the electronic nose device coupled with the multivariate methodology used in this experiment could clearly separate samples on the basis of muscle type, treatment and degree of WOF development (data not shown, see O'Sullivan et al., 2003b). The electronic nose data from sample sets 1 and 2

TABLE 5.2 Sensory Descriptive Terms and References Developed for Sensory Profiling

Sensory Term	Definition (With Appropriate Reference)
Odour	**Odour Reference**
1. Cardboard-O	Wet cardboard
2. Linseed oil-O	Warmed linseed oil/linseed oil-based paint
3. Rubber/sulphur-O	Warmed rubber/the white of a boiled egg
4. Nut-O	Crushed fresh hazel nuts
5. Green-O	Fresh green French beans
6. Fatty-O	Pig back fat (fresh, nonoxidised)
Taste	**Taste Reference**
7. Sweet-T	Sucrose 1 g/L aqueous solution
8. Salt-T	Sodium chloride 0.5 g/L aqueous solution
9. Sour-T	Citric acid monohydrate 0.3 g/L aqueous solution
10. Bitter-T	Quinine chloride 0.05 g/L aqueous solution
11. MSG/Umami-T	Monosodium glutamate 0.5 g/L aqueous solution
Flavour	**Flavour Reference**
12. Metallic-F/bloody-F	Ferrous sulphate 0.1 g/L aqueous solution
13. Fresh cooked pork-F	Oven cooked pork without browning
14. Rancid-F	Oxidised vegetable oil
15. Lactic acid/fresh sour-F	Natural yoghurt
16. Vegetable oil-F	Fresh vegetable oil
17. Piggy/Animal-F	Skatole 0.06 µg/mL refined vegetable oil
18. Fish-F	Fish stock in boiling water
19. Tinny-F	Stainless steel strip
20. Livery-F	Cooked beef liver
Aftertaste	**Aftertaste Reference**
21. Astringent-AT	Aluminium sulphate 0.02 g/L aqueous solution

Suffix to sensory terms indicates method of assessment by panellists; -AT, aftertaste; -O, odour; -F, flavour; -T, taste.
From O'Sullivan, M.G., Byrne, D.V., Jensen, M.T., Andersen, H.J. Vestergaard, J., 2003b. A comparison of warmed-over flavour in pork by sensory analysis, GC/MS and the electronic nose. Meat Science 65, 1125–1138.

TABLE 5.3 Gas Chromatography/Mass Spectrometry (GC/MS) Isolated Compounds

Compound	Kovats Value	Identification/Ions
Unknown	NA	43,45,47
Pentane	506	MS of AS
Pentanol	519	MS of AS
Dimethyl sulphide	573	MS of AS
Trimethylamine	586	MS of AS
Methyl-branched alkane	627	41,56,85
2-Methylpropanal	703	MS
2-Methylfuran	706	MS
Butanal	715	MS
2-Butanone	721	MS of AS
3-Methylhexane	725	MS
Benzene	731	MS
3-Methylbutanal	747	MS of AS
2-Ethylfurane	750	MS
Pentanal	782	MS of AS
Aliphatic alcohol	813	41,55,70
Aliphatic alcohol	824	41,55,70
Unknown	852	42,43,45
A octadiene, e.g., 1,3-octadiene	855	67,81,110
Hexanal	890	MS of AS
Nonane	905	MS of AS
Ethylbenzene	915	MS
A dimethylbenzene	925	MS
2-Butylfuran	938	MS
Butanoic acid	974	MS of AS
2-Heptanone	986	MS
Heptanal	991	MS of AS
4-Ethyl-1-octyn-3-ol	1032	MS
2-Pentylfuran	1036	MS of AS

TABLE 5.3 Gas Chromatography/Mass Spectrometry (GC/MS) Isolated Compounds—cont'd

Compound	Kovats Value	Identification/Ions
A methyl-branched alkane	1054	41,57,71
1-Octen-3-ol	1084	MS of AS
Benzaldehyde	1089	MS of AS
Octanal	1094	MS of AS
trans-2-Octenal	1179	MS of AS
Nonanal	1198	MS of AS

Ions, where the identification was ambiguous, the three most prominent ions in the mass spectra are given; *MS*, good fit with mass spectra in NIST library; *MS of AS*, fit with mass spectra of authentic standard; *NA*, not applicable.
From O'Sullivan, M.G., Byrne, D.V., Jensen, M.T., Andersen, H.J. Vestergaard, J., 2003b. A comparison of warmed-over flavour in pork by sensory analysis, GC/MS and the electronic nose. Meat Science 65, 1125–1138.

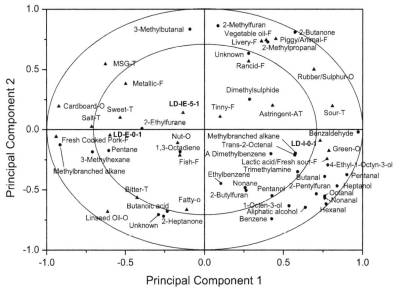

FIGURE 5.1 An overview of the variation found in the mean data from the ANOVA partial least squares regression correlation loadings plot for sample set 2, *M. longissimus dorsi* samples. Shown are the loadings of the *X*- and *Y*-variables for PC 1 versus 2. ▲ = sensory descriptor and sample, and • = GC/MS isolated compounds. The concentric circles represent 100% and 50% explained variance respectively. *From O'Sullivan, M.G., Byrne, D.V., Jensen, M.T., Andersen, H.J. Vestergaard, J., 2003b. A comparison of warmed-over flavour in pork by sensory analysis, GC/MS and the electronic nose. Meat Science 65, 1125–1138.*

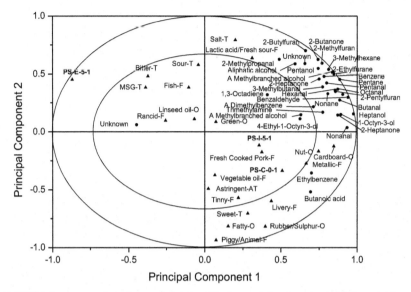

FIGURE 5.2   An overview of the variation found in the mean data from the ANOVA partial least squares regression correlation loadings plot for sample set 2, *M. psoas major* samples. Shown are the loadings of the *X*- and *Y*-variables for PC 1 versus 2. ▲ = sensory descriptor and sample, and ● = GC/MS isolated compounds. The concentric circles represent 100% and 50% explained variance respectively. *From O'Sullivan, M.G., Byrne, D.V., Jensen, M.T., Andersen, H.J. Vestergaard, J., 2003b. A comparison of warmed-over flavour in pork by sensory analysis, GC/MS and the electronic nose. Meat Science 65, 1125–1138.*

agreed with and correlated to sensory analysis. The electronic nose was effective in the determination of the oxidative state of the experimental samples and could be reproduced using similar samples measured using the same instrument, but in two separate laboratories and with a time separation of 11 months (O'Sullivan et al., 2003b). This case study demonstrates the versatility of multivariate data analysis in the analysis, validation and interpretation of different sensory and instrumental data streams (O'Sullivan et al., 2003b).

REFERENCES

Arnold, G.M., Williams, A.A., 1986. The use of generalised procrustes analysis in sensory analysis. In: Pigott, J.R. (Ed.), Statistical Procedures in Food Research. Elsevier, London, pp. 233–253.

Byrne, D.V., Bak, L.S., Bredie, W.L.P., Bertelsen, G., Martens, M., 1999. Development of a sensory vocabulary for warmed-over flavour 1: in porcine meat. Journal of Sensory Studies 14, 47–65.

Byrne, D.V., Bredie, W.L.P., Mottram, D.S., Martens, M., 2002. Sensory and chemical investigations on the effect of oven cooking on warmed-over flavour in chicken meat. Meat Science, 61, 127–139.

Byrne, D.V., O'Sullivan, M.G., Dijksterhuis, G.B., Bredie, W.L.P., Martens, M., 2001. Sensory panel consistency during development of a vocabulary for warmed-over flavour. Food Quality and Preference 12, 171—187.

Cavanagh, D., Kilcawley, K.N., O'Sullivan, M.G., Fitzgerald, G.F., McAuliffe, O., 2014. Assessment of wild non-dairy lactococcal strains for flavour diversification in a mini Gouda type cheese model. Food Research International 62, 432—440.

Dijkterhuis, G.B., 1994. Procrustes analysis in sensory-instrumental relations. Food Quality and Preference 5, 115—120.

Dijkterhuis, G.B., 1995. Multivariate data analysis in sensory and consumer science: an overview of developments. Trends in Food Science and Technology 6, 206—211.

Dijkterhuis, G.B., 1996. Procrustes analysis in sensory research. In: Næs, T., Risvik, E. (Eds.), Multivariate Analysis of Data in Sensory Science. Elsevier, Amsterdam, NL, pp. 185—219.

Esbensen, Midtgaard, Schönkopf, 1996. Multivariate Analysis in Practice. Wennbergs Trykker AS, Norway.

Fellendorf, S., O'Sullivan, M.G., Kerry, J.P., 2015. Impact of varying salt and fat levels on the physiochemical properties and sensory quality of white pudding sausages. Meat Science 103, 75—82.

Fellendorf, S., O'Sullivan, M.G., Kerry, J.P., 2016a. The reduction of salt and fat levels in black pudding and the effects on physiochemical and sensory properties. International Journal of Food Science and Technology (Online).

Fellendorf, S., O'Sullivan, M.G., Kerry, J.P., 2016b. Effect of using replacers on the physico-chemical properties and sensory quality of low salt and low fat white puddings. European Food Research and Technology (Online).

Fellendorf, S., O'Sullivan, M.G., Kerry, J.P., 2016c. Impact of using replacers on the physico-chemical properties and sensory quality of reduced salt and fat black pudding. Meat Science 113, 17—25.

Fellendorf, S., O'Sullivan, M.G., Kerry, J.P., 2016d. Impact on the physicochemical and sensory properties of salt reduced corned beef formulated with and without the use of salt replacers. Meat Science (submitted for publication).

Gower, J.C., 1975. Generalised procrustes analysis. Psychometrika 40, 33—50.

Guerrero, P., O'Sullivan, M.G., Kerry, J.P., de la Caba, K., 2015. Application of soy protein coatings and their effect on the quality and shelf-life stability of beef patties. Royal Society of Chemistry Advances 5, 8182—8189.

Hempel, A., O'Sullivan, M.G., Papkovsky, D., Kerry, J.P., 2013a. Use of optical oxygen sensors to monitor residual oxygen in pre- and post-pasteurised bottled beer and its effect on sensory attributes and product acceptability during simulated commercial storage. LWT-Food Science and Technology 50, 226—231.

Hempel, A., O'Sullivan, M.G., Papkovsky, D., Kerry, J.P., 2013b. Non-destructive and continuous monitoring of oxygen levels in modified atmosphere packaged ready-to-eat mixed salad products using optical oxygen sensors. Journal of Food Science 78, S1057—S1062.

Hempel, A., O'Sullivan, M.G., Papkovsky, D., Kerry, J.P., 2013c. Assessment and use of optical oxygen sensors as tools to assist in optimal product component selection for the development of packs of ready-to-eat mixed salads and for the non-destructive monitoring of in-pack oxygen levels using chilled storage. Foods 2, 213—224.

Hempel, A., O'Sullivan, M.G., Papkovsky, D., Kerry, J.P., 2013d. Use of smart packaging technologies for monitoring and extending the shelf-life quality of modified atmosphere packaged (MAP) bread: application of intelligent oxygen sensors and active ethanol emitters. European Food Research and Technology 237, 117—124.

Hurley, J.R., Cattell, R.B., 1962. The procrustes program: producing direct rotation to test a hypothesised factor structure. Behavioral Science 7 (2), 258–262.

Kennedy, J., Heymann, H., 2009. Projective mapping and descriptive analysis of milk and dark chocolates. Journal of Sensory Studies 24, 220–233.

King, M.C., Cliff, M.A., Hall, J.W., 1998. Comparison of projective mapping and sorting data collection and multivariate methodologies for identification of similarity-of-use of snack bars. Journal of Sensory Studies 13, 347–358.

MacFie, H., 2007. Preference mapping and food product development. In: MacFie, H. (Ed.), Consumer-led Product Development. Woodhead Publishing, Limited, Cambridge England.

Martens, H., Martens, M., 1999. Validation of PLS Regression models in sensory science by extended cross-validation. In: Tenenhause, M., Monineau, A. (Eds.), Les Methodes PLS. CISIA-CERESTA, France, pp. 149–182.

Martens, H., Martens, M., 2000. Modified Jack-knife estimation of parameter uncertainty in bilinear modelling by partial least squares regression (PLSR). Food Quality and Preference 11, 5–16.

Martens, H., Martens, M, 2001. In Multivariate Analysis of Quality. An Introduction. J. Wiley and Sons Ltd, Chichester, pp. 139–145 (Chapter 8).

Martens, H., Næs, T., 1989. Multivariate Calibration. John Wiley and Sons Ltd., London.

Martens, H., Martens, M., 2007. Chapter 21. The use of partial least squares methods in new food product development. In: Macfie, H. (Ed.), Consumer-led Food Product Development. Woodhead, Cambridge, UK, pp. 492–523.

Meilgaard, M.C., Civille, G.V., Carr, B.T., 1999. Sensory Evaluation Techniques, third ed. Academic Press, Boca Raton, Florida.

Meilgaard, M.C., Civille, G.V., Carr, B.T., 2007. Sensory Evaluation Techniques. CRC Press, Boca Raton, Florida.

Michon, C., O'Sullivan, M.G., Delahunty, C.M., Kerry, J.P., 2010a. Study on the influence of age, gender and familiarity with the product on the acceptance of vegetable soups. Food Quality and Preference 21, 478–488.

Michon, C., O'Sullivan, M.G., Sheehan, E., Delahunty, C.M., Kerry, J.P., 2010b. Investigation of the influence of age, gender and consumption habits on the liking for jam-filled cakes. Food Quality and Preference 21, 553–561.

Næs, T., Baardseth, P., Helgesen, H., Isaksson, T., 1996. Multivariate techniques in the analysis of meat quality. Meat Science 43, S111–S123.

Nestrud, M.A., Lawless, H.T., 2010. Perceptual mapping of apples and cheeses using projective mapping and sorting. Journal of Sensory Studies 25, 390–405.

O'Sullivan, M.G., Byrne, D.V., Stagsted, J., Andersen, H.J., Martens, M., 2002a. Sensory colour assessment of fresh meat from pigs supplemented with iron and vitamin E. Meat Science 60, 253–265.

O'Sullivan, M.G., Byrne, D.V., Martens, M., 2002b. Data analytical methodologies in the development of a vocabulary for evaluation of meat quality. Journal of Sensory Studies 17, 539–558.

O'Sullivan, M.G., Martens, H., Boberg Baech, S., Kristensen, L., Martens, M., 2002c. The reliability of naïve assessors in sensory evaluation visualised by pragmatical multivariate analysis. Journal of Food Quality 25, 395–408.

O'Sullivan, M.G., Kerry, J.P., Byrne, D.V., 2011a. Use of sensory science as a practical commercial tool in the development of consumer-led processed meat products. In: Kerry, J.P., Kerry, J.F. (Eds.), Processed Meats. Woodhead Publishing Ltd., United Kingdom.

O'Sullivan, M.G., Le Floch, S., Kerry, J.P., 2015. Meat preparation prior to cooking and consumer quality in modified atmosphere packed beef steak. Meat Science 101, 13–18.

O'Sullivan, M.G., Cruz, M., Kerry, J.P., 2011b. Evaluation of carbon dioxide flavour taint in modified atmosphere packed beef steaks. LWT-Food Science and Technology 44, 2193–2198.

O'Sullivan, M.G., Byrne, D.V., Nielsen, J.H., Andersen, H.J., Martens, M., 2003a. Sensory and chemical assessment of pork supplemented with iron and vitamin E. Meat Science 64, 175–189.

O'Sullivan, M.G., Byrne, D.V., Jensen, M.T., Andersen, H.J., Vestergaard, J., 2003b. A comparison of warmed-over flavour in pork by sensory analysis, GC/MS and the electronic nose. Meat Science 65, 1125–1138.

O'Sullivan, M.G., Byrne, D.V., Martens, M., 2003c. Evaluation of pork colour: sensory colour assessment using trained and untrained sensory panellists. Meat Science 63, 119–129.

O'Sullivan, M.G., Byrne, D.V., Martens, H., Gidskehaug, L.H., Andersen, H.J., Martens, M., 2003d. Evaluation of pork meat colour: prediction of visual sensory quality of meat from instrumental and computer vision methods of colour analysis. Meat Science 65, 909–918.

Oreskovich, D.C., Klein, B.P., Sutherland, J.W., 1991. Procrustes analysis and its applications to free choice and other sensory profiling. In: Lawless, H.T., Klein, B.P. (Eds.), Sensory Science Theory and Applications in Foods. Marcel Dekker, New York, pp. 353–394.

Risvik, E., McEwan, J.A., Rødbotten, M., 1997. Evaluation of sensory profiling and projective mapping data. Food Quality and Preference 8, 63–71.

Sæbøa, S., Almøya, T., Flatbergb, A., Aastveita, A.H., Martens, H., 2008. LPLS-regression: a method for prediction and classification under the influence of background information on predictor variables. Chemometrics and Intelligent Laboratory Systems 91, 121–132.

Tomic, O., Berget, I., Næs, T., 2015. A comparison of generalised procrustes analysis and multiple factor analysis for projective mapping data. Food Quality and Preference 43, 34–46.

Tobin, B.D., O'Sullivan, M.G., Hamill, R.M., Kerry, J.P., 2012a. Effect of varying salt and fat levels on the sensory quality of beef patties. Meat Science 4, 460–465.

Tobin, B.D., O'Sullivan, M.G., Hamill, R.M., Kerry, J.P., 2012b. Effect of varying salt and fat levels on the sensory and physiochemical quality of frankfurters. Meat Science 92, 659–666.

Tobin, B.D., O'Sullivan, M.G., Hamill, R.M., Kerry, J.P., 2013. The impact of salt and fat level variation on the physiochemical properties and sensory quality of pork breakfast sausages. Meat Science 93, 145–152.

van Kleef, E., Van Trijp, H.C.M., Luning, P., 2006. Internal versus external preference analysis: an exploratory study on end-user evaluation. Food Quality and Preference 17, 387–399.

Varela, P., Ares, G., 2012. Sensory profiling, the blurred line between sensory and consumer science. A review of novel methods for product characterization. Food Research International 48, 893–908.

Yarlagadda, A., Wilkinson, M.G., Ryan, S., Doolan, A.I., O'Sullivan, M.G., Kilcawley, K.N., 2014. Utilisation of a cell free extract of lactic acid bacteria entrapped in yeast to enhance flavour development in Cheddar cheese. International Journal of Dairy Science Technology 67, 21–30.

Yarlagadda, A., Wilkinson, M.G., O'Sullivan, M.G., Kilcawley, K.N., 2014. Utilisation of microfluidisation to enhance enzymatic and metabolic potential of lactococcal strains as adjuncts in Gouda type cheese. International Dairy Journal 38, 124–132.

Yusop, S.M., O'Sullivan, M.G., Kerry, J.F., Kerry, J.P., 2010. Effect of marinating time and low pH on marinade performance and sensory acceptability of poultry meat. Meat Science 85, 657–663.

Part II

Product Quality, Development and Optimisation

Chapter 6

Shelf Life and Sensory Quality of Foods and Beverages

INTRODUCTION

Food and beverage products have a defined shelf life that is governed by firstly microbiological safety and then product sensory quality including, in some circumstances, also rheological properties. According to Walker (1994), prior to determining the shelf life of a food, it is essential to determine which factor(s) limit the shelf life, which may cause physical, chemical or biological changes that result in a sensory change in the food. All products must be safe for human consumption as a default, but the sensory properties define quality which has a direct impact on consumer purchases. When a product comes to the end of its shelf life, undesirable characteristics may develop which can be microbiological, chemical or physical. Shelf life is defined in European legislation as the 'date of minimum durability' (FSAI, 2014a). The Codex Alimentarius defines shelf life as the period during which a food product maintains its microbiological safety and suitability at a specified storage temperature and where appropriate, under specified storage and handling conditions. Also, the UK Institute of Food Science and Technology (IFST) defines shelf life as 'the period of time during which the food product will remain safe, retain its desired sensory, chemical and microbiological characteristics, and comply with any label declaration of nutrition data' (IFST, 1993).

Microbiological safety can be determined by manufacturers by appropriately testing sufficient quantities of the product held at its normal storage temperature, which in many cases is ambient temperature, or it could be refrigerated or freezer temperatures for products requiring cold or frozen storage. Samples tested should reflect the normal batch to batch variation of the product. This variation should include variation to component ingredients, within specification fluctuations in the manufacturing process, and the sample should be contained in its specified packaging. Inadequate shelf life determination will lead to consumer dissatisfaction or complaints which will eventually affect the acceptance and sales of brand name products and at worst can lead to malnutrition or illness. For these reasons, food processors pay great attention to adequate storage stability and shelf life (Freitas and Costa, 2006).

A Handbook for Sensory and Consumer-Driven New Product Development.
http://dx.doi.org/10.1016/B978-0-08-100352-7.00006-3

Sensory analysis should also be performed in parallel with microbiological shelf life analysis to monitor the sensory profile of the product for potential deleterious sensory attribute changes. It is also important to note that the sensory analysis is only conducted on the product if it is deemed safe for human consumption by reference to the latent available microbiological data. The latter being the main information used for shelf life duration assignment. Sensory variables used during sensory shelf life testing could include the monitoring of specific sensory attributes which can be used as indices of sensory quality. These attributes may include the loss of particular fresh sensory notes in the case of bread and beer and the development of oxidised sensory notes in the case of beers or products containing fats such as meats or dairy products. The level of oxidation may also be measured accurately by chemical testing which is often done for foods containing polyunsaturated fats. A specific level of chemical indicator like malondialdehyde in the case of animal or plant fats or E-2-nonanal for beer can help determine oxidation end points.

For new products, the determination of product shelf life usually starts at the product development stage. However, irrespective of the stage a food product is at in its development, it is important to ensure that shelf life is considered at each stage and determined accurately using all available data (FSAI, 2014a). Product development involves defining the new product concept, followed by pilot scale production. The product is then initially assessed through affective sensory testing such as focus groups, sensory acceptance and preference tests using consumers. Descriptive sensory analysis may also be undertaken and preference mapping utilised to explore the sensory dimensions in conjunction with affective sensory data. Microbial evaluations should be included at the earliest possible stage, once the appropriate stage gate go decision indices have been met to estimate product shelf life. Consumer sensory evaluation or preference studies should also be undertaken to determine the commercial impact of the new product prior to the production of a finished product on a commercial scale for consumer purchase.

The procedure leading to the definition of the acceptance limit by the consumer strictly depends on the nature of the deteriorative reaction most limiting food quality. For all microbiologically stable foods with no health specific claim, sensory evaluation is the key factor for determining the acceptance limit, and hence shelf life (Manzocco and Lagazio, 2009). This is particularly the case with most alcoholic beverages due to the antimicrobial properties of ethanol, where spoilage or pathogen growth is prevented. Also, foods with very low water activities, which prevent microbial growth, such as biscuits or crackers will have shelf lives defined by sensory changes. These will usually involve textural changes such as softening or even hardening and will also depend on how the product is packaged and its exposure to temperature, atmospheric gasses and humidity.

It is important to note that shelf life can often be prolonged through packaging optimisation, and when a new market opens to an existing product, manufacturers must determine whether there is likely to be a change in storage conditions. The use of combined 'hurdle' strategies to prolong shelf life can also be implemented. This might include an antimicrobial process like a heat treatment such as retorting or high pressure processing in combination with an organic acid to synergistically retard microbial growth (Rodríguez-Calleja et al., 2012). A product made for ambient European retail conditions may not survive the prevailing hot or humid environment of a shipping container aboard a ship bound for Africa, Asia or South America. Also, the product may ultimately end up in air-conditioned supermarket conditions, but in transit could spend periods of time in unconditioned docksides, warehouses and depots. These likely storage or shipping conditions can be simulated using Binder type incubation cabinets (Fig. 6.1)

FIGURE 6.1   Diplays a Binder (BINDER Gmbh, Tuttlingen, Germany) climatic cabinet (KBF 720) which are used for accelerated storage testing. Temperature, humidity and light parameters can be controlled with ranges from −10 to 100°C without humidity and lighting, 10−90°C with humidity and 20−90°C with humidity and lighting. Photo stability testing can be undertaken with the doors incorporating Lux 4000 lighting. In the system presented, a remote data recording system has been incorporated. These wireless recorders communicate by radio frequency with the server PC at defined regular intervals and record many different parameters such as temperature and humidity in this case, but can also record additional measurements, depending on the application such as hygrometry measurements or pressure if required (O'Sullivan, 2011b, 2016).

to stress test the product in package during product optimisation and performing the usual battery of microbiological, sensory, physicochemical and rheological tests. Also, as a final component of product testing in the market place producers could include data loggers with suitable shipments. These devices can accumulate data on temperature and humidity of the prevailing environment the product encounters where seasonal variation should also be factored in. Only then can the producer determine if the existing format is suitable or requires optimisation.

Colour is often the first sensory modality the consumer encounters. This could be either without any package such as in the fishmonger's or butcher's retail cabinet, through transparent packaging or when the pack is first opened. Colour may be relatively stable for most products but where changes occur over shelf life they can be limiting factors for consumer purchase. Such changes are important for some foods such as fresh meat. One of the primary purchase criterions of fresh meat is meat colour. In red meats, consumers relate the bright-red colour to freshness, while discriminating against meat that has turned brown in colour (Hood and Riordan, 1973; Morrissey et al., 1994). This is more an issue with red meat due to the greater concentration of myoglobin present in the muscle compared to chicken, but poultry such as turkey, duck and ostrich have higher concentrations of myoglobin which can also make them susceptible to discolouration during retail display. Also, products such as dairy-based powders or butters and margarines can darken or become more yellow if oxidation occurs. Colour is easily monitored during shelf life testing using either simple sensory hedonic or descriptive tests for colour or by using portable measurement instruments such as the Minolta colorimeter which can measure the CIE or Hunter Lab colour measurement indices.

Rheological changes in foods and beverages can also be factors that require monitoring and will have an influence on shelf life determination. The texture of bread, confectionary, crackers and biscuits may become either softer or harder depending on composition, packaging and storage humidity and temperature. Also, a limitation to cream liqueur shelf life can be an increase in viscosity, phase separation and plugging of the neck of product bottles. These changes can again be monitored through sensory hedonic or descriptive analysis or using instruments such as the texture profile analyser for determining the texture of meat, bread, crackers or devices like the Brookfield rheometer for measuring the viscosity of liquids.

Comprehensive shelf life analysis essentially consists of storing a single large batch of product under normal conditions and to test, using suitable methods at various storage times (Lawless and Heymann, 2010). This chapter will discuss the microbiological, sensory methods as well as colour, flavour and textural changes and off-flavour development. Additionally, physicochemical methods employed for the shelf life determination of a selection of foods and beverages as well as other useful tests such as accelerated storage analysis will be discussed.

MICROBIAL LOADING

For food safety it is important to determine how long a product is safe to consume within the designed storage parameters. To achieve this the product is stored at the recommended storage temperature and conditions until the maximum legally allowable limit for numbers of defined microflora is reached and then manufactures subtract a further period of time to allow an additional safety margin by which the product can be consumed. This margin of safety should be determined and applied by the manufacturer after examining all reasonably foreseeable conditions of processing and use (FSAI, 2014a). With microbial growth it is often useful to consider safety and spoilage separately although the controlling factors for both may be identical (Walker, 1994). The safety of foodstuffs is mainly ensured by a preventive approach, such as implementation of good hygiene practice and application of procedures based on hazard analysis and critical control point principles (EC 2073/2005). Such practices ensure the absence of pathogenic microorganisms during production and minimise the microbial load within the regulatory guidelines.

The growth of spoilage or pathogenic microorganisms in a product will depend on a number of factors which include the prevailing hygienic conditions of the product during manufacture; the physiochemical composition of the product, such as moisture content, ethanol concentration, pH; the heat treatment used in manufacture such as pasteurisation; the packaging materials used and finally storage temperatures and conditions. Storage conditions may include defining the temperature of storage such as refrigeration temperatures or perhaps defining an upper temperature limit or exposure to light or humidity.

Food spoilage organisms can be readily identifiable by a change in appearance, off-odour development or perhaps changes in texture. The principal danger of pathogenic species of microorganism, such as *Salmonella* or *Listeria*, is the lack of sensory changes to the product after growth of these bacteria. Appearance, odour, flavour or texture sensory cues may be absent and thus do not indicate to the individual consuming the product that growth has occurred which presents a health risk.

Foodstuffs with the greatest risk of contamination include foods of animal origin and uncooked vegetables products. The species and population of microorganisms on meat are influenced by animal species; animal health, handling of live animals, slaughter practices; plant and personnel sanitation, carcass chilling; fabrication sanitation, type of packaging, storage time and storage temperature (Nottingham, 1982; Grau, 1986; McMillin, 2008). The European regulatory guidelines, EC 2073/2005, present comprehensive and specific tolerance limits for microbial loading by which meat and poultry production must comply (EU, 2005). For example, the aerobic colony count for cattle, sheep, goats and horse carcasses after dressing, but before chilling, must not exceed 5.0 log cfu/cm^2 daily mean log. For minced meat, aerobic colony counts should not exceed 5×10^6 cfu/g, and 500 cfu/g for *Escherichia coli* at the end of the manufacturing process. *Salmonella* must be absent in the carcasses of cattle, sheep, goats, horses and pigs after dressing but before chilling and absent in poultry carcasses of broilers and turkeys after chilling. The regulation also defines the sampling criteria which must be enforced for each specific meat and poultry product. The sampling rules for poultry carcases, for example, for *Salmonella* analyses, require a minimum of 15 carcases to be sampled at random during each sampling session and after chilling. A piece of approximately 10 g from neck skin should be obtained from each carcase. On each occasion the neck skin samples from three carcases need to be pooled before examination to form 5×25 g final samples. *Salmonella* must be absent from these pooled samples (EU, 2005).

The microbiological guideline for cooked meat, including cured products, by the International Commission on Microbiological Specifications for Foods (ICMSF, 2011) sets the acceptable limit as $< 10^5$ cfu/g of sample. It is well known that salt acts as a food preservative by reducing the water activity of food and therefore inhibits microbial growth (Fellendorf et al., 2016). Processed meat products may also contain nitrites and include bacon, bologna, corned beef, frankfurters, luncheon meats, ham, fermented sausages, shelf-stable canned meats, cured meats, perishable canned meats, cured meat (e.g., ham) and a variety of fish and poultry products (Pennington, 1998). The shelf life of meats can be extended significantly by curing as the nitrate/nitrite has a strong antibacterial effect, particularly to the growth of *Clostridium botulinum*. Nitrite is strongly inhibitory to anaerobic bacteria, most importantly *C. botulinum*, and contributes to control of other microorganisms, such as *Listeria monocytogenes* (Sebranek and Bacus, 2007).

Microbiological criteria have been set in legislation for various foods, including meat products as described above but, also ready-to-eat foods (FSAI, 2014b). Commission Regulation (EC 2073/2005) (EU, 2005) sets guideline limits on various microorganisms at point of sale (EU, 2005). Cheeses made from milk or whey that has undergone heat treatment should have a limit for *E. coli* of <100 cfu/g, precut fruit and vegetables (ready-to-eat) should have a limit for *E. coli* of 100–10^3 cfu/g and unpasteurised fruit and vegetable juices (ready-to-eat) should have cut-off level set for *E. coli* bacteria of $>10^3$ cfu/g. Similarly limits for coagulase-positive staphylococci at point of sale for shelled and shucked products of cooked crustaceans and molluscan shellfish should be set at $>10^4$ cfu/g. Cheeses made from milk that has undergone a lower heat treatment than pasteurisation and ripened cheeses made from milk or whey that has undergone pasteurisation or a stronger heat treatment should have limits of between 20 and 10^4 cfu/g.

SENSORY SHELF LIFE TESTING

Experimental Design and Sample Handling

As stated earlier, shelf life analysis essentially consists of storing a single large batch of product under normal conditions and to test, using suitable methods at various storage times. The sensory analysis involves testing the product as specific time points and can include hedonic or descriptive testing. One of the principle drawbacks of this method is that the panel (trained or naïve) may become aware that they are participating in a sensory shelf life test and expect that samples become more deteriorated as time passes by, which could lead to biased results (Lawless and Heymann, 2010). A more sophisticated approach to prevent assessor bias is to incorporate reverse storage design. Reversed storage design can be performed by staggering product times, so that all products with different storage times are evaluated on the same day (Lawless and Heymann, 2010). Byrne et al. (2003) undertook a similar strategy in a descriptive sensory profiling test of warmed-over flavour (WOF) in meat patties from carriers and noncarriers of the RN allele stored and tested on days 0, 1, 3 and 5. Each replicate was presented over 2 days to each panellist, eight samples per day. In all, 8 days of panel sessions of 1.5 h each were carried out. Presentation to individual panellists was in a randomised order. However, all storage days and cooking temperatures and genotypes were included on each day (Byrne et al., 2003). This was achieved by storing samples in reverse such that all end points (days 0, 1, 3 and 5) were reached on the analysis day. Similarly, O'Sullivan et al. (2003a) used a reversed design to assess WOF by a trained sensory panel ($n = 8$) for four treatments which were cooked and refrigerated at $4°C$ for up to 5 days. The day 5 samples were prepared first, followed by the day 3 samples, and so on, so that on the last day, the day 0 samples were prepared and then all samples can be presented to the panel. However, one of the drawbacks with reverse block design is that the sensory end points should be defined and known beforehand. To do this, then samples have to be analysed for deteriorative sensory changes in real time so that actual quality cut-off end points are reached or passed, but still are within microbiological safety limits. Simply including fresh control samples in the test sample set during real time analysis, and also informing the panel controls are present, could be a simple method to prevent assessor bias and correcting this potential problem.

Another way of avoiding bias by real time analysis is to freeze samples once certain time points have been reached. This is a routine procedure for butter analysis whereby samples are blast frozen (Fig. 6.2) at specific time points and then when all samples are present does the actual affective or descriptive sensory assessment take place. The blast freezing temperature can be set at $-35°C$ to rapidly freeze the sample. Also, freezing is a convenient solution when analysing food samples, but particularly cheese samples, or other fermented food products such as salami, as the freezing process stops

FIGURE 6.2 Diplays a Foster F35 Blast Freezer. The high-powered refrigeration system blasts cold air laterally over the product at high speed, extracting heat at an optimum rate, while maintaining food quality. Once the cycle is complete, the equipment switches into 'hold mode' to keep the food at the required temperature.

ageing or maturation at the required time point (O'Sullivan, 2016; personal communication). Essentially, samples are frozen very rapidly with blast freezing which in turn freezes the water in samples producing very small ice crystals which minimally effect the texture of samples once thawed. After blast freezing, samples can be stored in a normal freezer at $-20°C$. If samples were frozen normally, then larger ice crystals grow and form in the product which can rupture cells thus causing changes to texture on thawing that are sufficiently different from the initial textural quality. Blast freezing and then subsequent storage at $-20°C$ can also be used to store 'fresh' control samples used for comparison with subsequent aged samples for testing during sensory shelf life analysis.

SENSORY SHELF LIFE METHODS

Many of the methods covered in Chapter 1 of this book, on difference testing, including 'Paired Comparison', 'Duo Tri', 'Triangle', 'Tetrad', 2-AFC or

3-AFC could be potentially used as part of sensory shelf life testing. In each case the control sample could consist of the fresh product and the test sample(s) the product at various stages of shelf life. Ultimately the level at which sensory quality deteriorates to the extent that the sensory shelf life end point is reached must be set and known as part of the experimental design. Sensory shelf life is defined as the length of time during which the product does not significantly change its sensory characteristics and still corresponds to a product of high quality, or at least sufficient sensory quality.

Additionally, some of the other difference tests could be used including the methods 'In/Out', 'Ratings for degree of difference from a standard' and 'Weighting of differences from control' (Muñoz et al., 1992a,b,c; Lawless and Heymann, 1998a,b).

For the 'In/Out' method, shelf life test batches can be evaluated by a trained panel as being either within or outside the defined sensory specifications. Again, as previously discussed, the method can be limited as it does not provide any descriptive information that can be used to amend problems (Muñoz et al., 1992b).

The relative deviation from a fresh product could be used to determine differences in shelf life samples for the 'Ratings for degree of difference from a standard' method. This method is also limited, in that it does not provide any information regarding the source of differences compared to a fresh control sample (Muñoz et al., 1992c).

The final difference test 'Weighting of differences from control (Individual experts)' is similar to the degree of difference from a standard method but involves an even more complex judgement procedure on the part of panellists. This is because it is not only the differences that matter, but also how they are weighted in determining product quality (Lawless and Heymann, 1998a; Muñoz et al., 1992b).

Costell (2002) discusses in detail the very popular quality rating system called the Quality Index Method (QIM). This type of method requires the definition of specifications or quality standards and the selection of criteria to evaluate if products comply with the requirements of the quality standards. A highly trained assessor panel rates the quality of products using a scale in which points are defined in terms of the sensory characteristics that characterise the quality of each grade. QIM is based on the objective evaluation of the key sensory attributes of fish species using a scoring system that ranges from 0 to 3 with the lower the score denoting freshness (Costell, 2002). With this type of method, assessors must be able to recall the sensory characteristics of the ideal product and interpret descriptions that corresponding to each point of the scale as well as identifying and quantifying the common sensory defects that appear as a result of shelf life storage (Lawless and Heymann, 2010).

Descriptive analysis using trained panels could also be used during shelf life assessment, but the main sensory attributes effecting shelf life quality must be known and the intensity at which they individually or in combination deteriorate sufficiently to determine end points must also be established. Assessors are provided with a set of samples which include the different storage times and the sensory dimensions reflecting the spectrum of quality and are asked to generate the descriptors needed to describe differences amongst samples. As is usual with quantitative descriptive analysis, the assessors produce a consensus lexicon through open discussion with the panel leader and define the sensory attributes to be assessed (Lawless and Heymann, 2010). Usually, a sensory attribute or attributes that are determined to be critical to shelf life limitations are included, e.g., this could be rancid flavour or aroma with respect to oxidative deterioration, and the products are assessed by a trained panel whom then mark their respective assessments on 10 or 15 cm unstructured line scales (Stone and Sidel, 2004).

The sensory shelf life of the product is defined as the storage time at which overall quality, or the intensity of a specific sensory attribute, reaches a predetermined value or 'failure criterion', assuming that once the product has reached this point it is no longer saleable (Lawless and Heymann, 2010). Regulatory agencies do not usually monitor the sensory changes of food products throughout storage, and the maximum tolerable change in the sensory characteristics of a food product could not be determined based on regulations. Thus, in quality-based methods, the failure criteria for estimating sensory shelf life are arbitrarily selected by the researchers (Giménez et al., 2012). However, these failure criteria, or sensory cut-off limits, may not correlate to consumer rejection criteria and to some extent are simply plucked out of the air. To account for this, sensory shelf life testing should somehow be benchmarked to the consumer and reflect the real world scenario when quality is no longer adequate to satisfy consumer expectations.

The acceptability limit for shelf life is often chosen by company management on the basis of available experience of the product performance on the market or on the emulation of competitors. Despite being simple and inexpensive, such an approach risks critical overestimations or disadvantageous underestimation of the shelf life. This hazard is much more probable in the case of new foods, for which no previous experience is available (Manzocco et al., 2011). Affective sensory analysis using sensory acceptance tests is possibly the closest type of analysis to the consumer where hedonics for appearance, aroma, flavour, texture and overall acceptability is determined. Assessors are typically asked to mark on a category hedonic scale (e.g., 7 or 9 point) how they rate a presented product from dislike extremely to like extremely. This method is most appropriate as subjective liking can be determined for each sensory modality until the appropriate sensory cut-off level is achieved. This level is set as part of the experimental design and will depend on a number of variable factors that can also be used to set limits for the previously mentioned descriptive and difference tests. However, sensory acceptance tests are not a substitute for the more costly large-scale consumer tests so ultimately the final shelf life products, those near end points determined through cheaper and smaller scale affective sensory analysis, should be validated by appropriately conducted larger scale consumer testing. This is appropriate as sensory shelf life is usually determined by consumers who find that if the quality of a product is lower than accepted, they reject it and do not repurchase (Labuza and Schmidl, 1988). Ultimately, large scale consumer tests can be used to establish the hedonic cut-offs at which sensory quality is deemed no longer high enough and the end point for shelf life is reached. These tests can assist the product developer in assessing aged samples in a standardised and scientific manner using hedonics. However, the evaluation of consumer rejection as a function of storage time is a time-consuming, cumbersome and expensive process requiring large sample sizes, a large number of consumers as well as the application of appropriate statistical techniques (Manzocco et al., 2011). Therefore, once validated, the smaller scale affective sensory analysis can be utilised as part of routine shelf life analysis as the end points have been determined from the large scale consumer testing. The level at which shelf life is reached and the 'failure criterion' should thus incorporate relevant consumer data to establish suitable end point levels. This could again be somewhat arbitrary but using preference mapping, as discussed in Chapter 5, could help to establish suitable end points.

Preference mapping techniques involve statistically relating external information about perceived product characteristics to consumer preference or acceptance ratings to determine a product's characteristics and obtain a better understanding of what attributes of a product are driving preferences (Van Kleef et al., 2006; Meilgaard et al., 2007). Combined with modern multivariate data analytical software packages, the analysis of different modes of data, the subjective hedonic determinations of consumers to more objective measurements such as sensory profiling data (trained descriptive panel data) and physicochemical analysis (rheological, compositional, volatile flavour (GCMS), etc.) has become relatively easy. Consumer preference data based on shelf life assessment can be mapped on to other sensory affective data, sensory descriptive and physicochemical data to determine which products are the most accepted, and the most rejected with respect to consumer sensory quality, functionality and shelf life and can assist in determining accurate shelf life of products. In most shelf life studies a medium risk level (50% consumer rejection) is chosen as a reasonable acceptability limit but it has been suggested that lower percentages of consumer rejection could be much more reliable (Manzocco et al., 2011).

COLOUR AND SENSORY SHELF LIFE

For some food products, colour will be one of the principal sensory factors which will define shelf life, thus food manufacturers need to ensure that the appearance of the product will be stable over its required shelf life as consumers will perceive colour loss as a sign of quality deterioration (Sutherland et al., 1986). Colour has a close association with quality factors such as freshness, ripeness and desirability, and food safety, its prominent role is unquestionable for the consumer acceptability of food products (Wu and Sun, 2013). The measurement of colour allows the detection of certain anomalies or defects that food items may present (Leon et al., 2006). Colour fading can occur in soft drinks and thus stable colours must be able to retain suitable consumer sensory colour quality for many months. Colours can become oxidised by light and deteriorate over time. Similar oxidation can result in negative sensory colour changes. Carpenter et al. (2001) noted a strong association between colour preference and purchasing intent with consumers discriminating against beef that is not red (i.e., beef that is purple or brown). Meat colour or bloom develops through interaction with the air which permeates the gas permeable packaging materials used to overwrap meat, with meat discolouration or pigment oxidation, typically preceding lipid oxidation. At the point of sale, colour and colour stability are the most important attributes of meat quality, and various approaches have been used to meet consumer expectations that an attractive, bright-red colour indicates a long shelf life and good eating quality (Hood and Mead, 1993).

The packaging format can be crucial in prolonging shelf life. The colour of meat is extended by storage under modified atmosphere packaging (MAP) conditions (Kerry et al., 2000). For fresh meat this is particularly important and has a dramatic effect on shelf life, which will be discussed in greater detail in Chapter 11 of this book. For beverages, packaging in bottles with UV barrier protection or the inclusion of antioxidants like ascorbic acid or tocopherols during manufacture can help retard colour fade effects due to light oxidation.

Colour can be measured by sensory affective and descriptive testing or by using instruments such as the portable Minolta colorimeter. The use of structured line scaling, as presented, has been the convention in sensory evaluation of meat colour studies to date. However, unstructured line scaling may give a higher degree of discrimination between samples. O'Sullivan et al. (2002) used unstructured line scaling to evaluate the colour stability of pork chops from four dietary treatments (control, vitamin E, iron/vitamin E and iron) under standard commercial retail display conditions (Fig. 6.3) by a trained sensory panel. A higher degree of discrimination in visual assessment was determined relative to instrumental methods (Hunter L, a, b) as measured by a Minolta colorimeter. O'Sullivan et al. (2002) found that M. longissimus dorsi was more stable to oxidative deterioration compared to M. psoas major for all the experimental treatments as determined by sensory and instrumental assessment (Minolta colorimeter, L, a, b values). O'Sullivan et al. (2002) showed that trained panellists were able to differentiate four experimental treatment groups (control, vitamin E, iron/vitamin E and iron) on each day of a retail display study of uncooked samples of whole cuts of M. longissimus dorsi muscle. These authors found that the trained panel was more effective in

FIGURE 6.3 Retail refrigerated display cabinet used for shelf life assessment (O'Sullivan, 2011b, 2016).

evaluating the colour quality of samples than instrumental methods as determined by a portable colorimeter (Minolta). In another study, O'Sullivan et al. (2003b) investigated if training of sensory panellists for sensory visual assessment of meat gave a discriminative improvement in the subsequent evaluation compared to using an untrained sensory panel. In this study *M. longissimus dorsi* and *M. psoas major* minced pork patties from three dietary treatment groups (control, iron/vitamin E and iron) were packaged in polythene bags and placed in a retail refrigerated display cabinet (Fig. 6.3) at $5 \pm 1°C$, under fluorescent light (1000 Lux) for up to 5 days. Samples were subjected to visual colour evaluation by a trained sensory panel ($n = 8$) and an untrained panel ($n = 8$) on days 0, 1, 3 and 5 of retail display. These authors (O'Sullivan et al., 2003b) found that the signal to noise (S/N) ratio for assessors and replicates for the trained assessor group was better than those of the untrained assessor group indicating that their results were more reliable.

TEXTURE AND SENSORY SHELF LIFE

Texture can change over the shelf life of some products and can result in the formation of negative sensory consumer perception. Bread can become stale, biscuits hard or soft and cream liqueurs can become viscous. Tenderness can be evaluated by objective methods, instrumental or sensory descriptive (trained panels) or by subjective affective methods (sensory acceptance testing or consumer panel). Sensory methods, either analytical or affective, are expensive, difficult to organise and time-consuming (Platter et al., 2003). Thus, there have been many attempts to devise instrumental methods of assessing the force shearing, penetrating, biting, mincing, compressing and stretching of foods whose results are a prediction of tenderness ratings obtained by taste panels (Lawrie and Ledward, 2006). Shear tests measure the force to cut through fibres of cooked meat samples. They are the simplest and most common tests used to document food texture. Viscosity of cream liqueurs will increase over time and as such viscosity can be used as an index of shelf life. Eventually the viscosity will become a limiting factor to consumer acceptability. Viscosity can be quantified using a viscometer such as the Brookfield Rheometer. The viscosity of these fluid emulsions is an important characteristic because it influences the rate of creaming, the physical shelf life of the product and the organoleptic properties of the product (McClements, 1999). Texture analysis will be covered in detail in Chapter 8 of this book on instrumentation.

SHELF LIFE AND FLAVOUR PROFILE CHANGES

Flavour profile changes that effect sensory shelf life quality can be varied and diverse, but a large proportion of these that result in deteriorative sensory changes are caused by lipid oxidation. Lipid oxidation is the primary cause of rancidity during the storage of foods that contain polyunsaturated fats such as meat and meat products, either fresh, frozen (Buckley et al., 1989) or cooked as well as foods containing meat-based ingredients such as ready meals, soups or processed meat products. The oxidation of these polyunsaturated fatty acids causes the rapid development of meat rancidity (Kanner, 1994). Lipid oxidation is an autocatalytic chain reaction that occurs in the presence of O_2 which results in the development of rancid off-flavours. These off-flavours increase over the storage time of meats packed in O_2 environments such as in MAP and impact negatively on the oxidative stability of muscle lipids and lead to development of undesirable flavours (Rhee and Ziprin, 1987; Estevez and Cava, 2004). Additionally, one of the principle flavour defects to occur in pork products includes WOF, which is an objectionable flavour which becomes most noticeable when refrigerated cooked meat is reheated. These off-flavours can also occur in cooked poultry products such as chicken and turkey. It is a rapidly occurring phenomenon in contrast to the slowly developing oxidative rancidity which results from long-term storage of frozen raw meat or storage of vegetable oils at ambient temperature (St. Angelo and Bailey, 1987). O'Sullivan et al. (2002) found that cooked pork *M. psoas major* muscle was more susceptible to WOF development as determined by sensory profiling compared to *M. longissimus dorsi* for all treatments (Control, vitamin E, iron/vitamin E and Iron) in an animal feeding study constructed to determine the oxidative stability of these different muscles produced from animals fed diets in either prooxidant iron or antioxidant vitamin E.

   The development of oxidative reactions causes not only the loss of pleasant aroma compounds, in coffee products, but also the formations of off-flavours.

The occurrence of these oxidative reactions could also lead to the quality decay of instant coffee, coffee concentrates and drinks (Nicoli and Savonitto, 2005). The sensitivity of roasted coffee towards oxidation reactions is high due to the presence of a large number of strongly active volatile and nonvolatile compounds that easily react with oxygen (Manzocco et al., 2011). Similarly, oxidation plays a major role in beer ageing with the development, during storage, of the cardboard-flavoured component (E)-2-nonenal and its formation by lipid oxidation with other stale compounds (Vanderhaegen et al., 2006). The limiting factor to beer shelf life is the development of these stale, oxidised flavours such as E-2-nonenal, increases in the compound furfural, with a parallel reduction in fresh fruity or hop notes with beer ageing. In beers and wines, ethyl esters contribute to positive sensory attributes, for beer they denote freshness, and decrease during ageing. These positive flavour attributes of beer, such as fruity/estery and floral aroma compounds, tend to decrease in intensity during ageing (O'Sullivan, 2011a). On the other hand, ethyl esters contribute to the overall flavour balance of cream liqueurs, but at high concentrations, they can develop a very fruity flavour which can become the dominant flavour in the liqueur and increase in concentration as cream liqueurs age. Ethyl esters form due to the reaction of the alcohol and fatty acids and can result in the formation of excessive levels of short-chain fatty acids such as ethyl acetate, ethyl butanoate and ethyl hexanoate, which are the principal causes of this fruity defect. Sensory descriptors for fruity flavours may include 'pear-like, banana-like, pineapple-like, apple-like, strawberry-like, ester-like, ethereal or just fruity'. The level of fruitiness may become a limiting factor to consumer acceptance in aged liqueurs (O'Sullivan, 2011a).

CHEMICAL ANALYSIS

Lipid oxidation in fat-containing food products can lead to quality deterioration which has the potential of limiting shelf life from a flavour perspective, but also colour in the case of the brown metmyoglobin formation in fresh meat or the darkening of whole milk powder. In the case of cooked meat and poultry the product will be spoilt due to microbial action long before oxidation generally becomes an issue. However, it is necessary to establish the oxidative stability food products during shelf life testing and this will vary depending on the amount of polyunsaturated lipid present.

The method of Tarladgis et al. (1960) is regarded as the standard method for malonaldehyde (MDA) analysis. This method uses distillation and spectrophotometric analysis to quantify MDA in lipid-containing foods. MDA is a major carbonyl decomposition products of autoxidised, polyunsaturated lipid materials (Crawford et al., 1966). In the reaction, one molecule of MDA reacts with two molecules of 2-thiobarbituric acid (TBA) to form a MDA-TBA complex, which allows the resultant pink pigment to be quantitated spectrophotometrically. A more rapid method of quantifying MDA in foods has been described by Siu and Draper (1978) for a TCA extraction method for measurement of lipid oxidation.

Using either of the above methods, it is typical to monitor oxidative stability of fat-containing foods, in particular, meat and poultry high in polyunsaturated fat, over the course of the shelf life of the product under defined storage temperature and conditions. The design of the shelf life study might involve placing the product in a commercial retail display situation, like a display cabinet (Fig. 6.3) and tracking MDA formation periodically over time. The samples are placed in a retail refrigerated display cabinet at $5 \pm 1°C$, under a defined light source strength fluorescent light (e.g., 1000 Lux) for the storage period of the study (O'Sullivan et al., 2003b). Although there are no legal limits to the amount of MDA that can form in foods, above a certain level it may become objectionable to consumers and thus influence any repeat purchase. Thiobarbituric acid reactive substances (TBARS) have been correlated to sensory determined rancidity. Greene and Cumuze (1982) state the general population of meat consumers would not detect oxidation flavours until oxidation products reached levels of at least 2.0 mg/kg tissue. Similarly, Campo et al. (2006) reported that TBARS value of 2.28 could be considered as the limiting threshold for acceptability of oxidation in beef. This value indicates the point where the perception of rancidity overpowers beef flavour. Zakrys et al. (2008) found that sensory panellists preferred steaks stored in 50% O_2 during a shelf life experiment, even though oxidised flavours were detected which could be attributed to adaptation, or familiarity with, oxidised flavours in meat (Zakrys et al., 2008). Tarladgis et al. (1960) showed that threshold TBARS values of 0.5—1.0 for fresh ground pork were highly correlated (0.89) to intensity of rancid odour 1.0 by trained sensory panellists (Tarladgis et al., 1960). Whereas Jayasingh and Cornforth (2004) presented data which showed that consumers preferred cooked pork patties with TBARS of less than 0.5 compared with patties that had TBARS numbers greater than 1.4.

ACCELERATED STORAGE TESTS

Real-time shelf life studies have the potential of taking a very long time to complete and because of this producers utilise other methods to assist in calculating shelf life which can assist in getting the product to the market place (O'Sullivan, 2011b, 2016). Accelerated ageing traditionally involves use of temperature increases to speed reactions. The process of estimating ambient stability involves estimating the reaction rate at different temperatures, and then extrapolating to the desired temperature (Waterman and Adami, 2005). Accelerated storage tests are designed to reduce shelf life testing time by speeding up the deteriorative changes that occur in the product by exposure to factors such as high temperatures or humidity (O'Sullivan, 2011b, 2016). Accelerated tests involve a single test condition where it is assumed, for example, that if a product lasts 1 month at 40°C, it will last 4 months at ambient temperature, 20°C. Thus, it is assumed that the accelerating factor from 20 to 40°C is equal to 4. Often this value is a company tradition with unknown origins, other times it is estimated from data on a previous similar product (Nelson, 1990).

Generally, specialist storage cabinets, such as the Binder (Fig. 6.1), can be used for accelerated storage testing where temperature, humidity and light parameters can be accurately and carefully controlled. Mathematical modelling can be used to take into account the differences, linear or nonlinear between real-time shelf life and accelerated tests and can be applied to any deterioration process that has a valid kinetic model, which could be chemical, physical, biochemical or microbial (O'Sullivan, 2011b, 2016). Temperature is the most frequently used acceleration factor in the Arrhenius equation. For example, for nonhermetically sealed coffee products, storage temperature is critical for volatile release and the temperature dependence of volatile release can be well described by the Arrhenius equation from 4 to 40°C (Nicoli and Savonitto, 2005).

The kinetic model approach is the most common method for accelerated shelf life testing, which has been extensively described by Mizrahi (2000). The kinetic model approach involves the selection of the desired kinetically active factors for acceleration of the deterioration process to be measured, evaluating the parameters of the kinetic model and extrapolating the data to normal storage conditions. These extrapolated data for the kinetic model are then used to predict shelf life at actual storage conditions (Mizrahi, 2000).

There are certain problems that occur with accelerated shelf life testing. The most difficult problem is to find a simple mathematical relation able to describe the evolution of the chosen attribute as a function of storage time (Guerra et al., 2008) as the relationship to real time shelf life analysis is not always linear. Deteriorative mechanisms can prevail in the accelerative test that would not normally occur in the real time situation under normal storage conditions. Additionally, accelerated shelf life testing may be difficult to implement when no valid kinetic model is believed to exist for any accelerating kinetic factor, thus there is no test procedure available to make measurements. Also testing may not be practical in cases where too many parameters need to be measured and thus the analysis becomes overly complicated (Mizrahi, 2000). Also, there is no evidence, nor theoretical reason, for the effects of different parameters such as temperature, pH, aW or any other factor to be additive or multiplicative in accelerated shelf life models (Peleg, 2006).

Appropriate statistical methods must be used in accelerated shelf life testing which test the sensitivity of the model by a cross-validation method by using part of the data to verify the validity of the model. Guerra et al. (2008) state that given the number of arbitrary choices, the shelf life concept for microbiologically stable food products is more company or researcher driven than product or consumer dependent (Guerra et al., 2008). Also it is important to consider that many of the deterioration processes in foods that might be utilised for accelerated shelf life predictions do not follow any of the standard kinetic models and thus the assumption that the process or reaction must follow a particular kinetic is therefore unnecessary (Corradini and Peleg, 2007).

REFERENCES

Buckley, D.J., Gray, J.I., Asghar, A., Booren, A.M., Crackel, R.L., Price, J.F., Miller, E.R., 1989. Effects of dietary antioxidants and oxidised oil on membranol lipid stability pork product quality. Journal of Food Science 54, 1193—1197.

Byrne, D.V., O'Sullivan, M.G., Bredie, W.L.P., Martens, M., 2003. Descriptive sensory profiling and physical/chemical analyses of warmed-over flavour in meat patties from carriers and noncarriers of the RN(−) allele. Meat Science 63, 211—224.

Campo, M.M., Nute, G.R., Hughes, S.I., Enser, M., Wood, J.D., Richardson, R.I., 2006. Flavour perception of oxidation in beef. Meat Science 72, 303—311.

Carpenter, C.E., Cornforth, D.P., Whittier, D., 2001. Consumer preferences for beef color and packaging did not effect eating satisfaction. Meat Science 57, 359—363.

Corradini, M.G., Peleg, M., 2007. Shelf-life estimation from accelerated storage data. Trends in Food Science & Technology 18, 37—47.

Costell, E., 2002. A comparison of sensory methods in quality control. Food Quality and Preference 13, 341—353.

Crawford, D.L., Yu, T.C., Sinnhuber, R.O., 1966. Reaction mechanism, reaction of malonaldehyde with glycine. Journal of Agricultural and Food Chemistry 14, 184.

Estevez, M., Cava, R., 2004. Lipid and protein oxidation, release of iron from heme molecule and colour deterioration during refrigerated storage of liver pate. Meat Science 68, 551—558.

EU, 2005. EC 2073/2005, On Microbiological Criteria for Foodstuffs. Commission Regulation (EC) No 2073/2005. Microbiological criteria for foodstuffs. https://www.fsai.ie/uploadedFiles/Consol_Reg2073_2005.pdf.

Fellendorf, S., O'Sullivan, M.G., Kerry, J.P., 2016. Impact on the physicochemical and sensory properties of salt reduced corned beef formulated with and without the use of salt replacers. Meat Science (Submitted).

Freitas, M.A., Costa, J.C., 2006. Shelf life determination using sensory evaluation scores: a general Weibull modeling approach. Computers & Industrial Engineering 51, 652—670.

FSAI, 2014a. Guidance Note No. 18 Validation of Product Shelf-Life (Revision 2), ISBN 1-904465-33-1.

FSAI, 2014b. Food Safety Authority of Ireland. Guidance Note No. 3 Guidelines for the Interpretation of Results of Microbiological Testing of Ready-to-Eat Foods Placed on the Market (Revision 1), ISBN 0-9539183-5-1.

Giménez, A., Ares, F., Ares, G., 2012. Sensory shelf-life estimation: a review of current methodological approaches. Food Research International 49, 311—325.

Grau, F.H., 1986. Microbial ecology of meat and poultry. In: Pearson, A.M., Dutson, T.R. (Eds.), Advances in Meat Research, vol. 2, pp. 1—47. Westport.

Greene, B.E., Cumuze, T.H., 1982. Relationship between TBA numbers and inexperienced panellists' assessments of oxidized flavour in cooked beef. Journal of Food Science 47, 52–54.

Guerra, S., Lagazio, C., Manzocco, L., Barnabà, M., Cappuccio, R., 2008. Risks and pitfalls of sensory data analysis for shelf life prediction: data simulation applied to the case of coffee. LWT – Food Science and Technology 41, 2070–2078.

Hood, D.E., Mead, G.C., 1993. Modified atmosphere storage of fresh meat and poultry. In: Parry, R.T. (Ed.), Principles and Applications of Modified Atmosphere Packing of Food. Blackie Academic and Professional, London, pp. 269–298.

Hood, D.E., Riordan, E.B., 1973. Discoloration in pre-packed beef. Journal of Food Technology 8, 333–348.

ICMSF, 2011. In: Swanson, K.M.J. (Ed.), Microorganisms in Foods 8. Use of Data for Assessing Process Control and Product Acceptance, first ed. Springer, p. 400. http://doi.org/10.1007/978-1-4419-9374-8.

IFST, 1993. Shelf Life of Foods – Guidelines for Its Determination and Prediction. Institute of Food Science and Technology, London.

Jayasingh, P., Cornforth, D.P., 2004. Comparison of antioxidant effects of milk mineral, butylated hydroxytoluene and sodium tripolyphosphate in raw and cooked ground pork. Meat Science 66 (1), 83–89.

Kanner, J., 1994. Oxidative processes in meat and meat products: quality implications. Meat Science 36, 169–189.

Kerry, J.P., O'Sullivan, M.G., Buckley, D.J., Lynch, P.B., Morrissey, P.A., 2000. The effects of dietary α-tocopheryl acetate supplementation and modified atmosphere packaging (MAP) on the quality of lamb patties. Meat Science 56, 61–66.

Labuza, T.P., Schmidl, M.K., 1988. Use of sensory data in the shelf life testing of foods: principles and graphical methods for evaluation. Cereal Foods World 33, 193–206.

Lawless, H.T., Heymann, H., 1998a. Sensory evaluation in quality control. In: Lawless, H.T., Heymann, H. (Eds.), Sensory Evaluation of Food, Principles and Practices. Chapman and Hall, New York, pp. 548–584.

Lawless, H.T., Heymann, H., 1998b. Descriptive analysis. In: Lawless, H.T., Heymann, H. (Eds.), Sensory Evaluation of Food, Principles and Practices (pp. 117–138, pp. 341–378). Chapman and Hall, New York.

Lawless, H.T., Heymann, H., 2010. Sensory Evaluation of Food. Principles and Practices, second ed. Springer, New York.

Lawrie, R.A., Ledward, D.A., 2006. Lawrie's Meat Science. Woodhead Publishing Ltd, Cambridge, England.

Leon, K., Mery, D., Pedreschi, F., Leon, J., 2006. Color measurement in L*a*b* units from RGB digital images. Food Research International 39 (10), 1084–1091.

Manzocco, L., Lagazio, C., 2009. Coffee brew shelf life modelling by integration of acceptability and quality data. Food Quality and Preference 20, 24–29.

Manzocco, L., Calligaris, S., Nicoli, M.C., 2011. The stability and shelf life of coffee products. In: Kilcast, D., Subramaniam, P. (Eds.), Food and Beverage Shelf-Life and Stability. Woodhead Publishing Limited, Cambridge, UK.

McClements, D.J., 1999. Food Emulsions: Principles, Practices and Techniques. CRC Press, Boca Raton, FL.

McMillin, K.W., 2008. Where is MAP going? A review and future potential of modified atmosphere packaging for meat. Meat Science 80, 43–65.

Meilgaard, M.C., Civille, G.V., Carr, B.T., 2007. Sensory Evaluation Techniques. CRC Press, Boca Raton, Florida.

Mizrahi, S., 2000. CH5, Accelerated shelf-life tests. In: Kilcast, D., Subramaniam, P. (Eds.), The Stability and Shelf-Life of Food. Woodhead Publishing Limited, Cambridge, UK.

Morrissey, P.A., Buckley, D.J., Sheehy, P.J.A., Monaghan, F.J., 1994. Vitamin E and meat quality. Proceedings of the Nutrition Society 53, 289−295.

Muñoz, A.M., Civille, G.V., Carr, B.T., 1992a. Comprehensive descriptive method. Sensory Evaluation in Quality Control. Van Nostrand Reinhold, New York, pp. 55−82.

Muñoz, A.M., Civille, G.V., Carr, B.T., 1992b. "In/Out" method. In: Sensory Evaluation in Quality Control. Van Nostrand Reinhold, New York, pp. 140−167.

Muñoz, A.M., Civille, G.V., Carr, B.T., 1992c. Difference-from-control method (Degree of difference). In: Sensory Evaluation in Quality Control. Van Nostrand Reinhold, New York, pp. 168−205.

Nelson, W., 1990. Accelerated Testing. Statistical Models, Test Plans and Data Analyses. John Wiley & Sons, New York.

Nicoli, M.C., Savonitto, O., 2005. Physical and chemical changes of roasted coffee during storage. In: Viani, R., Illy, A. (Eds.), Espresso Coffee: The Science of Quality, second ed. Elsevier Academic Press, San Diego, CA, pp. 230−245.

Nottingham, P.M., 1982. Microbiology of carcass meats. In: Brown, M.H. (Ed.), Meat Microbiology. Applied Science Publishers, London, pp. 13−65.

O'Sullivan, M.G., Byrne, D.V., Stagsted, J., Andersen, H.J., Martens, M., 2002. Sensory colour assessment of fresh meat from pigs supplemented with iron and vitamin E. Meat Science 60, 253−265.

O'Sullivan, M.G., Byrne, D.V., Nielsen, J.H., Andersen, H.J., Martens, M., 2003a. Sensory and chemical assessment of pork supplemented with iron and vitamin E. Meat Science 64, 175−189.

O'Sullivan, M.G., Byrne, D.V., Martens, M., 2003b. Evaluation of pork colour: sensory colour assessment using trained and untrained sensory panellists. Meat Science 63, 119−129.

O'Sullivan, M.G., 2011a. CH4, Sensory shelf-life evaluation. In: Piggott, J.R. (Ed.), Alcoholic Beverages: Sensory Evaluation and Consumer Research. Woodhead Publishing Limited, Cambridge, UK.

O'Sullivan, M.G., 2011b. CH25, case studies: meat and poultry. In: Kilcast, D., Subramaniam, P. (Eds.), Food and Beverage Shelf-Life and Stability. Woodhead Publishing Limited, Cambridge, UK.

O'Sullivan, M.G., 2016. CH18. The stability and shelf life of meat and poultry. In: Subramaniam (Ed.), The Stability and Shelf Life of Food. Elsevier Academic Press, Oxford, UK, Ltd.

Peleg, M., 2006. Advanced Quantitative Microbiology for Food and Biosystems: Models for Predicting Growth and Inactivation. CRC Press, Boca Raton, FL.

Pennington, J.A.T., 1998. Dietary exposure models for nitrates and nitrites. Food Control 9, 385−395.

Platter, W.J., Tatum, J.D., Belk, K.E., Chapman, P.L., Scanga, J.A., Smith, G.C., 2003. Relationship of consumer sensory ratings, marbling score, and shear force value to consumer acceptance of beef strip loin steaks. Journal of Animal Science 81, 2741−2750.

Rhee, K.I., Ziprin, Y.A., 1987. Lipid oxidation in retail beef, pork and chicken muscles as affected by concentrations of heme pigments and nonheme iron and microsomal enzymic lipid peroxidation activity. Journal of Food Biochemistry 11, 1−15.

Rodríguez-Calleja, J.M., Cruz-Romero, M.C., O'Sullivan, M.G., Kerry, J.P., 2012. High-pressure-based hurdle strategy to extend the shelf-life of fresh chicken breast fillets. Food Control 25, 516—524.

Sebranek, J.G., Bacus, J.N., 2007. Cured meat products without direct addition of nitrate or nitrite: what are the issues? Meat Science 77, 136—147.

Siu, G.M., Draper, H.H., 1978. A survey of the malonaldehyde content of retail meats and fish. Journal of Food Science 43, 1147—1149.

St Angelo, A.J., Bailey, M.E., 1987. Warmed-Over Flavour of Meat. Academic Press, Florida, pp. vii—viii.

Stone, H., Sidel, J.L., 2004. Sensory Evaluation Practices. Elsevier Academic Press, London.

Sutherland, J.P., Varnum, A.H., Evans, M.G., 1986. A Color Atlas of Food Quality Control. CRC Press, Weert.

Tarladgis, B.G., Watts, B.M., Younathan, M.T., Dugan, L., 1960. A distillation method for the quantitative determination of malonaldehyde in foods. Journal of the American Oil Chemists Society 37, 44—49.

Van Kleef, E., Van Trijp, H.C.M., Luning, P., 2006. Internal versus external preference analysis: an exploratory study on end-user evaluation. Food Quality and Preference 17, 387—399.

Vanderhaegen, B., Neven, H., Verachtert, H., Derdelinckx, G., 2006. .The chemistry of beer aging — a critical review. Food Chemistry 95, 357—381.

Walker, S.J., 1994. The principles and practice of shelf life prediction of microorganisms. In: Man, C.M.D., Jones, A.A. (Eds.), Shelf Life Evaluation of Foods. Chapman and Hall, London, pp. 40—51.

Waterman, K.C., Adami, R.C., 2005. Accelerated aging: prediction of chemical stability of pharmaceuticals. International Journal of Pharmaceutics 293, 101—125.

Wu, D., Sun, D.W., 2013. Food colour measurement using computer vision. Instrumental Assessment of Food Sensory Quality 165—195.

Zakrys, P.I., Hogan, S.A., O'Sullivan, M.G., Allen, P., Kerry, J.P., 2008. Effects of oxygen concentration on sensory evaluation and quality indicators of beef muscle packed under modified atmosphere. Meat Science 79, 648—655.

Chapter 7

Packaging Technologies for Maintaining Sensory Quality

INTRODUCTION

Packaging is defined as the enclosure of products, items or packages in a wrapped pouch, bag, box, cup, tray, can, tube, bottle or other container to perform one or more of the following functions: containment, protection, preservation, communication, convenience, utility and performance. Packaging is such a massive topic that this book chapter will focus primarily on the protective function of packaging with respect to maintaining sensory food quality of foods. Food safety is mandatory and the default requirement with respect to food product packaging and only then followed by the maintaining of optimal sensory quality. Each food product category has its own unique challenges when it comes to packaging, and the maintenance of sensory quality presents quite different requirements to maintain safety and shelf life. The ways in which packaging can maintain sensory quality include the prevention of water migration, oxidation (pigment and lipid) and food spoilage. Deteriorative changes can be enzymatic, chemical, physical (moisture variation) and biological [microbiological and macrobiological (insects)] (Robertson, 2011). The main chemical reactions include oxidation and nonenzymatic browning. Microbial changes result from spoilage or pathogenic bacterial growth. Biochemical reactions occur as a result of enzymatic action resulting in browning, lipolysis and proteolysis. Finally, physical changes can result in a loss of optimal texture, emulsion stability, sedimentation, coalescence etc.

In order for any given food product to be commercially successful, consumer desires and demands must be addressed and met with respect to the sensory properties of such products, before other quality dimensions become relevant (Chambers and Bowers, 1993). For this reason it is paramount that the product package contains, protects and preserves sufficiently that a satisfactory sensory quality is maintained through the shelf life of the product. This chapter will address the major issues pertaining to the maintenance of optimal shelf life of food or beverage products through the control of packaging environment. Ultimately it is the packaging environment which will determine shelf

A Handbook for Sensory and Consumer-Driven New Product Development.
http://dx.doi.org/10.1016/B978-0-08-100352-7.00007-5

life to a large degree be that through gas manipulation or gas/moisture exclusion. For these reasons this chapter will focus on manipulation of packaging gases to optimise shelf life and the differing barrier materials which facilitate this manipulation.

PACKAGING MATERIALS

Traditionally, product packaging was quite simple with waxed paper used in the past and present for bread and aluminium inner and paper outer wrappers used for chocolate bars. The waxed paper retards moisture migration in bread, and the foil wrapper prevents grease (cocoa butter) migration to the paper outer wrapping with chocolate products. Even frozen foods, like fish fingers, were simply placed in cardboard boxes not so very long ago. Perhaps the great advantage of traditional packaging was its environmental friendliness. Today, plastics represent by far the commonest forms of packaging material and allow foods and beverages to be stored optimally for longer than perhaps their counterparts in the 1970s and 1980s, but with the sacrifice of being more challenging to dispose of.

A wide variety of materials are used in packaging, and primary packaging materials consist of one or more of the following materials: metals, glass, paper and plastic polymers (Robertson, 2006). Glass, metals and paper used in bottles, cans and cartons have given way to a plethora of different plastic laminates. However, paper-based cartons such as the tetrapac remain one of the leading formats worldwide for packaged liquid food and beverage products. Such plastic laminated cardboard provides a good barrier to water vapour and gases.

Plastic/metal laminate retort packages or pouches are taking over from traditional canning technologies. These pouches have a thin profile and a high ratio of surface area to volume and can be used to package anything from dog food to sous vide and ready meals right through to baby foods. Heat penetrates the food much more quickly when it only has to reach the inside of a half-inch-thick mass rather than a much larger mass in a round can. This results in a product with a more optimal sensory profile which looks better and has a reduced energy footprint during processing compared to traditional heat treatment methods. This is the type of laminate packaging that is replacing outdated tin can technology.

The advent of plastics or organic polymers has allowed the development of innovative packaging formats such as those mentioned above. However, plastic is not simply one compound but a term used to classify a wide variety of materials where the chemical nature of the polymer dictates stability to temperature, light, moisture and solvents and thus the degree of protection when used as a packaging material (Robertson, 2011).

Polyethylenes [low-density polyethylene (LDPE), high-density polyethylene (HDPE), low linear density polyethylene (LLDPE)] are thermoplastics that are some of the most common packaging materials. LDPE is a tough translucent flexible material with good water vapour but poor gas barrier properties. LLDPE has higher strength than LDPE with improved chemical and puncture resistance. HDPE is stiff and hard with good oil and grease resistance and is used in films and bottles. Polypropylene (PP) is also a thermoplastic with better barrier properties than the polyethylenes; it is translucent and is commonly used in closures and thin-walled containers (Robertson, 2006, 2011).

Polyvinyl chloride (PVC) is used for stretch wrapping of fresh meat in trays and can be used for shrink wrapping. Polyethylene terephthalate (PET) has superceded many applications where PVC used to be employed because of its high tensile strength, strong chemical resistance, light weight, temperature stability and elasticity and are commonly used in blow moulded bottles (Robertson, 2006).

Polyamides (PAs) or nylons have good strength and stiffness characteristics and are also good water vapour and gas barriers and have found application as a component of laminate pouches and bags in retort pouches as they also have good heat stability properties. It can also be used as a laminate layer in PET bottles and PA vacuum pack (VP) bags (Robertson, 2006, 2011).

MODIFIED ATMOSPHERE PACKAGING

The primary objective for packaging is to limit and delay both the growth of spoilage and pathogenic microorganisms and deteriorative chemical reactions through adequate containment. This is then followed by the maintenance of the optimal sensory and other quality characteristics of the product within a specified shelf life through adequate protection and preservation (O'Sullivan and Kerry, 2011). There are four categories of preservative packaging that can be used with foods. These are VPs, high oxygen modified atmosphere packs (high O_2 MAP), low oxygen modified atmosphere packs (low O_2 MAP) and controlled atmosphere packs (CAP) (Gill and Gill, 2005). Vacuum packaging has found application across a broad section of food products from cheese to meat, fish and even coffee products. High O_2 MAP treatments are principally applied to fresh red meat applications, but can also be used for products like salads. The MAP requirements for meat products are quite unique in that a high oxygen environment is sought because of the necessity to oxidise the pigment myoglobin to the very consumer appealing bright cherry-red colour. Low O_2 applications can also be applied to other meat products (cooked) and the requirements of products, like cheese and coffee, necessitate the removal of O_2 thus preventing oxidation and microbial growth with respect to cheese and oxidation of flavour volatiles with respect to coffee. It has uses for many food products and will be discussed in this section specifically relating to meat, cheese, bread, salads, seafood and coffee products.

In red meats, consumers relate the bright red colour to freshness, while discriminating against meat that has turned brown in colour (Hood and

Riordan, 1973; Morrissey et al., 1994). It is because of such sensory quality changes in fresh meat that so much attention has focused on developments within the area of packaging technologies, especially since the 1980's. MAP is one of the most popular refrigerated packaging formats for pork products, such as pork loin chops (O'Sullivan and Kerry, 2012). Additionally the bright cherry-red colour of fresh beef is used by consumers as an indicator of meat quality (Cassens et al., 1988; Kennedy et al., 2004), but fresh meat colour is short lived and surface discolouration that occurs during chilled and frozen storage is considered a sign of unwholesomeness and product deterioration. Thus, one of the primary objectives of meat producers has been the maintaining of optimal meat colour through the use of innovative packaging solutions such as modified atmosphere packaging (MAP). It is also likely that colour quality will be the limiting requirement for meat in package discolouring even before the limit of total bacterial counts have even been reached. Microbiological standards and guidelines give guidance on the types of microorganisms and their number that can be considered acceptable, unacceptable or unsafe in a food product. The following are the recommended microbiological limits for aerobic plate counts as applied to raw chilled meat: $M = 106$ or 6 log (CFU/g of meat) (acceptable limit) and $M = 107$ or 7 log (CFU/g of meat) (unacceptable limit) (ICMSF, 1986; European Commision, 2007). Concerning Enterobacteriaceae the recommended microbiological limits are: $M < 2.5$ log (CFU/g) or $M = 2.5$ log (CFU/g of meat).

The cutting and packaging of raw refrigerated meat in air-permeable overwrap packaging at individual stores has been gradually replaced by case-ready or centralised operations in many developed countries (McMillin, 1994). MAP is defined as 'A form of packaging involving the removal of air from the pack and its replacement with a single gas or mixture of gasses' (Parry, 1993).

The idea of MAP is by no means a new concept, it has its origins within the meat sector and perhaps for this reason it has found the widest adoption. In 1930, Killefer (1930) found that pork and lamb remained fresh for twice as long when stored in 100% carbon dioxide (CO_2) at $4-7°C$ compared to equivalent products stored in air and held at similar temperatures. Subsequently, Callow worked on the shelf life extension of bacon by packaging in CO_2-enriched atmospheres (Callow, 1932). A further commercial application was used in the 1930s to transport refrigerated beef carcasses from Australia and New Zealand in a CO_2-enriched environment (Floros and Matsos, 2005). MAP of retail-size meat packages was not introduced until the 1950s in the form of vacuum packaging (Floros and Matsos, 2005). It was not until 1981 that gas-flushed fresh meat in plastic trays was introduced by Marks and Spencer (Inns, 1987). It is now used ubiquitously across the meat industry for many different meat products. Modified atmosphere (MA) packs usually contain mixtures of two or three gases: O_2 (to enhance colour stability), CO_2 (to inhibit microbiological growth) and nitrogen (N_2, to maintain pack shape)

(Sorheim et al., 1999; Jakobsen and Bertelsen, 2000; Kerry et al., 2006). The fact that these gases promote the overall quality of fresh red meat has been well established (Gill, 1996). Meat colour is the first sensory modality that consumers encounter when purchasing both fresh and processed meat products. Only when the product has been brought home and cooked do sensory flavour and texture come in to play. For this reason the packaging methods by which meats are presented to the consumer have been optimised to maintain the optimum colour of meat and retard the development of unsightly discolourations. This, for the most part, has been achieved through the use of MAP (O'Sullivan and Kerry, 2011). High O_2 concentrations promote oxymyoglobin (OxyMb), the cherry-red form of myoglobin (O'Grady et al., 2000). To optimise shelf life, sensory quality and microbiological safety using MAP, the packaging system applied is highly product specific (Church and Parsons, 1995). By packaging beef in an MA and storage at low temperature, the shelf life can be prolonged considerably (Young, 1983; Gill and Penney, 1988). Beef and lamb are both red meats and share similar properties but considerable differences in shelf lives are apparent between them due to their relative susceptibility to chemical and microbial spoilage. In contrast to beef cuts, much of the surface of lamb is adipose tissue, which has a pH close to neutrality and has no significant respiratory activity (Robertson, 2006). The pH of beef is lower than that of lamb, thus making it less susceptible to microbial spoilage (Gill, 1989; Kerry et al., 2000).

The colour stability of meat products depends on variables such as muscle type, pH, storage temperature, oxygen availability and lighting type and intensity during display (Andersen and Skibstead, 1991). Additionally, different muscles exhibit different rates of OxyMb oxidation under controlled conditions (Ledward, 1985). The vast majority of meat products, have been and continue to be, offered in high oxygen pack formats (approximately 80% O_2) to maintain bloom, with at least 20% CO_2 to prevent selective microbial growth (Eilert, 2005). A patent in 1970 specified a range of O_2 and CO_2 concentrations suitable for MAP beef (Georgala and Davidson, 1970), which demonstrated that at least 60% O_2 is required to achieve a colour shelf life of 9 days and the patent claims that a mixture of 80% O_2 plus 20% CO_2 keeps meat red for up to 15 days at 4°C (Georgala and Davidson, 1970). Only O_2 and CO_2 have preservative effects (Gill and Gill, 2005). Okayama et al. (1995) also observed that MAP using a high level of O_2 (70–80%) preserved the bright red colour of fresh meat. In European countries such as Ireland, UK and France, beef steaks are commonly displayed in 70 mL O_2 and 30 mL CO_2 per 100 mL pack gas in MAP, whereas the concentrations used in the USA are 80 mL O_2 and 20 mL CO_2 per 100 mL pack gas (O'Sullivan et al., 2011). The packaging of muscle-based foods is necessary to ensure that such products reach the consumer in a condition which satisfies the demands of the consumer on a number of levels, namely, nutrition, quality, safety and convenience, as well as being capable of delivering a product shelf life that will endure the stresses of handling, transportation, storage, sale and consumer contact (O'Sullivan and Kerry, 2009). Thus, the packaging requirements for meat products are designed to optimise sensory quality and present highly appealing products to the consumer. The three sensory properties by which consumers most readily judge meat quality are appearance, texture and flavour (Liu et al., 1995). Colour perception plays a major role in consumer evaluation of meat quality (Lanari et al., 1995; Risvik, 1994) and is the sensory modality that has been the primary concern of meat packers. In the United States, there is also a high demand for fresh 'bone in' pork chops; however, colour deterioration of both muscle and bone limits marketing options (Lanari et al., 1995). Pork has a shorter shelf life than beef which may be due to a lower hygienic status during the production of pork. Also, the more rapid depletion of glycogen and glucose in pork compared with beef is of importance in dictating the shorter shelf life of pork (O'Sullivan and Kerry, 2012). The rate of discolouration of meat is believed to be related to the effectiveness of oxidation processes and enzymic reducing systems in controlling metmyoglobin levels in meat (Faustman and Cassens, 1989). Discolouration in retail meats during display conditions may occur as a combined function of muscle pigment oxidation (OxyMb to met-myoglobin) and lipid oxidation in membrane phospholipids (Sherbeck et al., 1995). High O_2 concentrations promote the OxyMb, cherry-red form of myoglobin (O'Grady et al., 2000) but may impact negatively on the oxidative stability of muscle lipids and lead to development of undesirable flavours (Rhee and Ziprin, 1987; Estevez and Cava, 2004). Also, the breakdown products of lipid oxidation have been associated with the development of off-flavours and off-odours and loss of colour in meat (Faustman and Cassens, 1989). Many researchers believe transition metals, notably iron, play a role in initiating lipid oxidation either directly (Harel and Kanner, 1985) or indirectly by facilitating the generation of other initiating species. Furthermore, MAP meat products held in high oxygen (O_2) atmospheres may result in protein oxidation which may have a negative effect on meat tenderness. Rowe et al. (2004) found that increased oxidation of muscle proteins early postmortem could have negative effects on meat tenderness. Recent studies have indicated that storage under high O_2 atmospheres can result in a decrease in beef tenderness due to protein oxidation (Torngren, 2003; Zakrys-Walwander et al., 2012). As such, the requirements for colour stability must be balanced against the deteriorative action of lipid oxidation. Protein oxidation as well as color and flavor changes in meat products are discussed in detail in Chapter 11 of this book.

Many attempts have been made to reduce pigment and lipid oxidation in meats through treatments with, for instance, antioxidants and MA packaging (O'Grady et al., 2006; Carpenter et al., 2007; O'Sullivan et al., 1998). MAP still remains one of the principle methods of maintaining and prolonging meat colour sensory quality.

Beef steaks are commonly displayed under high oxygen concentrations in MAP to promote colour stability (Zakrys et al., 2008). As stated earlier, MA packs usually contain mixtures of two or three gases: O_2 to enhance colour stability, CO_2 to inhibit growth of spoilage bacteria (Seideman and Durland, 1984) and the N_2 is used in MAP as an inert filler gas either to reduce the proportions of the other gases or to maintain pack shape (Bell and Bourke, 1996; Sorheim et al., 1999; Jakobsen and Bertelsen, 2000; Kerry et al., 2006). High O_2 concentrations promote the development of OxyMb, cherry-red form of myoglobin (O'Grady et al., 2000) but may impact negatively on the oxidative stability of muscle lipids and lead to development of undesirable flavours (Rhee and Ziprin, 1987; Estevez and Cava, 2004). Much work has been done on a distinctive off-flavour that develops rapidly in meat that has been precooked, chilled-stored and reheated. The term warmed-over flavour (WOF) has been adopted to identify this flavour deterioration (Renerre and Labadie, 1993). High oxygen levels within MAP also promote oxidation of muscle lipids over time (O'Grady et al., 1998). High O_2 MAP increases lipid oxidation in meat; Jakobsen and Bertelsen (2000) in beef; Lund et al. (2007a,b) in pork; and Kerry et al. (2000) in lamb. Oxidation of poly-unsaturated fatty acids not only causes the rapid development of meat rancidity but also affects the colour, the nutritional quality and the texture of beef and pork (Kanner, 1994; Lund et al., 2007a,b). While high O_2 levels promote the oxidation of lipids, it is the membrane phospholipids that are particularly susceptible to oxidation processes, thereby causing the rapid development of meat rancidity (Renerre, 1990). Overall high O_2 MAP will prolong the colour quality of packaged meat products but may negatively affect other sensory quality variables such as flavour, aroma and texture (O'Sullivan and Kerry, 2012) (Fig. 7.1).

FIGURE 7.1   Modified atmosphere packed gourmet beef burgers.

The very recent advent of vacuum skin packaging (VSP) has found widespread consumer appeal with respect to meat packaging because it is so convenient and looks good. With VSP, the meat is held under vacuum and appears purple, but nonetheless this format has been widely adopted by the consumer. This will be discussed in detail in the later section on vacuum packaging.

LOW O_2 MODIFIED ATMOSPHERE PACKAGING

Typically fresh red meats are stored in MAP containing 80% O_2:20% CO_2 (Georgala and Davidson, 1970) and cooked meats are stored in 70% N_2:30% CO_2 (Smiddy et al., 2002a,b,c). However, low O_2 MAP are generally packed with CO_2 (usually enough to dissolve into the product), and also N_2, while residual O_2 may be present or included during the packing process. Here, again the CO_2 acts as the preservative and the N_2 maintains the pack shape. For low O_2 MAP, carbon monoxide (CO) may also be used as a gas for prolonging meat colour integrity. CO is a colourless, odourless and tasteless gas. It is produced mainly through incomplete combustion of carbon-containing materials (Sørheim et al., 1997). Although a substantial increase in the shelf life of meat can be obtained by using various MAs, it is often limited by discolouration due to the oxidation of OxyMb to metmyoglobin. This discolouration can be prevented by the inclusion of a low level of CO in the gas mixture. Inclusion of 0.4% CO in conjunction with O_2 will not influence colour stability, metmyoglobin-reducing activity or O_2 consumption. This is likely the result of greater formation of OxyMb in atmospheres containing 20–80% O_2, which dominates or limits the ability of carboxymyoglobin (COMb) to form (Seyfert et al., 2007). COMb is more resistant to oxidation than OxyMb, owing to the stronger binding of CO to the iron-porphyrin site on the myoglobin molecule (Wolfe, 1980). However, one of the main consumer fears relating to the use of CO is the possible loss of quality due to a break in the cold chain causing deterioration in spite of its attractive appearance (Wilkinson et al., 2006). CO also results in the development of off-odours which may warn consumers of possible loss of quality (Knut and Nolet, 2006).

The storage life of chilled meat can be extended by packaging the product under CAP with N_2 or CO_2 (Gill and Molin, 1991). CAP do not change with respect to their atmospheres during storage and, as such, have very low gas permeabilities. O_2 scavengers may also be incorporated to remove any residual O_2 that may have been included during the manufacturing process.

Fresh baked bread is naturally a highly perishable product, with a typical shelf life of less than 7 days when stored under optimum conditions. This degree of perishability in bread-based products is hard to delay without causing negative sensory attributes (Hempel et al., 2013d). Typically and traditionally, fresh bread products are packaged using waxed paper or polyolefin-based plastic packaging. Since 2000, higher value, speciality type breads have been packaged using laminate constructions and packaged under MA (Hempel et al., 2013d). Bake-off technology not only reduces economic losses due to staling but also provides the possibility of fresh bread for consumers at any time (Lainez et al., 2008). However, the shelf life of part-baked products is very short and it is a problem to maintain their quality during prolonged storage (Karaoglu et al., 2005). A combination of various gases can be utilised in the MAP of breads, thereby creating an atmosphere capable of extending shelf life. Packaging under MA, commonly 100% CO_2, is important to optimise the preservation of part-baked bread (Khoshakhlagh et al., 2014). Carbon dioxide has the ability to selectively inhibit the growth of microorganisms, specifically yeasts and moulds, and so is the most popular gas used for bakery goods (Cauvain and Young, 2000). Also, bread products with low water activity are good candidates for MAP including baguettes, bagels, pita tortilla wraps and naan bread, where a high CO_2 atmosphere will prevent mould growth (Penicillium species are by far the most common), the principal cause of microbiological deterioration in such products. However, it is difficult to reduce O_2 content to a very low level in bakery product packs due to the porous interior of many of these products which tend to trap oxygen in such a way that it does not readily interchange with the gas which is flushed and flowing through the package at the point of air evacuation or displacement (Matz, 1989).

Consumer demand for freshness and convenience has led to the evolution and increased production of numerous varieties of minimally-processed vegetables presented in a wide range of packaging formats (Hempel et al., 2013b,c). The shelf-life of ready-to-eat (RTE) vegetable products or salads established by manufacturers is usually 7−14 days depending on the type of fresh produce selected, and is determined by loss in organoleptic qualities (Garcia-Ginemo and Zurera-Cosano, 1997). As a result of peeling, grating and shredding, produce will change from a relatively stable product with a shelf-life of several weeks or months to a perishable entity that possesses a very short shelf life, even as short as 1−3 days at chill temperatures (Hempel et al., 2013b,c). It is possible to achieve a shelf-life of 7−8 days at refrigeration temperatures (5°C), but for some markets this is not enough and a shelf-life of 2−3 weeks is sometimes necessary (Ahvenainen, 1997). MAP technology is thus now extensively used for extended storage of minimally processed fruit and vegetables, including fresh ready to eat (RTE) salad products. Oxygen (O_2), carbon dioxide (CO_2) and nitrogen (N_2) are the gases typically implemented in MAP of fresh produce, with O_2 in packs generally being employed between 1% and 5% in order to support product respiration but used at low enough concentration to discourage the proliferation of microbial spoilage in the form of bacteria and fungi (Hempel et al., 2013b,c). The use of high O_2 atmospheres (i.e., $>70\%$ O_2) have been used as an alternative technique to low O_2 equilibrium modified atmosphere (3% O_2) (Jacxsens et al., 2001).

The delicate flavour of coffee can deteriorate as important flavour volatiles oxidise in air. For this reason it is important that coffee packaging is impermeable to O_2. Typically roasted coffee will diffuse CO_2 for several hours after roasting which can displace any O_2 present in the package and facilitated by a built in nonreturn valve in the product packaging. Nitrogen can also be flushed through the packaging just prior to closure as a further reduction in any potential O_2 levels as even in very low concentrations (5–10%) oxidation of flavour volatiles can occur.

Cheese is another example of a foodstuff with varying packaging requirements.

MAP is used to pack cheddar cheese slices where the gas is usually 100% CO_2 or CO_2/N_2 combinations which prevent mould growth on the surface. Grated cheese can also be packed in similar gas mixtures, but due to the increased surface area of the product it is much more susceptible to lipid oxidation and thus a light absorbing layer must be incorporated by using a printable and transparent UV-protected and light resistant film.

Similarly cheese can be vacuum packed, where it is very stable for long periods. Also, cheeses can be packaged in packaging materials that allow the cheese to continue to mature, and thus allow CO_2 produced by resident microflora to escape through the package barrier. Carbon dioxide may originate from lactobacilli and secondary starters, such as propionic acid bacteria in various hard or semihard cheese varieties (Schneider et al., 2010). The latter example must allow CO_2 to escape. The transfer of carbon dioxide through the packaging material(s) is necessary to avoid bulging/blowing of the package (Kammerlehner, 2003). The plastic packaging material must thus reflect the barrier properties depending on the requirement of the particular cheese. Laminated cheese packaging materials may be composed of PA/PE, PE/PA/ PE, PET/PE, oriented polyamide (OPA)/PE, polystyrene/ethylene vinyl alcohol (PS/EVOH), PE/PE or PE/Barex. Aluminium foil laminates are virtually impervious to gas and water vapour while EVOH, PET, Barex and to a lesser extent, polyvinylidene chloride prevent O_2 diffusion (Schneider et al., 2010)

Soft cheeses like Brie, Camembert and Roquefort, because of their soft characteristics, are often packaged in trays or boxes to maintain product integrity. Suitable primary packaging materials usually consist of a coated paper or parchment and an outer layer frequently consisting of cellophane, OPP or PET films (Sturm, 1998). This outer layer is permeable to oxygen to allow the resident microflora to continue to respire but relatively impermeable to water vapour so that the water activity of the cheese remains in the range allowing the microflora to continue living.

Seafood is a highly perishable product but shelf life can be extended by up to three times using refrigeration and MAP packaging formats typically with a product to gas ration of 3:1. For white fish, CO_2 is effective in inhibiting the growth of common aerobic bacteria whereas O_2 assists in preserving the colour of the flesh and prevents the growth of the anaerobic *Clostridium botulinum* pathogen, with a typical gas mixture of 40% CO_2/30% N_2/30% O_2. Fish with high fat contents are typically packaged in a CO_2 atmosphere with N_2 as a filler to prevent pack collapse (40% CO_2/60% N_2). Prawns, lobster, crab, squid, mussels and cockles are usually packed in mixtures of 40% CO_2/30% N_2/30% O_2. Goulas et al. (2005) in a packaging study using refrigerated mussels (*Mytilus galloprovincialis*) found that the optimal format was MAP with 80%/20% CO_2/N_2 which extended shelf life to 14−15 days compared to aerobic, other MAP and vacuum pack formats.

VACUUM PACKAGING

Vacuum packaging was one of the earliest forms of MAP methods developed commercially and still is extensively used for products such as primal cuts of fresh red meat and cured meats (Parry, 1993), cheese (Schneider et al., 2010) and fish (Kaale et al., 2013) etc. Coffee beans are often vacuum packed to preserve freshness. Vacuum packaging involves the evacuation of air from the packs prior to sealing and is also extensively used for the packaging of pork products such as pork steak or fillet, bacon rashers, whole hams and cured or barbecue ribs to mention just a few products (O'Sullivan and Kerry, 2012). Vacuum packs are comprised of evacuated pouches or vacuum skin packs, in which a film of low gas permeability is closely applied to the surface of the product. Preservative effects are achieved by the development of an anaerobic environment within the pack (Gill and Gill, 2005; Bell et al., 1996). Under good vacuum conditions, the oxygen level is reduced to less than 1%. Due to the barrier properties of the film used, entry of oxygen from the outside is restricted (Parry, 1993; Robertson, 2006). In the case of meat products it is hoped that any residual O_2 in any remaining atmosphere or dissolved in the product will be removed by enzymatic reactions within the muscle tissue or through other chemical reactions with tissue components (Gill and Gill, 2005). In the case of vacuum-packaged meat, respiration of the meat will quickly consume the vast majority of residual O_2, replacing it with CO_2 which eventually increases to 10−20% within the package (Taylor, 1985; Parry, 1993; Gill, 1996). As the capacity of the muscle tissue for removing O_2 is limited, the amount of O_2 remaining in the pack at the time of closure must be very small if the product is to be effectively preserved (Gill and Gill, 2005).

It has been often stated that vacuum-packaged meat is unsuitable for the retail market because depletion of O_2 coupled with low O_2 permeability of the packaging film causes a change of meat colour from red to purple due to the conversion of OxyMb to deoxymyoglobin. These are not acceptable meat colours to the consumer (Allen et al., 1996; Parry, 1993). If fresh beef colour is not a bright cherry-red, the meat may be considered undesirable or even spoiled. Bright red beef outsells discoloured beef (20% surface met-myoglobin) by a ratio of 2:1 (Hood and Riordan, 1973). Colour is perhaps the most important sensory attribute of a food because if it is deemed unacceptable, the food will not be purchased and/or eaten, and consequently, all other sensory attributes lose significance (Clydesdale, 1978). American consumers have demonstrated a bias against the purchase of vacuum-packaged beef which displays the purple colour of deoxymyoglobin (Meischen et al., 1987). A further disadvantage is the accumulation of drip during prolonged storage of meat in vacuum packs (Jeremiah et al., 1992; Parry, 1993; Payne et al., 1997). However, this can partly be overcome by VSP (Fig. 7.2) using a film that fits very tightly to the meat surface, leaving little space for the accumulation of any fluid exudate (Hood and Mead, 1993). VSP involves production of a skin package in which the product is the forming mould. It was first introduced using an ionomer film, which softens on heating to such an extent that it can be draped over sharp objects without puncturing (Robertson, 2006). The meat is placed in trays, the upper packaging film is then heated and shrinks tightly around the meat and adheres to the trays when vacuum is drawn. One advantage of this is that it will produce almost no wrinkles in the packaging in which purge loss may collect (Vázquez et al., 2004). An advantage and disadvantage of this package is that it gives the product a unique look. The product shelf life can be 15–22 days depending on the cut which is even more so than traditional vacuum packaging probably due to the surface heat treatment during shrinking and the subsequent lower rate of microbial growth. Since the product is displayed in the myoglobin state, there is no loss of colour in the display case and oxidation issues are minimised with this type of package (Belcher, 2006). Lagerstedt et al. (2011) found no clear differences between skin-packed and vacuum-packed steaks for shear force and total loss, however, skin-packed steaks had lower purge loss which might be more appealing to the consumers in retail display. These authors also state that VSP offers a consumer friendly alternative for the retail display of meat and thus supercedes high O_2 MAP due to the many advantages it has to offer. Another advantage of VSP, put forward by consumers, is the lack of off-odour when the package is opened. With traditional vacuum packs odours may be detected

FIGURE 7.2 Vacuum skin packed mackerel fillets and an ILPRA (Vigevano, Italy) vacuum skin packaging machine.

when opening vacuum packs and have been described as sour, acid or cheesy (Dainty et al., 1979). Finally, as the product is conveniently vacuum packed the consumer can easily freeze the product if they so wish without having to repack, and due to O_2 exclusion the product can potentially last in the freezer for several months with minimal loss of quality. Vacuum packaging also prevents dehydration and evaporative water loss from the surface of the food and can minimise the effects of freezer burning (excessive hydration loss from the product surface) and drip loss on thawing (Fernández et al., 2010; Pornchai and Chitsiri, 2011).

Traditional rind cheeses like Edam, Gouda and Parmesan can be coated in wax to prevent further moisture losses once maturation has been reached. The wax can be composed of polymers/copolymers of ethylene, polyvinyl acetate, esters of maleic or fumaric acid (Schneider et al., 2010; Sturm, 1998; Strehle, 1997; Spreer, 2006). To facilitate production efficiency, cheese blocks can be dried at the surface after brining and then packed under medium vacuum (50−70 kPa) in bags which in turn may undergo heat shrinking at 85−92°C. This ensures a tight contact between surfaces/edges of the product and the packaging material (Sturm, 1998). Similarly, vacuum packaging is a very common format for retail packaging of many different types of cheese. The elimination of O_2 from the pack extends shelf life and retards the development of off-flavours due to lipid oxidation. Calcium lactate crystals sometimes forms on the surface of Cheddar cheese and may pose a significant expense to manufacturers due to consumer rejection of cheeses that prominently display the white crystalline deposits. This problem may be prevented by using a combination of smooth cheese surface plus very tight packaging which virtually eliminates crystal formation, presumably by eliminating available sites for nucleation (Rajbhandari and Kindstedt, 2014). Clearly, vacuum packaging can be an important format in the prevention of calcium lactate crystal formation.

ACTIVE AND INTELLIGENT PACKAGING

Active packaging has been defined as 'a type of packaging that changes the condition of the packaging to extend shelf life or improve safety or sensory properties while maintaining the quality of the food' (Quintavalla and Vicini, 2002). Active packaging systems can include oxygen scavengers, carbon dioxide scavengers and emitters, moisture control agents and antimicrobial packaging technologies (Kerry et al., 2006). Active packaging has the advantage of maintaining the preservative effects of various compounds (antimicrobial, antifungal or antioxidant), but without being in direct contact with the food product (O'Sullivan and Kerry, 2012). By incorporation of the preservative effects directly into packaging, preservation may be maintained which will compensate for any removal of preservative effect from actual food products (O'Sullivan and Kerry, 2012). Ahvenainen (2003) defined active packaging as packaging, which 'changes the condition of the packed food to extend shelf-life or to improve safety or sensory properties, while maintaining the quality of packaged food'. Packaging may be termed active when it performs some desired role in food preservation other than providing an inert barrier to external conditions (Hutton, 2003). Rooney (1995) defined an active package as a material that 'performs a role other than an inert barrier to the outside environment'. Looking to the consumers' demand for chemical preservative-free foods, food manufacturers are now using naturally occurring antimicrobials to sterilise and/or extend the shelf life of foods (Han, 2005). For example, the preservative effect of active packaging can substitute for the reduced preservative effects of salt or nitrate. By reducing the growth and spread of spoilage and pathogenic microorganisms in meat foodstuffs, antimicrobial packaging materials can inhibit or kill the microorganisms and thus extend the shelf life of perishable products and enhance the safety of packaged products (Han, 2005). Similarly, packaging films that release organic acids offer potential for reducing the effect of the growth of slime-forming bacteria on meat (Rooney and Han, 2005). The aim of active packaging is to increase the display life of contained products, while maintaining their quality, safety and sensory properties, without direct addition of the active agents to the product (Coma et al., 2008). Chemical preservatives can be employed in antimicrobial-releasing film systems, including organic acids and their salts (sorbates, benzoates and propionates), parabens, sulfites, nitrites, chlorides, phosphates, epoxides, alcohols, ozone, hydrogen peroxide, diethyl pyrocarbonate, antibiotics and bacteriocins (Ozdemir and Floros, 2004). The antimicrobial agent is incorporated into the packaging material by either spraying, coating, physical mixing or chemical binding (Berry, 2000). This is an important development, considering the consumer drive toward clean labelling of food products and the desire to limit the use of food additives (O'Sullivan and Kerry, 2009).

The most prevalent form of active packaging in the food industry is based on oxygen scavenging. Oxygen scavengers operate on the principal of binding-free oxygen present in packages after sealing to make it unavailable for the growth of aerobic microorganisms. Additionally, this also prevents lipid oxidation and retards the formation of oxidised off-flavours. Oxygen scavenging technologies utilise one or more of the following concepts: iron powder oxidation, ascorbic acid oxidation, photosensitive dye oxidation, enzymatic oxidation (e.g., glucose oxidase and alcohol oxidase), unsaturated fatty acids (e.g., oleic or linolenic acid) rice extract or immobilised yeast on a solid substrate (Floros et al., 1997; Gill and McGinnis, 2003).

Carbon dioxide scavengers or absorbers (sachets), consisting of either calcium hydroxide and sodium hydroxide, or potassium hydroxide, calcium oxide and silica gel, may be used to remove carbon dioxide during storage to prevent bursting of the package (Kerry et al., 2006). Carbon dioxide emitters function on the basis of suppressing microbial growth. The release of carbon dioxide operates similar to O_2 scavenging. Since the permeability of carbon dioxide is 3—5 times higher than that of oxygen in most plastic films, it must be continuously produced to maintain the desired concentration within the package (Ozdemir and Floros, 2004).

Active packaging also refers to the incorporation of additives into packaging systems with the aim of maintaining or extending product quality and shelf life. Antimicrobial packaging includes systems such as adding a sachet into the package, dispersing bioactive agents in the packaging, coating bioactive agents on the surface of the packaging material or utilising antimicrobial macromolecules with film-forming properties or edible matrices (Coma, 2008). Several categories of antimicrobials have been tested for antimicrobial packaging applications and include: organic acids, fungicides, bacteriocins, proteins, inorganic gases, silver substitute zeolite and enzymes (Han, 2000; Suppakul et al., 2003; Cutter, 2002). Antimicrobial food packaging materials have to extend the lag phase and reduce the growth phase of microorganisms to extend shelf life and to maintain product quality and safety (Han, 2000; Quintavalla and Vicini, 2002). Among the functional substances which are thought most suitable for their incorporation in the package wall are phytochemicals, vitamins, nanofibres and prebiotics (Lopez-Rubio et al., 2006). Also, antimicrobial substances incorporated into packaging materials can control microbial contamination by reducing the growth rate and maximum growth population and/or extending the lag phase of the target microorganism, or by inactivating microorganisms by contact (Quintavalla and Vicini, 2002).

It has been suggested that the addition of ethanol to modify the atmosphere in MAP bakery foods could increase shelf life (Matz, 1989). Ethanol has been demonstrated to extend the shelf life of packaged bread and other baked products, where ethanol vapours have been shown to be effective in controlling Aspergillus and Penicillium species and various spoilage yeasts (Brody, 2001). Ethanol emitters release absorbed or encapsulated ethanol from a sachet placed within the product package prior to closure and are particularly effective against moulds, but can also inhibit yeasts and bacteria, particularly for bakery products (Day, 2008). Different commercial sources of alcohol (whisky, brandy) are used in many premium bakery products, not only to inhibit mould growth but also to impart unique flavours associated with such alcohol-based products (Cauvain Young, 2000).They have found widespread use in Japan to extend the mould-free shelf life of cakes and other high moisture bakery products by up to 2000% (Rooney, 2005; Day, 2003). Hempel et al. (2013d) in a study on speciality bread found that ethanol emitters could extend shelf life without the need of additional MA gas. Acceptable limits for microbial quality were maintained for 16 days when packaged in air using ethanol emitters with no negative effect on product sensory quality (Hempel et al., 2013d).

Rodríguez-Calleja et al. (2012) investigated the combined effects of high hydrostatic pressure, a commercial liquid antimicrobial edible coating consisting of lactic and acetic acid, sodium diacetate, pectin and water ('articoat DPL'), and MAP on the shelf life of chicken breast fillets. These authors found that the sensory and microbiological data obtained showed that the combined treatment of mild high pressure at 300 MPa and application of the commercial liquid antimicrobial edible coating 'articoat DPLTM' exhibited a strongly synergetic interaction, extending the shelf life of skinless chicken breast fillets up to 4 weeks. This confirms the potential utility of the hurdle strategy for improving the shelf life of raw poultry meat using an edible antimicrobial agent.

Active packaging refers to packages having active functions beyond the inert passive containment and protection of the product (Soroka, 2008). Intelligent and smart packaging, on the other hand, refers usually to the ability to sense or measure an attribute of the product, the inner atmosphere of the package or the shipping environment. Packages integrated with a sensor or an indicator such as time–temperature indicators, gas indicators and biosensors are regarded as intelligent packaging (IP) systems because they inform the consumer about the kinetic changes related to the quality of the food or the environment that is contained within the package (Yucel, 2016). One of the most effective forms of IP is the O_2 sensor. The use of a nondestructive method of O_2 detection to monitor O_2 levels in different forms of food packaging provides information as to the package environmental conditions of a food product throughout storage and shelf life (Hempel et al., 2013b,c). Problems associated with packaging and containment failures can be instantly observed post packaging using this technology (Hempel et al., 2013a). Optical oxygen sensors are thus a novel and effective way of nondestructively detecting and monitoring oxygen (Papkovsky, 1995, 2004) within product packages. The use of a reversible optical sensor incorporating a phosphorescent dye that is quenched by molecular oxygen has been well documented in food packaging applications. A colour change thus indicates when the level of O_2 has changed. The use of optical sensors has been previously shown to be a valuable and novel tool in determining the excess levels of O_2 in low O_2 packaging formats due to packaging containment failures brought about predominantly by the physical damage to packaging materials during the pack-forming process (Hempel et al., 2012). The reversible nature of the optical sensors demonstrates a novel and effective way of continuously monitoring fluctuations in O_2 levels within packs, from O_2 ingress within packs to O_2 consumption by microbial, biochemical and chemical processes (Hempel et al., 2013a; O'Mahoney et al., 2006; Smiddy et al., 2002a,b,c). These sensors have found applications in MAP cheese (O'Mahoney et al., 2006), vacuum-packed cheese (Hempel et al., 2012), MAP and vacuum-packed beef (Smiddy et al., 2002a,b,c), cooked meats (Smiddy et al., 2002a,b,c), MAP and vacuum-packed chicken (Smiddy et al., 2002a,b,c), as well as sous vide products (Papkovsky, 2004), bottled beer (Hempel et al., 2013a) and RTE salads (Hempel et al., 2013b,c). Hempel et al. (2013b,c) demonstrated the effectiveness of O_2 sensors in detecting packaging faults and containment failures in commercial MAP bread products.

The case study in Chapter 13 demonstrates an intelligent oxygen sensor system for beer freshness (Hempel et al., 2013a). Additionally, Chapter 14 demonstrates the use of smart packaging technologies for monitoring and extending the shelf life quality of MA packaged bread with the application of intelligent oxygen sensors and active ethanol emitters (Hempel et al., 2013d).

Future Trends in Packaging

The popularity of retort pouch product formats is increasing as the appeal for the traditional can is stagnating or falling with some sources predicting that canned desserts and sauces will completely disappear by 2020 (Holter, 2011). While offering the advantages of a long shelf life and low prices, canned food suffers from a poor consumer perception of healthiness and taste quality (O'Halloran, 2013). Consumers are increasingly turning to fresh, chilled-processed and frozen-processed food varieties.

High O_2 MAP is routinely employed in the meat industry for many different meat products. Whereas, low O_2 packaging systems have been readily available in the United States, but not as widely implemented as the high oxygen counterparts (O'Sullivan and Kerry, 2008). Vacuum packaging continues to be in many cases, the most cost-effective packaging strategy. As the food industry moves toward central processing that employs MAP and VSP, processors may need to overcome consumer preference for fresh meat that is bright red in colour and packaged with the traditional PVC overwrap (O'Sullivan and Kerry, 2008). VSP steaks will bloom sufficiently when exposed to air.

On pack opening, it has been reported that VSP meat will have a better colour stability than conventional VP meat (Li et al., 2012). It is also reported that some consumers were willing to pay more for VSP meat than for various types of gas-packaged meat (Aaslyng et al., 2010). A relatively recent innovation in vacuum packaging has been the evolution of shrinkable films in use with horizontal form-fill-seal machinery (Salvage and Lipsky, 2004). This packaging format uses a styrene or PP tray and a barrier film that can form around the product to reduce the amount of purge from coming out of the product. An additional web of film or a header can also be added for prepricing and prelabelling (Belcher, 2006). Nevertheless, it is encouraging that the initial perceptions of quality will likely not bias eating satisfaction once a decision to purchase is made and the meat is taken home, thereby hastening the acceptance of the newer packaging technologies (Carpenter et al., 2001; O'Sullivan and Kerry, 2012).

Processing and packaging technologies which are accepted by the market and adopted by the industry will have to become more efficient, consistent and leaner in activity if future global challenges are to be met as the consumer has become very much more discerning with respect to the origins of the food they consume (O'Sullivan and Kerry, 2008). Consumer demands for more environmentally friendly packaging and more natural products will also create increased demand for packaging from biodegradable and renewable resources (Cutter, 2006). Packaging materials are principally made from plastics including polyethylene. The materials used for food packaging today consist of a variety of petroleum-derived plastic materials (Weber et al., 2002). The volatility of oil prices has a direct effect on the cost of traditional petrochemical-based packaging materials (O'Sullivan and Kerry, 2012). Also, the environmental considerations of disposing of traditional packaging after use has become centre stage with respect to green solutions to modern living (O'Sullivan and Kerry, 2012). The increased costs of petroleum will continue to drive the demands for biobased packaging materials (Cutter, 2006; Marsh and Bugusu, 2007; O'Sullivan and Kerry, 2009, 2012). However a variety of biobased materials have been shown to prevent, drip loss, reduce lipid oxidation and improve flavour attributes, as well as enhancing the handling properties, colour retention, and microbial stability of foods (Cutter, 2006). Polymers obtained from renewable sources, widely used in the food industry, have been tested as raw materials in the production of a new type of more sustainable packaging with specific functionality, such as the control of moisture content, gases and the migration of food additives and/or nutrients (Martins Sousa et al., 2016). Of the polymers from renewable sources, starch, the largest constituent of cereal grains, including polished rice, is used to make biodegradable films (Liu et al., 2005). Other polymers made from renewable resources include polysaccharides such as alginates, carrageenans, chitosan/chitin, pectin and starch (O'Sullivan and Kerry, 2012). Proteins can also be used such as casein, whey, collagen, gelatin, corn, soy and wheat, as well as lipids including fats, waxes and oils (Cutter and Sumner, 2002; Cutter, 2006). Food-based biopolymer materials have also been successfully used to produce edible packaging and thus there is tremendous potential for edible packaging to be applied within the food industry. One of the oldest and most widespread examples of this is the use of gelatin in the manufacture of sausage casings (Liu et al., 2005).

REFERENCES

Aaslyng, M.D., Tørngren, M.A., Madsen, N.T., 2010. Scandinavian consumer preference for beef steaks packed with or without oxygen. Meat Science 85, 519–524.

Ahvenainen, R., 2003. Active and intelligent packaging: an introduction. In: Ahvenainen, R. (Ed.), Novel Food Packaging Techniques. Woodhead Publishing Ltd., Cambridge, UK, pp. 5–21.

Ahvenainen, R., 1997. New approaches in improving the shelf life of minimally processed fruit and vegetables. Trends Food Science and Technology 7, 179–187.

Allen, P., Doherty, A.M., Buckley, D.J., Kerry, J., O'Grady, M.N., Monahan, F.J., 1996. Effect of oxygen scavengers and vitamin E supplementation on colour stability of MAP beef. In: 42nd International Congress of Meat Science and Technology, pp. 88–89.

Andersen, H.J., Skibsted, L.H., 1991. Oxidative stability of frozen pork patties-effect of light and added salt. Journal of Food Science 56, 1182–1184.

Belcher, J.N., 2006. Industrial packaging developments for the global meat market. Meat Science 74, 143–148.

Bell, R.G., Bourke, B.J., 1996. Recent developments in packaging of meat and meat products. In: Proceedings of the International Developments in Process Efficiency and Quality in the Meat Industry, pp. 99–119. Dublin Castle, Ireland.

Bell, R.G., Penney, N., Moorhead, S.M., 1996. The retail display life of steaks prepared from chill stored vacuum and carbon dioxide-packed sub-primal beef cuts. Meat Science 42 (2), 165–178.

Berry, D., August 2000. Packaging's Role. Brief Article. Dairy Foods.

Brody, A.L., 2001. Antimicrobial packaging. In: Active Packaging for Food Applications. Technomic Publishing Company, PA, USA, pp. 157–159.

Callow, E.H., 1932. Gas storage of pork and bacon. Part 1.Preliminary experiments. Journal of the Society of Chemical Industry 51, 116–119.

Carpenter, C.E., Cornforth, D.P., Whittier, D., 2001. Consumer preferences for beef color and packaging did not effect eating satisfaction. Meat Science 57, 359–363.

Carpenter, R., O'Grady, M.N., O'Callaghan, Y.C., O'Brien, N.M., Kerry, J.P., 2007. Evaluation of the antioxidant potential of grape seed and bearberry extracts in raw and cooked pork. Meat Science 76, 604–610.

Cassens, R.G., Faustman, C., Jimenez-Colmenero, F., 1988. Modern developments in research on colour of meat. In: Krol, B., Van Roon, P., Houben, J. (Eds.), Trends in Modern Meat Technology 2. Pudoc, Wageningen, The Netherlands, p. 2.

Cauvain, S.P., Young, L.S., 2000. Strategies for extending bakery product shelf life. In: Bakery Food Manufacture and Quality. Blackwell Science Publishing, Cornwall, UK, pp. 201–203.

Chambers IV, E., Bowers, J., 1993. Consumer perception of sensory quality in muscle foods: sensory characteristics of meat influence consumer decisions. Food Technology 47, 116–120.

Church, I.J., Parsons, A.L., 1995. Modified atmosphere packaging technology: review. Journal of the Science of Food and Agriculture 67, 143–152.

Clydesdale, F.M., 1978. Colorimetry – methodology and applications. CRC Critical Reviews in Food Science and Nutrition 10, 243–301.

Coma, V., 2008. Bioactive packaging technologies for extended shelf life of meat-based products. Meat Science 78, 90–103.

Cutter, C.N., 2006. Opportunities for bio-based packaging technologies to improve the quality and safety of fresh and further processed muscle foods. Meat Science 74, 131–142.

Cutter, C.N., Sumner, S.S., 2002. Application of edible coatings on muscle foods. In: Gennadios, A. (Ed.), Protein-based Films and Coatings. CRC Press, Boca Raton, FL, pp. 467–484.

Dainty, R.H., Shaw, B.G., Harding, C.D., Michainie, S., 1979. The spoilage of vacuum packed beef by cold tolerant bacteria. In: Russell, A.D., Fuller, R. (Eds.), Cold Tolerant Microbes in Spoilage and the Environment. Academic Press, London and New York, pp. 83–100.

Day, B.P.F., 2003. Active packaging. In: Coles, R., McDowell, D., Kirwan, M. (Eds.), Food Packaging Technologies. CRC Prss, Boca Raton, FL., USA, pp. 282–302.

Day, B.P.F., 2008. CH1, active packaging of food. In: Kerry, J.P., Butler, P. (Eds.), Smart Packaging Technologies. John Wiley & Sons, Chichester, West Sussex, UK, pp. 1–18.

Eilert, S.J., 2005. New packaging technologies for the 21st century. Meat Science 71, 122–127.

Estevez, M., Cava, R., 2004. Lipid and protein oxidation, release of iron from heme molecule and colour deterioration during refrigerated storage of liver pate. Meat Science 68, 551–558.

European Commission, 2007. Commission regulation (EC) 1447/2007 of 5 December 2007 on amending Regulation (EC) 2073/2005 on microbiological criteria on foodstuffs. Official Journal of the European Union. L 322/12-L322/29, Brussels.

Faustman, C., Cassens, R.G., 1989. Strategies for improving fresh meat colour. In: Proceedings, 35th International Congress of Meat Science and Technology, pp. 446–453. Copenhagen, Denmark.

Fernández, K., Aspé, E., Roeckel, M., 2010. Scaling up parameters for shelf-life extension of Atlantic salmon (*Salmo salar*) fillets using superchilling and modified atmosphere packaging. Food Control 21 (6), 857–862.

Floros, J.D., Matsos, H.I., 2005. Chapter 10. Introduction to modified atmosphere packaging. In: Han, J.H. (Ed.), Innovations in Food Packaging, pp. 159–171.

Floros, J.D., Dock, L.L., Han, J.H., 1997. Active packaging technologies and applications. Food Cosmetics and Drug Packaging 20, 10–17.

Georgala, D.L., Davidson, C.M., 1970. Food package. British Patent 1,199,998.

Garcia-Ginemo, R.M., Zurera-Cosano, G., 1997. Determination of ready to eat vegetable salad shelf life. International Journal of Food Microbiology 36, 31–38.

Gill, C.O., McGinnis, J.C., 2003. The use of oxygen scavengers to prevent the transient discolouration of ground beef packaged under controlled, oxygen-depleted atmospheres. Meat Science 41, 19–27.

Gill, C.O., Penney, N., 1988. The effect of the initial gas volume to meat weight ratio on the storage life of chilled beef packaged under carbon dioxide. Meat Science 22 (1), 53–63.

Gill, C.O., 1989. Packaging for prolonged chill storage: the Captech process. British Food Journal 91 (7), 11–15.

Gill, C.O., Molin, G., 1991. Modified atmospheres and vacuum packaging. In: Rusel, N.J., Gould, G.W. (Eds.), Food Preservatives. Blackie and Sons Ltd., Glasgow, p. 172.

Gill, C.O., 1996. Extending the storage life of raw chilled meats. Meat Science 43 (Suppl.), S99–S109.

Gill, A.O., Gill, C.O., 2005. Chapter 13. Preservative packaging for fresh meats, poultry and fin fish. In: Han, J.H. (Ed.), Innovations in Food Packaging, pp. 204–220.

Goulas, A.E., Chouliara, I., Nessi, E., Kontominas, M.G., Savvaidis, I.N., 2005. Microbiological, biochemical and sensory assessment of mussels (*Mytilus galloprovincialis*) stored under modified atmosphere packaging. Journal of Applied Microbiology 98 (3), 752–760.

Han, J.H., 2000. Antimicrobial food packaging. Food Technology 54 (3), 56–65.

Han, J.H., 2005. Antimicrobial packaging systems. In: Han, J.H. (Ed.), Innovations in Food Packaging. Elsevier Academic Press, Amsterdam, pp. 81–107.

Harel, S., Kanner, J., 1985. Muscle membranal lipid peroxidation initiated by H_2O_2-activated metmyoglobin. Journal of Agriculture and Food Chemistry 33, 1188–1192.

Hempel, A.W., Gillanders, R.N., Papkovsky, D.B., Kerry, J.P., 2012. Detection of cheese packaging containment failures using reversible optical oxygen sensors. International Journal of Dairy Technology 65 (3), 456–460.

Hempel, A., O'Sullivan, M.G., Papkovsky, D., Kerry, J.P., 2013a. Use of optical oxygen sensors to monitor residual oxygen in pre- and post-pasteurised bottled beer and its effect on sensory attributes and product acceptability during simulated commercial storage. LWT − Food Science and Technology 50, 226–231.

Hempel, A., O'Sullivan, M.G., Papkovsky, D., Kerry, J.P., 2013b. Non-destructive and continuous monitoring of oxygen levels in modified atmosphere packaged ready-to-eat mixed salad products using optical oxygen sensors. Journal of Food Science 78, S1057–S1062.

Hempel, A., O'Sullivan, M.G., Papkovsky, D., Kerry, J.P., 2013c. Assessment and use of optical oxygen sensors as tools to assist in optimal product component selection for the development of packs of ready-to-eat mixed salads and for the non-destructive monitoring of in-pack oxygen levels using chilled storage. Foods 2, 213−224.

Hempel, A., O'Sullivan, M.G., Papkovsky, D., Kerry, J.P., 2013d. Use of smart packaging technologies for monitoring and extending the shelf-life quality of modified atmosphere packaged (MAP) bread: application of intelligent oxygen sensors and active ethanol emitters. European Food Research and Technology 237, 117−124.

Holter, G., March 2011. The Decline and Fall of Canned Food. WWW.foodmanufacture.co.uk.

Hood, D.E., Riordan, E.B., 1973. Discoloration in pre-packed beef. Journal of Food Technology 8, 333−348.

Hood, D.E., Mead, G.C., 1993. Modified atmosphere storage of fresh meat and poultry. In: Parry, R.T. (Ed.), Principles and Applications of Modified Atmosphere Packing of Food. Blackie Academic and Professional, London, pp. 269−298.

Hutton, T., 2003. Food Packaging: An Introduction. Key Topics in Food Science and Technology − Number 7. Campden and Chorleywood Food Research Association Group, Chipping Campden, Gloucestershire, UK, p. 108.

Inns, R., 1987. Modified atmosphere packaging. In: Paine, F.A. (Ed.), Modern Processing, Packaging and Distribution Systems for Food, vol. 4, pp. 36−51. Blackie, Glasgow, UK.

ICMSF (International Commission on Microbiological Specifications of Food), 1986. Microorganisms in Foods 2. Sampling for Microbiological Analysis: Principles and Specific Applications, second ed.

Jakobsen, M., Bertelsen, G., 2000. Colour stability and lipid oxidation of fresh beef. Development of a response surface model for predicting the effects of temperature, storage time, and modified atmosphere composition. Meat Science 54, 49−57.

Jacxsens, L., Devlieghere, F., Van der Steen, C., Debevere, J., 2001. Effect of high oxygen modified atmosphere on microbial growth and sensorial qualities of fresh cut produce. International Journal Food Microbiology 71, 197−210.

Jeremiah, L.E., Gill, C.O., Penney, N., 1992. The effect on pork storage life of oxygen contamination in nominally anoxic packagings. Journal of Muscle Foods 3, 263−281.

Kaale, L.D., Eikevik, T.M., Bardal, T., Kjorsvik, E., 2013. A study of the ice crystals in vacuum-packed salmon fillets (*Salmon salar*) during superchilling process and following storage. Journal of Food Engineering 115, 20−25.

Kanner, J., 1994. Oxidative processes in meat and meat products: quality implications. Meat Science 36, 169−189.

Karaoglu, M.M., Kotancilar, H.G., Gurses, M., 2005. Microbiological characteristics of part-baked white pan bread during storage. International Journal of Food Properties 8 (2), 355−365.

Kammerlehner, J., 2003. Käsetechnologie. Freisinger Künstlerpresse, Freising.

Kennedy, C., Buckley, D.J., Kerry, J.P., 2004. Display life of sheep meats retail packaged under atmospheres of various volumes and compositions. Meat Science 68, 649−658.

Kerry, J.P., O'Grady, M.N., Hogan, S.A., 2006. Past, current and potential utilisation of active and intelligent packaging systems for meat and muscle-based products: a review. Meat Science 74, 113−130.

Kerry, J.P., O'Sullivan, M.G., Buckley, D.J., Lynch, P.B., Morrissey, P.A., 2000. The effects of dietary α-tocopheryl acetate supplementation and modified atmosphere packaging (MAP) on the quality of lamb patties. Meat Science 56, 61−66.

Khoshakhlagh, K., Hamdami, N., Shahedi, M., Le-bail, A., 2014. Heat and mass transfer modeling during storage of part-baked Sangak traditional flat bread in MAP. Journal of Food Engineering 140, 52−59.

Killefer, D.H., 1930. Carbon dioxide preservation of meat and fish. Industry Engineering Chemistry 22, 140−143.

Knut, F., Nolet, G., 2006. Envasado con CO: Una nueva tecnología de envasado sin oxígeno para la industria cárnica de la Unión Europea. Eurocarne 143, 195−199.

Lagerstedt, Å., Ahnström, M.L., Lundström, K., 2011. Vacuum skin pack of beef − A consumer friendly alternative. Meat Science 88, 391−396.

Lainez, E., Vergara, F., Barcenas, M.E., 2008. Quality and microbial stability of partially baked bread during refrigerated storage. Journal of Food Engineering 89, 414−418.

Lanari, M.C., Schaefer, D.M., Scheller, K.K., 1995. Dietary vitamin E supplementation and discoloration of pork bone and muscle following modified atmosphere packaging. Meat Science 41, 237−250.

Ledward, D.A., 1985. Post-slaughter influences on the formation of metmyoglobin in beef muscles. Meat Science 15, 149−171.

Li, X., Lindahl, G., Zamaratskaia, G., Lundström, K., 2012. Influence of vacuum skin packaging on color stability of beef longissimus lumborum compared with vacuum and high-oxygen modified atmosphere packaging. Meat Science 92, 604−609.

Lopez-Rubio, A., Gavara, R., Lagaron, J.M., 2006. Bioactive packaging: turning foods into healthier foods through biomaterials. Trends in Food Science and Technology 17, 567−575.

Lund, M.N., Hviid, M.S., Skibsted, L.H., 2007a. The combined effect of antioxidants and modified atmosphere packaging on protein and lipid oxidation in beef patties during chill storage. Meat Science 76, 226−233.

Lund, M.N., Lametsch, R., Hviid, M.S., Jensen, O.N., Skibsted, L.H., 2007b. High-oxygen packaging atmosphere influences protein oxidation and tenderness of porcine longissimus dorsi during chill storage. Meat Science 77, 295−303.

Liu, L., Kerry, J.F., Kerry, J.P., 2005. Section of optimum extrusion technology parameters in the manufacture of edible/biodegradable packaging films derived from food-based polymers. Journal of Food Agriculture Environment 3, 51−58.

Liu, Q., Lanari, M.C., Schaefer, D.M., 1995. A review of dietary vitamin E supplementation for improvement of beef quality. Journal of Animal Science 73, 3131−3140.

Marsh, K., Bugusu, B., 2007. Food packaging Roles, materials, and environmental issues. Food Science 72 (3), 39−55.

Martins Sousa, G., Yamashita, F., Soares Soares Júnior, M., 2016. Application of biodegradable films made from rice flour, poly(butylene adipate-co-terephthalate), glycerol and potassium sorbate in the preservation of fresh food pastas. LWT − Food Science and Technology 65, 39−45.

Matz, S.A., 1989. Modified atmosphere packaging. In: Bakery Technology: Packaging, Nutrition, Product Development and quality Assurance. Elsevier Science Publishers, Essex, UK, pp. 150−152.

McMillin, K.W., 1994. Gas-exchange systems for fresh meat in modified atmosphere packaging. In: Brody, A.L. (Ed.), Modified Atmosphere Food Packaging. Institute of Packaging Professionals, Herndon, Virginia, pp. 85−102.

Meischen, H.W., Hu€man, D.L., Davis, G.W., 1987. Branded beef-product of tomorrow today. Proceedings of the Reciprocal Meat Conference 40, 37−46.

Morrissey, P.A., Buckley, D.J., Sheehy, P.J.A., Monaghan, F.J., 1994. Vitamin E and meat quality. Proceedings of the Nutrition Society 53, 289−295.

O'Grady, M.N., Monahan, F.J., Bailey, J., Allen, P., Buckley, D.J., Keane, M.G., 1998. Colour-stabilising effect of muscle vitamin E in minced beef stored in high oxygen packs. Meat Science 50, 73—80.

O'Grady, M.N., Monahan, F.J., Burke, R.M., Allen, P., 2000. The effect of oxygen level and exogenous α-tocopherol on the oxidative stability of minced beef in modified atmosphere packs. Meat Science 55, 39—45.

O'Grady, M.N., Maher, M., Troy, D.J., Moloney, A.P., Kerry, J.P., 2006. An assessment of dietary supplementation with tea catechins and rosemary extract on the quality of fresh beef. Meat Science 73, 132—143.

O'Halloran, S., July, 2013. US Canned Foods Market Declines. Food Engineering. http://www.foodengineeringmag.com/articles/90980-us-canned-foods-market-declines.

Okayama, T., Muguruma, M., Murakami, S., Yamada, H., 1995. Studies on modified atmosphere packaging of thin sliced beef.1. Effect of two modified atmosphere packaging systems on pH value, microbial-growth, metmyoglobin formation and lipid oxidation of thin sliced beef. Journal of the Japanese Society for Food Science and Technology-Nippon Shokuhin Kagaku Kogaku Kaishi 42, 498—504.

O'Mahoney, F., O'Riordan, T.C., Papkovskaia, N., Kerry, J.P., Papkovsky, D.B., 2006. Non-destructive assessment of oxygen levels in industrial modified atmosphere packaged cheddar cheese. Food Control 17 (4), 286—292.

O'Sullivan, M.G., Kerry, J.P., 2012. Chapter 4, Packaging of (fresh and frozen) pork. In: Nollet, L.M.L. (Ed.), Handbook of Meat, Poultry and Seafood Quality. Wiley-Blackwell Publishing Ltd., Oxford, UK.

O'Sullivan, M.G., Cruz, M., Kerry, J.P., 2011. Evaluation of carbon dioxide flavour taint in modified atmosphere packed beef steaks. LWT — Food Science and Technology 44, 2193—2198.

O'Sullivan, M.G., Kerry, J.P., 2011. Sensory quality of fresh and processed meats. In: Kerry, J.P., Ledward, D.A. (Eds.), Improving the Sensory and Nutritional Quality of Fresh and Processed Meats. Woodhead Publishing Limited, Cambridge, UK.

O'Sullivan, M.G., Kerry, J.P., 2008. Chapter 30, Sensory and quality properties of packaged meat. In: Kerry, J.P., Ledward, D.A. (Eds.), Improving the Sensory and Nutritional Quality of Fresh Meat. Woodhead Publishing Limited, Cambridge, UK.

O'Sullivan, M.G., Kerry, J.P., 2009. Chapter 13, Meat packaging. In: Toldrá, F. (Ed.), Handbook of Meat Processing. John Wiley & Sons, Chichester, West Sussex, UK, pp. 211—230.

O'Sullivan, M.G., Kerry, J.P., Buckley, D.J., Lynch, P.B., Morrissey, P.A., 1998. The effect of dietary vitamin E supplementation on quality aspects of porcine muscles. Irish Journal of Agriculture and Food Research 37, 227—235.

Ozdemir, M., Floros, J.D., 2004. Active food packaging technologies. Critical Reviews in Food Science and Nutrition 44, 185—193.

Papkovsky, D.B., 1995. New oxygen sensors and their application to bio sensing. Sensors and Actuators B Chemical 29, 213—218.

Papkovsky, D.B., 2004. Methods in optical oxygen sensing: protocols and critical analyses. Methods Enzymology 381, 715—735.

Parry, R.T., 1993. Introduction. In: Parry, R.T. (Ed.), Principles and Applications of Modified Atmosphere Packaging of Food. Blackie Academic and Professional, New York, p. 3.

Payne, S.R., Durham, C.J., Scott, S.M., Penney, N., Bell, R.G., Devine, C.E., 1997. The effects of rigor temperature, electrical stimulation, storage duration and packaging systems on drip loss in beef. In: Proceedings of the 43rd International Congress of Meat Science and Technology, pp. 592—593. Auckland, (Gl-22).

Pornchai, R., Chitsiri, R., 2011. Vacuum packaging. In: Sun, D.-W. (Ed.), Handbook of Frozen Food Processing and Packaging, second ed. CRC Press, NewYork, pp. 861–874.

Quintavalla, S., Vicini, L., 2002. Antimicrobial food packaging in meat industry. Meat Science 62, 373–380.

Rajbhandari, P., Kindstedt, P.S., 2014. Surface roughness and packaging tightness affect calcium lactate crystallization on Cheddar cheese. Journal of Dairy Science 97, 1885–1892.

Renerre, M., 1990. Review: factors involved in the discoloration of beef meat. International Journal of Food Science and Technology 25, 613–630.

Renerre, M., Labadie, J., 1993. Fresh meat packaging and meat quality. In: Proceedings of the 39th Int. Cong. of Meat Science, and Tech, p. 361. Calgary, Canada.

Risvik, E., 1994. Sensory properties and preferences. Meat Science 36, 67–77.

Rhee, K.I., Ziprin, Y.A., 1987. Lipid oxidation in retail beef, pork and chicken muscles as affected by concentrations of heme pigments and nonheme iron and microsomal enzymic lipid per-oxidation activity. Journal of Food Biochemistry 11, 1–15.

Robertson, G.L., 2006. Food Packaging Principles and Practice, 600 Broken sound Parkway NW, Suite 300, Boca Roton, FL33487–2742, second ed. CRC press, Taylor and Francis Group, pp. 286–309.

Robertson, G.L., 2011. Chapter 7, Packaging and food and beverage shelf life. In: Kilcast, D., Subramaniam, P. (Eds.), Food and Beverage Shelf-life and Stability. Woodhead Publishing Limited, Cambridge, UK.

Rodríguez-Calleja, J.M., Cruz-Romero, M., O'Sullivan, M.G., Kerry, J.P., 2012. High-pressure-based hurdle strategy to extend the shelf-life of fresh chicken breast fillets. Food Control 25, 516–524.

Rooney, M.L., 1995. Overview of active food packaging. In: Rooney, M.L. (Ed.), Active Food Packaging. Blackie Academic & Professional, London, pp. 1–37.

Rooney, M., Han, J.H., 2005. Introduction to active food packaging technologies. In: Han, J.H. (Ed.), Innovations in Food Packaging. Elsevier Academic Press, Amsterdam, pp. 63–69, 81–107.

Rowe, L.J., Maddock, K.R., Lonergan, S.M., Huff-Lonergan, E., 2004. Influence of early post-mortem protein oxidation on beef quality. Journal of Animal Science 82, 785–793.

Salvage, B., Lipsky, J., 2004. Focus on Packaging and Process. The National Provisioner, pp. 64–79.

Schneider, Y., Kluge, C., Weiß, U., Rohm, H., 2010. Chapter 12, Packaging materials and equipment. In: Law, B.A., Tamime, A.Y. (Eds.), Technology of Cheesemaking, second ed. Wiley-Blackwell, John Wiley & Sons, Chichester, West Sussex, UK, pp. 413–439.

Seideman, S.C., Durland, P.R., 1984. The utilization of modified atmosphere packaging for fresh meat: a review. Journal of Food Quality 6, 239–252.

Seyfert, M., Mancini, R.A., Hunt, M.C., Tang, J., Faustman, C., 2007. Influence of carbon monoxide in package atmospheres containing oxygen on colour, reducing activity, and oxygen consumption of five bovine muscles. Meat Science 75 (3), 432–442.

Sherbeck, J.A., Wulf, D.M., Morgan, J.B., Tatum, J.D., Smith, G.C., Williams, S.N., 1995. Dietary supplementation of vitamin E to feedlot cattle affects retail display properties. Journal of Food Science 60, 250–252.

Soroka, W., 2008. Illustrated Glossary of Packaging Terms. Institute of Packaging Professionals, p. 3.

Smiddy, M., Fitzgerald, M., Kerry, J.P., Papkovsky, D.B., O'Sullivan, C.K., Guilbault, G.G., 2002a. Use of oxygen sensors to nondestructively measure the oxygen content in modified atmosphere and vacuum packed beef: impact of oxygen content on lipid oxidation. Meat Science 61 (3), 285–290.

Smiddy, M., Papkovsky, D., Kerry, J.P., 2002b. Evaluation of oxygen content in commercial modified packs of processed cooked meats. Food Research International 35 (6), 571−575.

Smiddy, M., Papkovskaia, N., Papkovsky, D.B., Kerry, J.P., 2002c. Use of oxygen sensors for the non-destructive measurement of the oxygen content in modified atmosphere and vacuum packs of cooked chicken patties: impact of oxygen content on lipid oxidation. Food Research International 35, 577−584.

Sørheim, O., Aune, T., Nesbakken, T., September 1997. Technological, hygienic and toxicological aspects of carbon monoxide used in modified-atmosphere packaging of meat. Trends in Food Science and Technology 8 (9), 307−312.

Sorheim, O., Nissen, H., Nesbakken, T., 1999. The storage life of beef and pork packaged in an atmosphere with low carbon monoxide and high carbon dioxide. Meat Science 52, 157−164.

Spreer, E., 2006. Technologie der Milchverarbeitung. Behr's Verlag, Hamburg.

Strehle, G., 1997. Verpacken von Lebensmitteln. Behr's Verlag, Hamburg.

Sturm, W., 1998. Verpackung Milchwirtschaftlicher Lebensmittel, Edition. IMQ, Kempten.

Suppakul, P., Miltz, J., Sonneveld, K., Bigger, S.W., 2003. Active packaging technologies with an emphasis on antimicrobial packaging and its applications. Journal of Food Science 68 (2), 408−420.

Taylor, A.A., 1985. Packaging Fresh Meat. In: Lawrie, R.A. (Ed.), Developments in Meat Science, vol. 3. Elsevier Applied Science Publishers, Essex, England (Chapter 4).

Torngren, M.A., September 2003. Effect of packaging method on colour and eating quality of beef loin steaks. In: 49th International Congress of Meat Science and Technology, pp. 495−496. Brazil.

Vázquez, B.I., Carriera, L., Franco, C., Fente, C., Cepeda, A., Barros-Velázquez, J., 2004. Shelf life extension of beef retail cuts subjected to an advanced vacuum skin packaging system. European Food Research and Technology 218, 118−122.

Weber, C.J., Haugaard, V., Festersen, R., Bertelsen, G., 2002. Production and applications of biobased packaging materials for the food industry. Food Additives and Contaminants 19, 172−177.

Wilkinson, B.H.P., Janz, J.A.M., Morel, P.C.H., Purchas, R.W., Hendriks, W.H., 2006. The effect of modified atmosphere packaging with carbon monoxide on the storage quality of master packaged fresh pork. Meat Science 75 (4), 605−610.

Wolfe, S.K., 1980. Use of CO- and CO_2 enriched atmospheres for meats, fish, and produce. Food Technology 34 (3), 55−63.

Young, L.L., Reviere, R.D., Cole, A.B., 1983. Fresh red meats: a place to apply modified atmospheres. Food Technology 42, 65−69.

Yucel, U., 2016. Intelligent Packaging. Reference Module in Food Science. Elsevier, ISBN 978-0-08-100596-5.

Zakrys, P.I., Hogan, S.A., O'Sullivan, M.G., Allan, P., Kerry, J.P., 2008. Effects of oxygen concentration on the sensory evaluation and quality indicators of beef muscle packed under modified atmosphere. Meat Science 79, 648−655.

Zakrys-Waliwander, P.I., O'Sullivan,, M.G., O'Neill, E.E., Kerry, J.P., 2012. The effects of high oxygen modified atmosphere packaging on protein oxidation of bovine. *M. longissimus dorsi* muscle during chilled storage. Food Chemistry 2, 527−532.

Chapter 8

Instrumental Assessment of the Sensory Quality of Food and Beverage Products

INTRODUCTION

Colour, aroma, taste, flavour and texture are all important factors affecting food and beverage quality. As these sensory modalities are assessed by human responses, it is important that reproducible and reliable methods are available to accurately quantify them (O'Sullivan and Kerry, 2008). Although sensory evaluation is routinely used in the industry to evaluate the quality of foods, it can be expensive and time-consuming. However, it must be noted that sensory analysis is still an optimum approach when undertaking such analysis. Alternatively, instrumental techniques are widely used to assess quality attributes and determine changes in quality (Kong and Singh, 2011; Dijksterhuis, 1995). Sensory-based instrumental methods are often employed to measure sensory changes that occur in food products. The idea behind such methods is that sensory perceptions have chemical and physical counterparts in the substance under investigation (Dijksterhuis, 1995). Instrumental methods are usually quick and directly correlated to a sensory-based criterion. Instrumental and sensory limits can be assessed using survival analysis results to determine what instrumental limits correspond to the appropriate sensory limit of acceptability (O'Sullivan and Kerry, 2013). A considerable amount of research has been undertaken to investigate the suitability of advanced sensor technology to simulate human sensory responses. The development of valid and relevant instrumental methods in concert with dynamic sensory methods has allowed for a more comprehensive analysis of human perception (Ross, 2009). Chemometrics, which uses multivariate data analysis to interpret sensory-based instrumental data, is an important research area which ultimately will determine the extent to which sensory-based instrumental technology is applied. This book chapter will discuss sensory-based instrumental methods for colour assessment based on colorimetry and computer vision technology, texture systems (texture, rheology) and odour-based instrumental methods (gas chromatography—mass spectrometry (GC/MS) and the electronic nose (EN))

A Handbook for Sensory and Consumer-Driven New Product Development.
http://dx.doi.org/10.1016/B978-0-08-100352-7.00008-7

and flavour-based systems (the electronic tongue (ET)). Near infrared (NIR) and Fourier transform infrared (FTIR) spectroscopic methods will also be discussed.

INSTRUMENTAL METHODS OF COLOUR ANALYSIS

The appearance of food and beverage products is generally the first sensory modality that the consumer encounters and for this reason product colour can be vital. As stated in the earlier chapter on shelf life, consumers will perceive colour loss as a sign of quality deterioration and thus, for certain foods and beverages where colour changes can occur, it will be one of the principal sensory factors which will define quality and shelf life (Sutherland et al., 1986). Colours can fade in products, like soft drinks, or become oxidised by light over time to develop negative sensory colour characteristics. Colour is one of the main quality parameters of wines and has an important influence on the overall acceptability by consumers (Martin et al., 2007). For standard beer analysis, EBC (European Brewing Convention) colour and bitterness units can be assessed by photometry (Lachenmeier, 2007). To ensure food conformity to consumer expectations, it is critical for the food processing industry to develop effective colour inspection systems to measure the colour information of food products during processing operations and storage periods (Wu and Sun, 2013).

Of primary importance to sensory quality is the colour of fresh meat products where the cherry red colour of oxymyoglobin is oxidised to the grey-brown pigment of metmyoglobin. Oxymyoglobin is a haeme protein in which iron exists in the ferrous form (Fe^{+2}), unlike metmyoglobin that possesses the ferric form (Fe^{+3}). The conversion of the ferrous to the ferric form is a result of oxidation (Liu et al., 1995). At the point of sale, colour and colour stability are the most important attributes of meat quality and various approaches have been used to meet consumer expectations that an attractive, bright-red colour indicates a long shelf life and good eating quality (Hood and Mead, 1993). Consumers relate this bright-red colour to meat that is fresh, while, discriminating against meat that has turned brown in colour, which they consider unsightly. Thus, it is important to quantify the colour quality, particular of red meat products such as beef, lamb and pork (O'Sullivan, 2011b). Instrumental methods have been developed which can accurately measure the colour characteristics of food products and have been available within the food and beverage industries for many years. The Minolta colorimeter (Minolta Camera Co. Ltd, Osaka, Japan, Fig. 8.1) is widely used by both the industry and research institutes to measure meat colour quality. This small portable hand-held device can use different measurement coordinates, but the principal one used for meats are the CIE, L (white to black), a (green to red) and b (blue to yellow) values. The $L^*a^*b^*$, or CIE Lab, colour space is an international standard for colour measurements, adopted by the Commission Internationale d'Eclairage (CIE) in 1976 (CIE, 1986) and is very close to human perception

FIGURE 8.1   A Minolta colorimeter (Minolta Camera Co. Ltd, Osaka, Japan, Fig. 8.1) used for quantitative food and beverage colour measurements.

of colour (Wu and Sun, 2013). O'Sullivan et al. (2002) used a Minolta colorimeter to assess the colour sensory quality of retail displayed pork patties and found that the RED sensory term, as utilised by a trained sensory panel, was significantly ($p < .01$) positively correlated to a values determined by the Minolta colorimeter. However, these point-to-point colorimeters can have certain drawbacks (O'Sullivan and Kerry, 2013). Huselegge et al. (2001) analysed 56,000 veal carcasses and found significant differences between Minolta CR300 devices and explained this to some extent by the fact that the individual Minolta instruments were operated by two different persons, who may not have placed the instrument on the same site.

Computer vision methods are a relatively recent innovation to the instrumental colour measurement area (O'Sullivan and Kerry, 2013). Increased requirements for quality by consumers require the colour evaluation of food products to be more rapid, objective and quantifiable. Therefore, the food industry has paid numerous efforts for a long time to measure and control the colour of their products (Wu and Sun, 2013). Advantages of computer vision systems are noninvasive, objective, consistent and rapid estimations of meat palatability (Jackman et al., 2009). Computer vision is a rapid, objective and economic inspection technique that could provide a detailed characterisation of colour uniformity at pixel-based level (Wu and Sun, 2013). Papadakis et al. (2000) used a high-resolution digital camera to capture images of different types of microwavable pizza bases and used computer software (Photoshop, Adobe Systems Inc., California, USA) to obtain colour parameters from digital images. These authors found this system to be versatile, affordable and easy to use and indicated that it was a technique that can be applied to many other foods besides pizza. O'Sullivan et al. (2003a) used a high-resolution digital camera and a custom algorithm (MATLAB, Massachusetts, USA) to determine the colour parameters of various fresh pork meat patties and found that the digital camera-derived a values correlated to a greater extent to the sensory RED term compared to the Minolta colorimeter for both a trained and an untrained colour sensory evaluation of two different muscles, M. *longissimus dorsi* and M. *psoas major*. Jackman et al. (2010) used computer vision methodology to correlate with consumer panel palatability data and found that it is possible for consumer opinion of beef likeability to be accurately modelled by using image colour, marbling and surface texture features. Meat is not a homogenous material and variations in pigment concentration and therefore development of metmyoglobin occurs on the surface of meat during display in commercial retail conditions. Thus, meat colour is difficult to assess because of this colour variation over a meat cut, even from within the same muscle (AMSA, 1991). Taking a picture of the entire surface of a meat sample can provide a more representative colour profile compared to the point-to-point measurements of traditional colorimeters (O'Sullivan et al., 2003a).

O'Sullivan et al. (2003a) found that instrumental a values and R values (RGB colour measurement index, red, green and blue) as determined by a digital camera correlated with RED and negatively with the BROWN sensory terms in an experiment designed to correlate instrumental methods of colour analysis to sensory assessment as performed by a trained and an untrained sensory panel. Similarly, digital camera-derived b values correlated to a greater degree to the BROWN sensory term for both assessor groups and muscles compared to Minolta derived b values. The digital camera-derived instrumental measurements had a high affinity for their respective sensory descriptors and instrumental measurements taken with the digital camera and were more highly correlated to and predictive of the sensory terms particularly for the RED, BROWN and L value descriptors when compared to Minolta colorimeter correlations. This is perhaps due to the fact that the camera took measurements over the entire surface of samples resulting in a more representative measurement, whereas the Minolta colorimeter took point-to-point measurements, which missed out on some relevant information (O'Sullivan et al., 2003a). These results agree with those of Lu et al. (2000) who found that computer vision can be used for predicting sensory colour scores of pork loin muscle. The sensory colour scores were nonlinearly related to the colour features extracted from the loin images. Both statistical and neural models resulted in satisfactory prediction, though the neural net model was better. J.peg (Joint Photographic Experts Group, a standard for compressing digital photographic images) images taken by the digital camera allow direct conversion of the resultant data via an algorithm (MATLAB, Massachusetts, USA) to any of a number of colour measurement systems, i.e., Hunter Lab, CIE Lab, RGB, XYZ, etc. Also, only one picture need be taken as opposed to the multiple single-point observations required by a colorimeter to obtain a representative colour profile (O'Sullivan et al., 2003a).

Computer vision technology has also been applied to many food products including pork (Lu et al., 2000; O'Sullivan et al., 2003a), beef (Jackman et al., 2010; Zheng et al., 2006), fish (Quevedo et al., 2010), ham (Valous et al., 2009), fruits (Balaban et al., 2008), pizza (Papadakis et al., 2000), cheese (Everard et al., 2007) and wheat (Zapotoczny and Majewska, 2010).

INSTRUMENTAL METHODS FOR MEASURING TEXTURE

Texture is a fundamental sensory quality aspect of many foods, including beverages, and is a quality attribute that is closely related to the structural and mechanical properties of a product. Firmness is important for fresh fruits, including apples where it is measured by the standard destructive penetrometric test. This type of test, or puncture test, is often used to measure the force required for a probe to penetrate into a food to a specified depth and involves both compression and shearing of a food sample and is analogous to the biting of a food item in the mouth (Lu, 2013). Also, tenderness has often been described as the most important factor in terms of high eating quality of meats, especially in beef (O'Sullivan and Kerry, 2013; Zakrys et al., 2008, 2009; Zakrys-Waliwander et al., 2010; Zakrys-Waliwander et al., 2011a,b).

Force (F), deformation (D) and time (t) are three basic variables used in studying the mechanical properties of foods. The external force, measured in Newtons (N) or deformation relationship for most food materials is dependent on time or loading rate (Lu, 2013).

Tenderness can be evaluated by objective methods, instrumental or sensory (trained panels) or by subjective methods (consumer panel). Sensory methods, either analytical or affective, are expensive, difficult to organise and time-consuming (Platter et al., 2003). Thus, there have been many attempts to devise instrumental methods of assessing the force shearing, penetrating, biting, mincing, compressing and stretching of foods whose results are a prediction of tenderness ratings obtained by taste panels (Lawrie and Ledward, 2006). Texture analysis of meat products involves the uniaxial compression of food samples between two plates and shear tests measure the force to cut through fibres of cooked samples. They are the simplest and most common tests used to document raw and cooked meat texture (O'Sullivan and Kerry, 2013). Consumers can distinguish between tender and tough beef steaks (Aaslyng, 2009). In addition to an acceptable flavour, the consumer desires meat to be palatable and consequently, meat tenderness is another critical determinant used by the consumer to determine meat quality (O'Sullivan and Kerry, 2013). In fact, it has been demonstrated that the consumer would be willing to pay a higher price in the marketplace for beef as long as it is guaranteed tender (Miller et al., 2001).

One drawback of texture analysis is that, information obtained from shearing devices that perform in a similar way may not be interchangeable (Lyon and Lyon, 1998). The mechanical process of mastication has been simulated using texture profile analysis (TPA, Fig. 8.2). This objective method measures the compression force of a probe and the related textural parameters of a test food during two cycles of deformation (Caine et al., 2003). The main advantage of TPA is that one can assess many variates with a double compression cycle. Variates that can be assessed with this analysis are: hardness, springiness, cohesiveness, adhesiveness, resiliency, fracturability, gumminess, chewiness, etc. In meat the variates assessed are hardness, springiness and cohesiveness; the three altogether permit the calculation of chewiness (Ruiz de Huidobro et al., 2001). However, the most widespread method normally used as an indicator of meat sensory hardness (tenderness) is the Warner—Bratzler (WB) shear test and almost the sole methodology used in

FIGURE 8.2   The mechanical process of mastication has been simulated using, texture profile analysis. Presented is a TA.XT2 Texture Analyser (Stable Micro Systems, England).

raw meat (Bratzler, 1932; Warner, 1928), and which is referred to in most papers (Culioli, 1995). The basic concept and design of the WB shear device (Bratzler, 1932) have been subject to modification and improvement over the years. Yet, the familiar blade, with its triangular hole in the middle, remains one of the most widely used devices to provide measurements of meat texture quality. Other machines built to measure rheological parameters of foods and materials usually include a version of the WB blade as a basic attachment to the system (Lyon and Lyon, 1998).

(Shackelford et al. 1991, 1997) state that WB shear force is an imprecise predictor of beef tenderness characteristics determined by sensory-trained panellists. However, according to Destefanis et al. (2008) and also McMillin (2008), WB shear values of less than 42.87 N and greater than 52.68 N allow classification of tough and tender beef in a sufficiently reliable way to be highly related to consumer tenderness perception (Destefanis et al., 2008; McMillin, 2008). The most widely used method for measuring meat texture is the single blade shear test of the WB type (Culioli, 1995). However, results obtained from using this method can be variable. This variability depends on many factors, such as muscle type, sample preparation, cooking method, shear apparatus, measurement procedure and panel type (Destefanis et al., 2008).

Miller et al. (2001) reported that consumer perceptions of beef flavour and juiciness have a greater impact on consumer overall acceptability levels of New York strip steaks as the WB shear force and toughness levels increase. In short, as beef steaks become tougher, flavour and juiciness have a greater effect on consumer satisfaction. Zakrys et al. (2008) observed that WB shear force values had positive correlations to O_2 levels in modified atmosphere packed *M. longissimus dorsi* muscle samples, displaying that all samples appeared to become less tender with increasing O_2 level during the 15 days storage. It appears that samples packed with 50% and 80% O_2 were tougher than low O_2-treated samples.

The width of the blades and the position of the triangle; the speed of the test; the shape, mass and orientation of the test sample are important to interpretation of the results of shear tests. Yet, these very specific conditions are often omitted from reports on testing protocols (Lyon and Lyon, 1998). Thus, it is sometimes difficult to replicate a previous worker's studies due to this lack of information.

Food rheology is the study of deformation and flow of foods under well-defined conditions and has been shown to be closely correlated with food texture (Bourne, 2002). For liquids, texture is described as viscosity. Emulsion-based alcoholic beverages such as cream liqueurs become more viscous over time and this can be a limiting factor with respect to quality and shelf life thus requiring monitoring using rheological measurement. Some of the measurements that can be used to assess cream liqueur shelf life include droplet size distributions and viscosity (O'Sullivan, 2011a). Droplet size distributions can be determined by laser diffraction using a Malvern Mastersizer

S (Malvern instruments, Malvern, Worcestershire, UK), which facilitates the measurement of particles in the size range of 0.050−880 Lm (Heffernan et al., 2009). A viscometer is an instrument used to measure the viscosity of a fluid. Typically, rotational viscometers are used to quantify the viscosity of cream liqueurs, such as the 'Cup and bob' type, which measure the torque required to rotate the bob in a fluid at a known speed, where this measured torque is a function of the viscosity of that fluid (O'Sullivan, 2011a). Also, viscosity of cream liqueurs can be determined using a viscometer such as Carrimed CSL 100 rheometer (Carri-Med Ltd, Dorking, Surrey, UK) (O'Sullivan, 2011a). This is generally performed at a constant shear rate at a specific temperature, 20°C (Power, 1996; Lynch and Mulvihill, 1997). Also, the Brookfield Synchro-Lectric viscometer utilises a series of cylindrical spindles and horizontal disks which are rotated at fixed speeds while the torque required to overcome the viscous drag of the fluid is recorded. Conversion tables are available to convert this into Newtonian viscosity (McKenna and Lyng, 2013).

However, sensory viscosity may not be well predicted by rheological measurements. Janhøj et al. (2008) investigated the relationships between rheological and sensory attributes of acidified milk drinks and found that creaminess appeared to be largely determined by sensory viscosity (viscosity as perceived by the consumer), but the science of relating sensory and rheological properties is hampered by poor physical definition of the sensory terms.

GAS CHROMATOGRAPHY/MASS SPECTROMETRY

Gas chromatography (GC) is the most useful method for the analysis of volatile compounds. Initially a sample is injected into an inlet and volatilised in the injector and swept onto a column by a carrier gas. Individual volatile compounds are separated by their vapour pressure, interaction (or lack of) with the column stationary phase and usually by a temperature gradient applied to the column (Kilcawley, 2015). GC is routinely used in the analysis of characteristic food odours and can be carried out by human assessment and headspace/direct GC/MS (Fig. 8.3) (Grigioni et al., 2000). For a standard spirit drink analysis, higher alcohols and other volatile compounds are determined using GC (Lachenmeier, 2007). In GC, the volatiles are separated in a column as they are forced along by a carrier gas, and the analytes are labelled and quantified by a detector. Utilisation of a flame ionisation detector or mass spectrometry (MS) are the most often used detectors (Kong and Singh, 2011) MS detectors operate by using ionisation energy (electron or chemical) to fragment molecules exiting a GC column in a gas stream. The fragments are preselected based on their mass-to-charge ratio (m/z) and detected as ion profiles generated which are compared to an online reference mass spectral library which in conjunction with retention indices of standards are used to positively identify the compound (Kilcawley, 2015).

FIGURE 8.3   Shimadzu 8030 TQMS (Triple Quadrupole Mass Spectrometer) with Combipal autosampler (CTC analytics, Basel, Switzerland).

GC-olfactometry is a third method which utilises trained panellists, by means of a sniffer port interface, to qualitatively and quantitatively evaluate the odour for each analyte leaving the chromatographic column, in order to understand the sensory properties of a certain compound at a given concentration (Plutowska and Wardencki, 2008; Kilcawley, 2015).

GC/MS, which utilises a mass spectrometer as a detector after volatile separation using the column, can yield information on the concentration of volatiles present in a sample. These volatiles can be flavour or odour in origin and can be used to quantify the flavour characteristics of foods and beverages or they can be chemicals which denote off-flavour such as oxidation compounds. GC/MS is routinely used to identify chemical markers that can be used as indices of sensory quality. With respect to food products, much work has undertaken in order to identify chemical markers of oxidation that can be directly correlated to human sensory responses. GC is used to monitor volatile lipid oxidation products of food and is commonly used to quantify the secondary oxidation products including aldehydes, ketones, alcohols, short carboxylic acids and hydrocarbons (Kong and Singh, 2011) and thus is a useful tool in shelf life analysis. Hexanal has a distinctive odour described as being green or grassy (Gasser and Grosch, 1988) and is a secondary breakdown product formed during the oxidation of linoleic acid (C18:2). It has been identified and used to evaluate the oxidative state and correlated as an index to sensory scores of pork (Shahidi et al., 1987), beef (Drumm and Spanier, 1991) and also chicken (Byrne et al., 2002). O'Sullivan et al. (2003b) found that the sensory descriptor *Green-O* (O = odour) as determined by a trained sensory panel covaried with hexanal levels obtained by GC/MS for warmed-over flavour (WOF) samples of cooked porcine *M. longissimus dorsi* and *M. psoas major* muscles. Also, pentanal and 2,4-decadienal could be used as marker compounds to follow the development of WOF and its associated rancid flavours in cooked meats (St. Angelo et al., 1987). Siegmund and

Pfannhauser (1999) determined that 2,4-decadienal was the most potent odorant and increased significantly with storage time of cooked chicken meat. The sensory properties of the two isomers (*EZ* and *EE*) of 2,4-decadienal were also described as fatty and fried and thus were sensory markers and not just a chemical index. These authors concluded that an increase of the influence of these compounds on the aroma of stored chicken meat would result in a more intense chicken aroma, and also in the undesired WOF of the meat. O'Sullivan et al. (2003b) found that WOF samples of cooked porcine *M. longissimus dorsi* and *M. psoas major* covaried with the sensory descriptor *Green-O* and the GC/MS identified compounds pentanal, 2-pentylfuran, octanal, nonanal and 1-octen-3-ol. Other authors have found these compounds to be indicative of lipid oxidation (Siegmund and Pfannhauser, 1999; Byrne et al., 2002). O'Sullivan et al. (2003b) also showed that the compounds 1-octen-3-ol and 2-pentylfuran covaried with the sensory oxidative descriptors *Rancid-F* (F = flavour), *Fish-F* and *Rubber/Sulphur-Like-O* when samples of *M. longissimus dorsi* and *M. psoas major* with varying degrees of WOF were analysed. These results agree with those of Siegmund and Pfannhauser (1999) who found that the relative concentrations of the lipid oxidation products 1-octen-3-ol and 2-pentylfuran increased in cooked chill stored chicken meat as storage time increased.

GC/MS is often used in the analysis of meat, poultry and fish products with respect to fatty acid composition. The intramuscular fatty acid composition of the monogastric animals, and in particular, the triacylglycerols are a reflection of the dietary fatty acids, while in ruminants, the biohydrogenation in the rumen (i.e., saturation of the dietary unsaturated fatty acids (UFA)) is responsible for the smaller variations in intramuscular fatty acid composition (Raes et al., 2004). Poultry and pork muscle typically have higher levels of polyunsaturated fatty acids (PUFA) than lamb or beef. Pork muscle has more linoleic acid than beef or lamb which contributes to the higher PUFA:saturated fatty acid (SFA) ratio. However, beef and lamb commonly have more favourable n6:n3 fatty acids ratios than pork (Wood and Enser, 1997). Fat content and fatty acid composition of meat are of major importance for consumers due to their importance for meat quality and nutritional value (Wood et al., 2004). It is well known that the fatty acid composition of meat from ruminants differs from meat from nonruminants. PUFA:SFA ratio is lower in beef than in pork due to ruminal biohydrogenation. However, beef has a lower n 6: n 3 fatty acid ratio than pork, which is considered beneficial to human health (Enser et al., 1996). Swine feeding has been formulated with a higher content of natural sources of UFA, such as n-3 series or conjugated linoleic acid (CLA), due to human health concerns (Boselli et al., 2008). Also, diets enriched with vegetable oils (such as sunflower oil, soybean oil or corn oil) that contain an elevated UFA percentage should result in healthier products for consumers (Mitchaothai et al., 2007). Beneficial effects of n-3 fatty acids have been shown in the secondary prevention of coronary heart disease, hypertension, type 2 diabetes and, in some patients with renal disease, rheumatoid arthritis, ulcerative colitis, Crohn's disease, and chronic obstructive pulmonary disease (Simopoulos, 1999). Also, the fatty acid composition of beef has been widely studied and fatty acids such as n-3 and n-6 PUFA as well as CLA are especially interesting to researchers or consumers considering their potential relationship to human health, but also fatty acid composition plays an important role with respect to beef eating quality (Jiang et al., 2010). High n-3 PUFA levels in grass-fed beef compared to traditional concentrate-fed beef (Vatansever et al., 2000) can manifest in fishy flavours in the resulting meat. Nuernberg et al. (2005), who investigated the effects of a grass-based and a concentrate feeding system on meat quality characteristics and fatty acid composition of longissimus muscle in different cattle breeds, found that fishy off-flavours were significantly higher and overall flavour liking scores were significantly lower in meat from grass-finished cattle with increased 18:1 trans-isomers and, notably, CLAcis-9, *trans*-11. On the other hand (C18:1n9) is associated with favourable beef palatability attributes (Dryden and Marchello, 1970).

Holm et al. (2012) showed that the GC/MS-derived microbial metabolites 2- and 3-methylbutanal, 2- and 3-methylbutanol, acetoin and diacetyl were closely related to the changes of the sensory attributes meaty and sour&old odour in saveloy, a Danish cooked sliced pork sausage or luncheon meat-type product. Thus, these aroma compounds could be used as chemical markers for the sensory shelf life of sliced saveloy (Holm et al., 2012).

Volatile compounds are principally responsible for specific cheese aroma and flavour and can be quantified by various GC methods. Key aromatic compounds are derived from the metabolism of carbohydrates (lactose and citrate) by lactic acid bacteria resulting in acetate, diacetyl (2,3-butanedione), acetaldehyde, acetoin (3-hydroxy 2-butanone), ethanol, 2,3-butanediol, propionate and lactate. Additionally, short chain free fatty acids (butyrate, caproate, caprylate and caprate) are the product of lipolysis and are both volatile and odour active. Many cheese aroma compounds are also formed from the metabolism of amino acids (leucine, isoleucine and valine) (Kilcawley, 2015).

Ethyl esters form due to the reaction of alcohol and fatty acids and can result in the formation of excessive levels of short chain fatty acids such as ethyl acetate, ethyl butanoate and ethyl hexanoate, which are the principal causes of a fruity defect in cream liqueurs. The level of fruitiness may become a limiting factor to consumer acceptance in aged liqueurs (O'Sullivan, 2011a). Typically, the levels of these ethyl esters can be monitored and quantified using GC or GC/MS and liquid extraction of methylated esters or using solid-phase micro-extraction. Effectively, flavour partition can be measured under static equilibrium headspace conditions but does not always relate well to the release profile that is observed in vivo during consumption of a product (Doyen et al., 2001).

The usefulness of GC/MS is obvious, but as a technique it has certain drawbacks such as high operating costs and method can be time-consuming (Pryzbylski and Eskin, 1995). However, the EN may provide a practical advantage over other methods and may have an application in an online/at-line capacity for the quality determination of meat products (O'Sullivan et al., 2003b).

THE ELECTRONIC NOSE

The EN is an artificial olfactory system based on GC volatile analysis which can detect and recognise a wide spectrum of odour patterns, and determine the odour intensity of mixtures of a variety of volatile oil degradation compounds (Kong and Singh, 2011). In contrast to the well-known analytical and sensory techniques that have been used for the analysis of flavour compounds, the EN does not give any information about the compounds causing the investigated aroma; neither the identity of the compounds nor their sensory properties (O'Sullivan and Kerry, 2013). An EN can function as a rapid and nondestructive tool for online flavour characterisation. The application of EN in the food industry has been increasing due to its rapidity, cost-effectiveness, objectivity and simplicity (Kong and Singh, 2011). Using the EN, the aroma is judged by the so-called 'aroma pattern', which should be characteristic to the investigated substrate (Siegmund and Pfannhauser, 1999). Thus, in food production plants, there is a growing demand for online/at-line measurement of sensory relevant quality criteria. One promising technology in this capacity is the application of an EN (Hansen et al., 2005). The EN may provide a practical advantage over other methods and may have an application in an online/at-line capacity for the quality determination of food products (O'Sullivan et al., 2003b). In order to classify samples, an EN combines the response profiles of the various sensors, which react to different types of volatile compounds in the sample gas (Rajamäki et al., 2006). The EN is an array of chemical gas sensors with a broad and partly overlapping selectivity for measurement of volatile compounds within the headspace over a sample combined with computerised multivariate statistical data processing tools (Gardner and Bartlett, 1994). The sensor array of an EN has a large information potential and will give a unique overall pattern of the volatiles. In principle, both the EN and the human nose operate by sensing simultaneously a high number of components giving rise to a specific response pattern (Haugen and Kvaal, 1998). However, the EN has both large differences in sensitivity and selectivity from the human nose (Haugen and Kvaal, 1998). The EN has been assessed in the analysis of a large variety of different food and beverage products and has found application in the analysis of meat, poultry and seafood including: Eklöv et al. (1998), fermented sausage; Rajamäki et al. (2006), chicken, Ólafsdóttir et al. (1997); O'Connell et al. (2001), fish, Hansen et al. (2005), meatloaf; Tikk et al. (2008), meatballs; Panigrahi et al. (2006), spoilage beef strip loin and in the WOF analysis of various meat products (e.g., Siegmund and Pfannhauser, 1999; chicken; Grigioni et al., (2000); beef; O'Sullivan et al., 2003b; pork). The EN has also been used for wine (Garcia et al., 2006), fruits (Infante et al., 2008) and cheese (Limbo et al., 2009).

The above-mentioned EN instruments all differed in the individual sensor array set-up, numbers of sensors and the data analysis used to process the characteristic signature data of the various meat types. The data analytical techniques employed included AromaScan, neural networks, principal component regression and partial least squares regression (PLSR) (O'Sullivan et al., 2003b). One of the principal drawbacks to date of ENs is the large variation in data acquisition. In order to compare results over time (weeks, months or years), it is required that these instruments give the same signal when identical samples are being measured over time. However, due to dynamic processes taking place in the sensors over time, the signal from a sensor array may vary significantly (Haugen and Kvaal, 1998). O'Sullivan et al. (2003b) found that the EN device used in a WOF experiment could clearly separate samples on the basis of muscle type, treatment and degree of WOF development. Also, the EN data from two separate sample sets analysed in different laboratories and with a time separation of 11 months agreed with the sensory analysis and the device used in this experiment was effective in determination of the oxidative state of the samples analysed. O'Sullivan et al. (2003b) used level correction, a multivariate data analytical tool, to normalise the data from sample sets analysed at a 11-month interval prior to subsequent PLSR and this allowed their direct comparison.

Another potential drawback of the EN is that sensors have a limited life, they must be replaced after some time and new sensors from another batch will differ in performance (Haugen and Kvaal, 1998). One potential method of solving these fundamental problems is to use a reliable data analytical tool to correct for variations over time, possibly using a reference sample and secondly to use an EN in which the sensors do not require replacement. This problem was overcome by O'Sullivan et al. (2003b) who employed an EN in the aforementioned WOF experiment, which used sensor technology (MGD-1, Environics Ltd, Finland), which had an advantage over existing EN devices in that the sensors do not wear because the molecules measured do not come in direct contact with the sensors. The method of sensor operation is based on the principle of ion mobility and ionisation of gas molecules. The clusters formed through ion—molecule reactions are brought into different electrical fields perpendicular to the sample flow and detected. In effect, the sensors used will not wear out, thus avoiding the requirement for replacement and the variations in sensor batch manufacture. This displayed the potential effectiveness of the EN as an objective online/at-line quality control monitoring device.

If an EN is to be used in quality assurance and quality control programs for raw materials and/or end products, there is a need to calibrate it against sensory assessment in order to determine the relevance of the measurements

(Hansen et al., 2005). However, the EN has both large differences in sensitivity and selectivity from the human nose (Haugen and Kvaal, 1998). Tikk et al. (2008) concluded that a significant, positive correlation existed between the EN gas sensor signals, the WOF-associated sensory attributes and the levels of secondary lipid oxidation products for pork meatballs, a very popular Danish dish. This also supports the potential of EN technology as a potential future quality control tool in the meat industry. Hansen et al. (2005) demonstrated that an EN could predict the sensory quality of porcine meatloaf, based on measuring the volatiles in either the raw materials or the meatloaf produced from those raw materials. They further stipulated that a strategy involving an operational and standardised methodology and vocabulary for in-house sensory evaluation of the raw materials was essential if the EN was to be calibrated properly and used online in the future. GC/MS can provide a better understanding of what an EN is measuring when both analyses are considered together and how they are related to the perceived sensory attributes as measured by sensory analysis (Hansen et al., 2005).

The Electronic Tongue

ETs consist of an array of electrochemical sensors interfaced with advanced data processing tools which are able to interpret the complex electrochemical signals developed to detect taste and olfaction in foods (Winquist et al., 1999; Kong and Singh, 2011). The ET can be an analytical tool including an array of nonspecific, poorly selective chemical sensors with partial specificity (cross selectivity), coupled with chemometric processing, for recognising the qualitative and quantitative composition of multispecies solutions (Ha et al., 2015). The data processing algorithms involved mainly include principal component analysis which is mostly used in identification/classification for qualitative purposes, partial least squares which is mainly used in multidetermination for quantitative purposes and artificial neural networks (ANN) which is a powerful parallel computing method, especially suitable for nonlinear sensor signals and extremely related to human pattern recognition (Ha et al., 2015). One of the most widely used methods for sample classification in ET systems are these ANN, formed by algorithms inspired in biological neural systems (Gil-Sánchez et al., 2015). ETs work by mimicking the human tongue and can differentiate as well as quantify tastes such as sourness, saltiness, bitterness, sweetness and umami (Kong and Singh, 2011). ET technology is mainly based on potentiometric, voltametric, ion-selective field-effect transistor, piezoelectric and optical sensors with pattern recognition tools for data processing (Wardencki et al., 2013). Liquid solutions can be sampled without alteration and the ET will provide a characteristic fingerprint, similar to the EN, for the sample tested with the detection thresholds of sensors similar or better than those of human receptors. The data from the ET can be calibrated against those from a trained sensory panel. The ET can easily 'taste' raw substances, semi-raw products and also new entities that are not yet allowed for human consumption (Wardencki et al., 2013). EN and ET technologies are sometimes combined for food flavour detection, simulating the coexistence of the two sensory functions in humans (Li et al., 2006; Kong and Singh, 2011). ET technologies are still in an early stage of development and can be limited by sensor drift which can occur as the sensors age and accumulate surface contamination (Ha et al., 2015).

To date, ETs have found application in wine (Cetó et al., 2014a,b), beer (Gutiérrez et al., 2013), dairy protein (Newman et al., 2014), milk (Dias et al., 2009), olive oil (Apetrei et al., 2010) and tomato (Beullens et al., 2008) products.

NEAR INFRARED AND FOURIER TRANSFORM INFRARED SPECTROSCOPY

Identification of compounds in food by IR spectroscopy is based on the property of molecules to absorb infrared light and experience a wide variety of vibrational motion characteristics of their composition. When coupled with chemometric data analysis techniques, NIR spectroscopy is a rapid technique that possesses potential selectivity for screening products for qualitative attributes (Kong and Singh, 2011). Since the 1980s, NIR reflectance spectroscopy has been proved to be one of the most efficient and advanced tools for the estimation of quality attributes in food products (Prieto et al., 2009; O'Sullivan and Kerry, 2013). NIR spectroscopy is a sensitive, fast and nondestructive analytical technique with simplicity in sample preparation allowing a simultaneous assessment of numerous food properties (Osborne et al., 1993). The NIR technique is widely accepted as one of the most promising on/in-line process control techniques detecting fat, moisture and protein content in meats, fruit and vegetables, grain and grain products, milk and dairy products, beverages and other products (Huang et al., 2008) and for determining authenticity of food products such as fats and oils, soluble coffee, green coffee and fruits (Kong and Singh, 2011). NIR spectroscopy has been successfully applied to the quantitative determination of major constituents (moisture, fat and protein) in meat and meat products including: beef (Eichinger and Beck, 1992; Tøgersen et al., 2003; Alomar et al., 2003; Prevolnik et al., 2005; Prieto et al., 2006), pork (Tøgersen et al., 1999; Brøndum et al., 2000; Ortiz-Somovilla et al., 2007) and poultry meat (Valdes and Summers, 1986; Renden et al., 1986; Abeni and Bergoglio, 2001; Berzaghi et al., 2005; Cozzolino and Murray, 2002; McDevitt et al., 2005) and mutton (Viljoen et al., 2007). Near infrared spectroscopy (NIRS) is an analytical technique that uses a source producing light of known wavelength pattern (usually 800–2500 nm) that enables the aquisition of a complete picture of the organic composition of the analysed substance/material (Van Kempen, 2001). It is based on the principle that different chemical bonds in organic matter absorb or emit light of different wavelengths when the sample is irradiated (Prevolnik et al., 2004).

FTIR spectroscopy differs to IR in that light is guided through an interferometer to generate modulated light and measures all infrared wavelengths simultaneously, rather than individual wavelengths of the radiation. The interferogram is then converted into a conventional spectrum using the Fourier transform algorithm (Kong and singh, 2011). Lachenmeier (2007) investigated FTIR in combination with PLSR as a complete multicomponent screening method for spirit drinks and beer in the context of official food control. They found that, the majority of samples were classified as being in conformance with legal and quality requirements and that FTIR spectroscopy is faster and only requires a simple sample preparation compared to conventional methods. FTIR has also found applications in distilled liquors and wines (Cocciardi et al., 2005; Moreira et al., 2002; Fernandez and Agosin, 2007; Villiers et al., 2012). FTIR (Fig. 8.4) spectroscopy has substantial potential as a quantitative control method in the food industry (Van de Voort et al., 1992) and has also been used to monitor shelf life of Crescenza cheese (Cattaneo et al., 2005) and ricotta cheese (Sinelli et al., 2005). Anjos et al. (2015) used FTIR with attenuated total reflectance (FTIR-ATR) and a PLSR model for the prediction of sugar content in honey samples. For lipid measurement, it has the advantage of easy sample preparation, rapid measurements and there is no use of chemicals in contrast to traditional solvent methods accompanied by chromatographic techniques (Flåtten et al., 2005). FTIR analysis is rapid, noninvasive, requires minimum sample pretreatment or specific consumables or reagents, and in conjunction with attenuated total reflectance (ATR) technology permits users to collect full spectra in a few seconds, allowing simultaneous assessment of numerous meat properties (Ammor et al., 2009; Ellis et al., 2004, 2002). For producers and manufacturers, a rapid method to

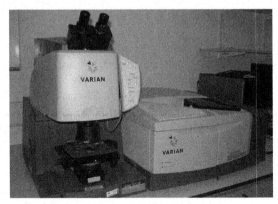

FIGURE 8.4   Fourier transform infrared (FTIR) spectroscopy. Presented is an Agilent 670IR with 620 microscope.

measure fat qualities would be useful and FTIR spectroscopy has proven to be a powerful tool in food research (Ripoche and Guillard, 2001). Flåtten et al. (2005) measured C22:5 and C22:6 marine fatty acids in pork fat with FTIR. Their results show that marine fatty acids and general fatty acid composition in pork fat can be measured with FT-MIR spectroscopy with good precision. Classification of the samples on the basis of these measurements gives the opportunity for useful implementations of the method in commercial situations, with less labour and time required than alternative chromatographic methods (Flåtten et al., 2005).

The data from NIR and FTIR analysis are processed using an appropriate chemometric or multivariate data analytical method. Essentially, results from test analysis are compared chemometrically to a database of similar known samples or standards (O'Sullivan and Kerry, 2013). The model then best fits the current results to those that are available and thus present the concentration of the known variable of interest. Of course a suitable database of known compounds in specific samples must be available, either through the instrument vendor or created by the scientists undertaking the work in order to derive any significant data from NIR or FTIR (O'Sullivan and Kerry, 2013).

Adulteration in meat using FTIR-ATR and multivariate analysis has been studied and quantified by Al-Jowder et al. (1999) and Al-Jowder et al. (2002), who identified adulteration of beef meat with heart, tripe, kidney and liver. Similarly, Meza-Márquez et al. (2010) used Mid-FTIR-ATR spectroscopy with multivariate analysis (soft independent modelling class of analogies SIMCA) to successfully detect and quantify the adulteration of minced lean beef with horse meat, textured soy protein or the addition fat beef trimmings in minced beef. At the same time, the model was capable of simultaneously determining the composition (water, protein, fat, ash and glycogen) of the meat samples with a 99% confidence limit (Meza-Márqucz ct al., 2010).

REFERENCES

Aaslyng, M.D., 2009. Improving the Sensory Quality and Nutritional Quality of Fresh Meat, Chapter I, Trends in Meat Consumption and the Need for Fresh Meat and Meat Products of Improved Quality. Danish Meat Research Institute, Denmark.

Abeni, F., Bergoglio, G., 2001. Characterization of different strains of broiler chicken by carcass measurements, chemical and physical parameters and NIRS on breast muscle. Meat Science 57, 133–137.

Al-Jowder, O., Defernez, M., Kemsley, E., Wilson, R.H., 1999. Mid-infrared spectroscopy and chemometrics for the authentication of meat products. Journal of Agricultural and Food Chemistry 47, 3210–3218.

Al-Jowder, O., Kemsley, E.K., Wilson, R.H., 2002. Detection of adulteration in cooked meat products by mid-infrared spectroscopy. Journal of Agricultural and Food Chemistry 50, 1325–1329.

Alomar, D., Gallo, C., Castaneda, M., Fuchslocher, R., 2003. Chemical and discriminant analysis of bovine meat by near infrared reflectance spectroscopy (NIRS). Meat Science 63, 441–450.

AMSA, 1991. American Meat Science Association Committee on Quidelines for Meat Color Evaluation. Contribution No. 91-545-A.

Ammor, S.A., Argyri, A., Nychas, G.J.E., 2009. Rapid monitoring of the spoilage of minced beef stored under conventionally and active packaging conditions using Fourier transform infrared spectroscopy in tandem with chemometrics. Meat Science 81, 507–514.

Anjos, O., Campos, M.C., Ruiz, P.C., Antunes, P., 2015. Application of FTIR-ATR spectroscopy to the quantification of sugar in honey. Food Chemistry 169, 218–223.

Apetrei, C., Apetrei, I.M., Villanueva, S., de Saja, J.A., Gutierrez-Rosales, F., Rodriguez-Mendez, M.L., 2010. Combination of an e-nose, an e-tongue and an e-eye for the characterisation of olive oils with different degree of bitterness. Analytica Chimica Acta 663 (1), 91.

Balaban, M.O., Aparicio, J., Zotarelli, M., Sims, C., 2008. Quantifying nonhomogeneous colors in agricultural materials. Part II: Comparison of machine vision and sensory panel evaluations. Journal of Food Science 73 (9), S438–S442.

Bratzler, L.J., 1932. Measuring the Tenderness of Meat by Means of a Mechanical Shear (Master of Science thesis). Kansas State College (KA), USA.

Berzaghi, P., Dalla Zotte, A., Jansson, L.M., Andrighetto, I., 2005. Near-infrared reflectance spectroscopy as a method to predict chemical composition of breast meat and discriminate between different n-3 feeding sources. Poultry Science 84, 128–136.

Byrne, D.V., Bredie, W.L.P., Mottram, D.S., Martens, M., 2002. Sensory and chemical investigations on the effect of oven cooking on warmed-over flavour in chicken meat. Meat Science 61, 127–139.

Bourne, M.C., 2002. Sensory methods of texture and viscosity measurement. In: Food Texture and Viscosity, second ed. Academic Press, London and New York, pp. 257–291.

Boselli, E., Pacetti, D., Lucci, P., Di Lecce, P., Frega, N., 2008. Supplementation with high-oleic sunflower oil and α-tocopheryl acetate: effects on meat pork lipids. European Journal of Food Science and Technology 110, 381–391.

Brøndum, J., Munck, L., Henckel, P., Karlsson, A., Tornberg, E., Engelsen, S.B., 2000. Prediction of water-holding capacity and composition of porcine meat by comparative spectroscopy. Meat Science 55, 177–185.

Beullens, K., Mészáro, P., Vermeir, S., Kirsanov, D., Legin, A., Buysens, S., Cap, N., Nicolaï, B.M., Lammertyn, J., 2008. Analysis of tomato taste using two types of electronic tongues. Sensors and Actuators B: Chemical 131, 10–17.

Caine, W.R., Aalhus, J.L., Best, D.R., Dugan, M.E.R., Jeremiah, L.E., 2003. Relationship of texture profile analysis and Warner–Bratzler shear force with sensory characteristics of beef rib steaks. Meat Science 64, 333–339.

Cattaneo, T.M.P., Giardina, C., Sinelli, N., Riva, M., Giangiacomo, R., 2005. Application of FT-NIR and FT-IR spectroscopy to study the shelf-life of Crescenza cheese. International Dairy Journal 15, 693–700.

Cetó, X., Apetrei, C., del Valle, M., Rodriguez-Mendez, M.L., 2014a. Evaluation of red wines antioxidant capacity by means of a voltammetric e-tongue with an optimized sensor array. Electrochimica Acta 120, 180–186.

Cetó, X., Capdevila, J., Puig-Pujol, A., del Valle, M., 2014b. Cava wine authentication employing a voltammetric electronic tongue. Electroanalysis. http://dx.doi.org/10.1002/elan.201400057.

CIE, 1986. Colorimetry, second ed. CIE Central Bureau Kegelgasse 27 A-1030, Wien, Austria.

Cocciardi, R.A., Ismail, A.A., Sedman, J., 2005. Investigation of the potential utility of single-bounce attenuated total reflectance Fourier transform infrared spectroscopy in the analysis of distilled liquors and wines. Journal of Agriculture and Food Chemistry 53 (8), 2803–2809.

Cozzolino, D., Murray, I., 2002. Effect of sample presentation and animal muscle species on the analysis of meat by near infrared reflectance spectroscopy. Journal of Near Infrared Spectroscopy 10, 37–44.

Culioli, J., 1995. Meat tenderness: mechanical assessment. In: Ouali, A., Demeyer, D.I., Smulders, F.J.M. (Eds.), Expression of Tissue Proteinases and Regulation of Protein Degradation as Related to Meat Quality. ECCEAMST, Utrecht, pp. 239–266.

Destefanis, G., Brugiapaglia, A., Barge, M.T., Dal Molin, E., 2008. Relationship between beef consumer tenderness perception and Warner-Bratzler shear force. Meat Science 78, 153–156.

Dias, L.A., Peres, A.M., Veloso, A.C.A., Reis, F.S., Vilas-boas, M., Machado, A.A.S.C., 2009. An electronic tongue taste evaluation: identification of goat milk adulteration with bovine milk. Sensors and Actuators B: Chemical 136, 209–217.

Dijksterhuis, G.B., 1995. Multivariate data analysis in sensory and consumer science: an overview of developments. Trends in Food Science & Technology (6), 206–211.

Doyen, K., Carey, M., Linforth, R.S.T., Marin, M., Taylor, A.J., 2001. Volatile release from an emulsion: headspace and in-mouth studies. Journal of Agriculture and Food Chemistry 49, 804–810.

Drumm, T.D., Spanier, A.M., 1991. Changes in the lipid content of autoxidation and sulphur-containing compounds in cooked beef during storage. Journal of Agriculture and Food Chemistry 49, 336–343.

Dryden, F.D., Marchello, J.A., 1970. Influence of total lipid and fatty acid composition upon the palatability of three bovine muscles. Journal of Animal Science 31, 36–41.

Eichinger, H., Beck, G., 1992. Possibilities for improving breeding value estimation of meat quality in cattle by using the near-infrared measurement technique. Archiv für Tierzucht 35, 41–50.

Eklöv, T., Johansson, G., Winquist, F., Lundström, I., 1998. Monitoring sausage fermentation using an electronic nose. Journal of Science and Food Agriculture 76, 525–532.

Enser, M., Hallet, K.G., Hewett, B., Fursey, G.A., Wood, J.D., 1996. Fatty acid content and composition of English beef, lamb, and pork at retail. Meat Science 42, 443–456.

Ellis, D.I., Broadhurst, D., Goodacre, R., 2004. Rapid and quantitative detection of the microbial spoilage of beef by Fourier transform infrared spectroscopy and machinelearning. Analytica Chimica Acta 514, 193–201.

Ellis, D.I., Broadhurst, D., Kell, D.B., Rowland, J.J., Goodacre, R., 2002. Rapid and quantitative detection of the microbial spoilage of meat by Fourier transform infrared spectroscopy and machine learning. Applied and Environmental Microbiology 68, 2822–2828.

Everard, C.D., O'Callaghan, D.J., Fagan, C.C., O'Donnell, C.P., Castillo, M., Payne, F.A., 2007. Computer vision and colour measurement techniques for inline monitoring of cheese curd syneresis. Journal of Dairy Science 90, 3162–3170.

Flåtten, A., Bryhni, E.A., Kohler, A., Egelandsdal, B., Isaksson, T., 2005. Determination of C22:5 and C22:6 marine fatty acids in pork fat with Fourier transform mid-infrared spectroscopy. Meat Science 69 (2005), 433–440.

Fernandez, K., Agosin, E., 2007. Quantitative analysis of red wine tannins using Fourier-transform mid-infrared spectroscopy. Journal of Agriculture and Food Chemistry 55, 7294–7300.

Garcia, M., Aleixandre, M., Gutierrez, J., Horrillo, M.C., 2006. Electronic nose for wine discrimination. Sensors and Actuators B: Chemical 113 (2), 911–916.

Gardner, J.W., Bartlett, P.N., 1994. A brief history of electronic noses. Sensors and Actuators B: Chemical 18, 211–220.

Gasser, U., Grosch, W., 1988. Flavour deterioration of soya bean oil: identification of intense odour compounds during flavour reversion. Fett Wissenschaft Technologie 90, 332–336.

Gil-Sánchez, L., Garrigues, J., Garcia-Breijo, E., Grau, R., Aliño, M., Baigts, D., Barat, J.M., 2015. Artificial neural networks (Fuzzy ARTMAP) analysis of the data obtained with an electronic tongue applied to a ham-curing process with different salt formulations. Applied Soft Computing 30, 421–429.

Grigioni, G.M., Margaria, C.A., Pensel, N.A., Sánchez, G., Vaudagna, S.R., 2000. Warmed-over flavour in low temperature-long time processed meat by an "electronic nose". Meat Science 56, 221–228.

Gutiérrez, J.M., Haddi, Z., Amari, A., Bouchikhi, B., Mimendia, A., Cetó, X., del Valle, M., 2013. Hybrid electronic tongue based on multisensory data fusion fordiscrimination of beers. Sensors and Actuators B: Chemical 177, 989–996.

Ha, D., Sun, Q., Su, K., Wan, H., Li, H., Xu, N., Sun, F., Zhuang, L., Hu, N., Wang, P., 2015. Recent achievements in electronic tongue and bioelectronic tongue as taste sensors. Sensors and Actuators B: Chemical 207, 1136–1146.

Haugen, J.E., Kvaal, K., 1998. Electronic nose and artificial neural network. Meat Science 49, S273–S286.

Hood, D.E., Mead, G.C., 1993. Modified atmosphere storage of fresh meat and poultry. In: Parry, R.T. (Ed.), Principles and Applications of Modified Atmosphere Packing of Food. Blackie Academic and Professional, London, pp. 269–298.

Holm, E.S., Schäfer, A., Skov, T., Koch, A.G., Petersen, M.A., 2012. Identification of chemical markers for the sensory shelf-life of saveloy. Meat Science 90 (2012), 314–322.

Hansen, T., Pedersen, M.A., Byrne, D.V., 2005. Sensory based quality control utilising an electronic nose and GC-MS analyses to predict end-product quality from raw materials. Meat Science 69, 621–634.

Heffernan, S.P., Kelly, A.L., Mulvihill, D.M., 2009. High-pressure-homogenised cream liqueurs: emulsification and stabilization efficiency. Journal of Food Engineering 95, 525–531.

Huang, H., Yu, H., Xu, H., Ying, Y., 2008. Near infrared spectroscopy for on/in-line monitoring of quality in foods and beverages: a review. Journal of Food Engineering 87, 303–313.

Huselegge, B., Engel, B., Buist, W., Merkus, G.S.M., Klont, R.E., 2001. Instrumental colour classification of veal carcasses. Meat Science 57, 191–195.

Infante, R., Farcuh, M., Meneses, C., 2008. Monitoring the sensorial quality and aroma through an electronic nose in peaches during cold storage. Journal of the Science of Food and Agriculture 88, 2073–2078.

Jackman, P., Sun, D.-W., Du, C.-J., Allen, P., 2009. Prediction of beef eating qualities from colour, marbling and wavelet surface texture features using homogenous carcass treatment. Pattern Recognition 42 (5), 751–763.

Jackman, P., Sun, D.W., Allen, P., Brandon, K., White, A.M., 2010. Correlation of consumer assessment of *longissimus dorsi* beef palatability with image colour, marbling and surface texture features. Meat Science 84, 564–568.

Janhøj, T., Bom Frøst, M., Ipsen, R., 2008. Sensory and rheological characterization of acidified milk drinks. Food Hydrocolloids 22, 798–806.

Jiang, T., Busboom, J.R., Nelson, M.L., O'Fallon, J., Ringkob, T.P., Joos, D., Piper, K., 2010. Effect of sampling fat location and cooking on fatty acid composition of beef steaks. Meat Science 84, 86–92.

Kilcawley, K., 2015. Cheese Flavour. second ed. of Fundamentals of Cheese Science.

Kong, F., Singh, R.P., 2011. Advances in instrumental methods to determine food quality deterioration. In: Kilcast, D., Subramaniam, P. (Eds.), The Stability and Shelf-Life of Food. Woodhead Publishing Limited, Cambridge, UK.

Lachenmeier, D.W., 2007. Rapid quality control of spirit drinks and beer using multivariate data analysis of Fourier transform infrared spectra. Food Chemistry 101, 825–832.

Lawrie, R.A., Ledward, D.A., 2006. Lawrie's Meat Science. Woodhead Publishing Ltd, Cambridge, England.

Li, Z., Wang, N., Vigneault, C., 2006. Electronic nose and electronic tongue in food production and processing. Stewart Postharvest Review 2, 1—5.

Liu, Q., Lanari, M.C., Schaefer, D.M., 1995. A review of dietary vitamin E supplementation for improvement of beef quality. Journal of Animal Science 73, 3131—3140.

Limbo, S., Sinelli, N., Torri, L., Riva, M., 2009. Freshness decay and shelf life predictive modelling of European sea bass (*Dicentrarchus labrax*) applying chemical methods and electronic nose. LWT, Food Science and Technology 42, 977—984.

Lu, J., Tan, J., Shatadal, P., Gerrard, D.E., 2000. Evaluation of pork color by using computer vision. Meat Science 56, 57—60.

Lu, R., 2013. Principles of solid food texture analysis. Instrumental Assessment of Food Sensory Quality 103—128.

Lynch, A.G., Mulvihill, D.M., 1997. Effect of sodium caseinate on the stability of cream liqueurs. International Journal of Dairy Technology 50, 1—7.

Lyon, B.G., Lyon, C.E., 1998. Assessment of three devices used in shear tests of cooked breast meat. Poultry Science 77, 1585—1590.

McKenna, B.M., Lyng, J.G., 2013. Principles of food viscosity analysis. Instrumental Assessment of Food Sensory Quality 130—162.

Martin, M.L.G.M., Ji, W., Luo, R., Hutchings, J., Heredia, F.J., 2007. Measuring colour appearance of red wines. Food Quality and Preference 18, 862—871.

Meza-Márquez, O.G., Gallardo-Velázquez, T., Osorio-Revilla, G., 2010. Application of mid-infrared spectroscopy with multivariate analysis and soft independent modeling of class analogies (SIMCA) for the detection of adulterants in minced beef. Meat Science 86, 511—519.

McDevitt, R.M., Gavin, A.J., Andres, S., Murray, I., 2005. The ability of visible and near infrared reflectance spectroscopy (NIRS) to predict the chemical composition of ground chicken carcasses and to discriminate between carcasses from different genotypes. Journal of Near Infrared Spectroscopy 13, 109—117.

McMillin, K.W., 2008. Where is MAP going? A review and future potential of modified atmosphere packaging for meat. Meat Science 80, 43—65.

Miller, M.F., Carr, M.A., Ramsey, C.B., Crockett, K.L., Hoover, L.C., 2001. Consumer thresholds for establishing the value of beef tenderness. Journal of Animal Science 79, 3062—3068.

Mitchaothai, J., Yuangklang, C., Wittayakun, S., Vasupen, K., Wongsutthavas, S., Srenanul, P., et al., 2007. Effect of dietary fat type on meat quality and fatty acid composition of various tissues in growing-finishing swine. Meat Science 76, 95—101.

Moreira, J.L., Marcos, A.M., Barros, P., 2002. Analysis of Portuguese wine by Fourier transform infrared spectrometry. Ciência e Técnica Vitivinícola/Journal of Viticulture and Enology 17, 27—33.

Newman, J., Harbourne, N., O'Riordan, D., Jacquier, J.C., O'Sullivan, M., 2014. Comparison of a trained sensory panel and an electronic tongue in the assessment of bitter dairy protein hydrolysates. Journal of Food Engineering 128, 127—131.

Nuernberg, K., Wood, J.D., Scollan, N.D., Richardson, R.I., Nute, G.R., Nuernberg, G., et al., 2005. Effect of a grass-based and a concentrate feeding system on meat quality characteristics and fatty acid composition of longissimus muscle in different cattle breeds. Livestock Production Science 94 (2), 137—147.

O'Connell, M., Valdora, G., Peltzer, G., Negri, R.M., 2001. A practical approach for fish freshness determinations using a portable electronic nose. Sensors and Actuators B: Chemical 80, 149—154.

Ólafsdóttir, G., Martindóttir, E., Jónsson, E.H., 1997. Rapid gas sensor measurements to determine spoilage of capelin (*Mallotus villosus*). Journal of Agriculture and Food Chemistry 45, 2654–2659.

Ortiz-Somovilla, V., Espana-Espana, F., Gaitan-Jurado, A.J., Perez-Aparicio, J., De Pedro-Sanz, E.J., 2007. Proximate analysis of homogenized and minced mass of pork sausages by NIRS. Food Chemistry 101, 1031–1040.

O'Sullivan, M.G., Byrne, D.V., Stagsted, J., Andersen, H.J., Martens, M., 2002. Sensory colour assessment of fresh meat from pigs supplemented with iron and vitamin E. Meat Science 60, 253–265.

O'Sullivan, M.G., Byrne, D.V., Martens, H., Gidskehaug, L.H., Andersen, H.J., Martens, M., 2003a. Evaluation of pork meat colour: prediction of visual sensory quality of meat from instrumental and computer vision methods of colour analysis. Meat Science 65, 909–918.

O'Sullivan, M.G., Byrne, D.V., Jensen, M.T., Andersen, H.J., Vestergaard, J., 2003b. A comparison of warmed-over flavour in pork by sensory analysis, GC/MS and the electronic nose. Meat Science 65, 1125–1138.

O'Sullivan, M.G., Kerry, J.P., 2008. Ch 27, Sensory evaluation of fresh meat. In: Kerry, J.P., Ledward, D.A. (Eds.), Improving the Sensory and Nutritional Quality of Fresh Meat. Woodhead Publishing Ltd, Cambridge, UK.

O'Sullivan, M.G., 2011a. Ch 4, Sensory shelf-life evaluation. In: Piggott, J.R. (Ed.), Alcoholic Beverages: Sensory Evaluation and Consumer Research. Woodhead Publishing Ltd, Cambridge, UK.

O'Sullivan, M.G., 2011b. Ch 25, Case studies: meat and poultry. In: Kilcast, D., Subramaniam, P. (Eds.), Food and Beverage Shelf-life and Stability. Woodhead Publishing Ltd, Cambridge, UK.

O'Sullivan, M.G., Kerry, J.P., 2013. In: Kilcast, D. (Ed.), Ch 10, Instrumental Assessment of the Sensory Quality of Meat, Poultry and Fish. Woodhead Publishing Ltd, Cambridge, UK.

Osborne, B.G., Fearn, T., Hindle, P.H., 1993. Near Infrared Spectroscopy in Food Analysis. Longman Scientific and Technical, Harlow, Essex, UK.

Panigrahi, S., Balasubramanian, S., Gub, H., Logue, C.M., Marchello, M., 2006. Design and development of a metal oxide based electronic nose for spoilage classification of beef. Sensors and Actuators B: Chemical 119, 2–14.

Papadakis, S.E., Abdul-Malek, S., Emery-Kamdem, R., Yam, K.L., 2000. A versatile and inexpensive technique for measuring colour of foods. Food Technology 54, 48–51.

Platter, W.J., Tatum, J.D., Belk, K.E., Chapman, P.L., Scanga, J.A., Smith, G.C., 2003. Relationship of consumer sensory ratings, marbling score, and shear force value to consumer acceptance of beef strip loin steaks. Journal of Animal Science 81, 2741–2750.

Plutowska, B., Wardencki, W., 2008. Application of gas chromatography-olfactometry (GC-O) in analysis and quality assessment of alcoholic beverages ± a review. Food Chemistry 107 (1), 449–463.

Power, P.C., 1996. The Formulation, Testing and Stability of 16% Fat Cream Liqueurs (Ph.D. thesis). National University of Ireland, Cork.

Prieto, N., Andres, S., Giraldez, F.J., Mantecon, A.R., Lavin, P., 2006. Potential use of near infrared reflectance spectroscopy (NIRS) for the estimation of chemical composition of oxen meat samples. Meat Science 74, 487–496.

Prieto, N., Roehe, R., Lavín, P., Batten, G., Andrés, S., 2009. Application of near infrared reflectance spectroscopy to predict meat and meat products quality: a review. Meat Science 83, 175–186.

Prevolnik, M., Čandek-Potokar, M., Škorjanc, D., 2004. Ability of NIR spectroscopy to predict meat chemical composition and quality: a review. Czechoslovak Journal of Animal Science 49, 500−510.

Prevolnik, M., Čanddek-Potokar, M., Škorjanc, D., Velikonja-Bolta, Š., Škrlep, M., Žnidaršič, T., et al., 2005. Predicting intramuscular fat content in pork and beef by near infrared spectroscopy. Journal of Near Infrared Spectroscopy 13, 77−85.

Pryzbylski, R., Eskin, N.A.M., 1995. Methods to measure volatile compounds and the flavour significance of volatile compounds. In: Warner, K., Eskin, N.A.M. (Eds.), Methods to Assess Quality and Stability of Oils and Fat-Containing Foods. AOCS Press, Illinois, pp. 107−133.

Quevedo, R.A., Aguilera, J.M., Pedreschi, F., 2010. Color of salmon fillets by computer vision and sensory panel. Food and Bioprocess Technology 3 (5), 637−643.

Raes, K., De Smet, S., Demeyer, D., 2004. Effect of dietary fatty acids on incorporation of long chain polyunsaturated fatty acids and conjugated linoleic acid in lamb, beef and pork meat: a review. Animal Feed Science and Technology 113, 199−221.

Rajamäki, T., Alakomi, H.L., Ritvanen, T., Skyttä, E., Smolander, M., Ahvenainen, R., 2006. Application of an electronic nose for quality assessment of modified atmosphere packaged poultry meat. Food Control 17, 5−13.

Renden, J.A., Oates, S.S., Reed, R.B., 1986. Determination of body fat and moisture in dwarf hens with near infrared reflectance spectroscopy. Poultry Science 65, 1539−1541.

Ripoche, A., Guillard, A.S., 2001. Determination of fatty acid composition of pork fat by Fourier transform infrared (FTIR) spectroscopy. Meat Science 58, 299−304.

Ross, C.F., 2009. Sensory science at the human-machine interface. Trends in Food Science & Technology 1−10.

Ruiz de Huidobro, F., Cañeque, V., Lauzurica, S., Velasco, S., Pérez, C., Onega, E., 2001. Sensory characterization of meat texture in sucking lambs. Methodology. Investigació n Agraria: Producción y Sanidad Animales 16 (2), 223−234.

Shackelford, S.D., Morgan, J.B., Cross, H.R., Savell, J.W., 1991. Identification of threshold levels for Warner−Bratzler shear force in beef top loin steaks. Journal of Muscle Foods 2, 289−296.

Shackelford, S.D., Wheeler, T.L., Koohmaraie, M., 1997. Tenderness classification of beef: 1. Evaluation of beef longissimus shear force at 1 or 2 days post mortem as a predictor of aged beef tenderness. Journal of Animal Science 75, 2417−2422.

Shahidi, F., Yun, J., Rubin, L.J., Wood, D.F., 1987. The hexanal content as an indicator of oxidative stability and flavour acceptability in cooked ground pork. Canadian Institute of Food Science and Technology Journal 20, 104−106.

Siegmund, B., Pfannhauser, W., 1999. Changes of the volatile fraction of cooked chicken meat during chill storing: results obtained by the electronic nose in comparison to GC-MS and GC olfactometry. Zeitschrift Für Lebensmittel-Untersuchung und forchung A 208, 336−341.

Simopoulos, A.P., 1999. Essential fatty acids in health and chronic disease. American Journal of Clinical Nutrition 70 (3), 560S−569S.

Sinelli, N., Barzaghi, S., Giardina, C., Cattaneo, T.M.P., 2005. A preliminary study using Fourier transform near infrared spectroscopy to monitor the shelf-life of packed industrial ricotta cheese. Journal of Near Infrared Spectroscopy 13, 293−300.

St. Angelo, A.J., Vercellotti, J.R., Legendre, M.G., Vennett, C.H., Kuan, J.W., James Jr., C., Dupuy, H.P., 1987. Chemical and instrumental analysis of warmed-over flavour in beef. Journal of Food Science 52, 1163−1168.

Sutherland, J.P., Varnum, A.H., Evans, M.G., 1986. A Color Atlas of Food Quality Control. CRC Press, Weert.

Tikk, K., Haugen, J.E., Andersen, H.J., Aaslyng, M.D., 2008. Monitoring of warmed-over flavour in pork using the electronic nose – correlation to sensory attributes and secondary lipid oxidation products. Meat Science 80 (4), 1254–1263.

Tøgersen, G., Isaksson, T., Nilsen, B.N., Bakker, E.A., Hildrum, K.I., 1999. On-line NIR analysis of fat, water and protein in industrial scale ground meat batches. Meat Science 51, 97–102.

Tøgersen, G., Arnesen, J.F., Nilsen, B.N., Hildrum, K.I., 2003. On-line prediction of chemical composition of semi-frozen ground beef by non-invasive NIR spectroscopy. Meat Science 63, 515–523.

Valous, N.A., Mendoza, F., Sun, D.W., Allen, P., 2009. Colour calibration of a laboratory computer vision system for quality evaluation of pre-sliced hams. Meat Science 81 (1), 132–141.

Van de Voort, F.F., Sedan, J., Emo, G., Ismail, A.A., 1992. Rapid and direct iodine value and saponification number determination of fats and oils by attenuated total reflectance/Fourier transform infrared spectroscopy. Journal of the Oil Chemistry Society 69, 1118–1123.

Van Kempen, L., 2001. Infrared technology in animal production. World's Poultry Science Journal 57, 29–48.

Valdes, E.V., Summers, J.D., 1986. Determination of crude protein in carcass and breast muscle samples of poultry by near infrared reflectance spectroscopy. Poultry Science 65, 485–490.

Vatansever, L., Kurt, E., Enser, M., Nute, G.R., Scollan, N.D., Wood, J.D., et al., 2000. Shelf life and eating quality of beef from cattle of different breeds given diets differing in n-3 poly-unsaturated fatty acid composition. Animal Science 71, 471–482.

Viljoen, M., Hoffman, L.C., Brand, T.S., 2007. Prediction of the chemical composition of mutton with near infrared reflectance spectroscopy. Small Ruminant Research 69, 88–94.

Villiers, A., Alberts, P., Tredoux, A.G., Nieuwoudt, H.H., 2012. Analytical techniques for wine analysis: an African perspective; a review. Analytica Chimica Acta 730, 2–23.

Wardencki, W., Chmiel, T., Dymerski, T., 2013. Gas chromatography-olfactometry (GC-O), electronic noses (e noses) and electronic tongues (e-tongues) for in vivo food flavour measurement. In: Kilcast, D., Subramaniam, P. (Eds.), The Stability and Shelf-Life of Food. Woodhead Publishing Ltd, Cambridge, UK.

Warner, K.F., 1928. Progress report of the mechanical test for tenderness of meat. Proceedings of the American Society of Animal Production 21, 114.

Winquist, F., Lundström, I., Wide, P., 1999. The combination of an electronic tongue and an electronic nose. Sensors and Actuators B: Chemical 58 (1–3), 512.

Wood, J.D., Enser, M., 1997. Factors influencing fatty acids in meat and the role of antioxidants in improving meat quality. British Journal of Nutrition 78 (Suppl. 1), S49–S60.

Wood, J.D., Richardson, R.I., Nute, G.R., Fisher, A.V., Campo, M.M., Kasapidou, E., et al., 2004. Effects of fatty acids on meat quality: a review. Meat Science 66 (1), 21–32.

Wu, D., Sun, D.W., 2013. Food colour measurement using computer vision. Instrumental Assessment of Food Sensory Quality 165–195.

Zakrys, P.I., Hogan, S.A., O'Sullivan, M.G., Allen, P., Kerry, J.P., 2008. Effects of oxygen concentration on sensory evaluation and quality indicators of beef muscle packed under modified atmosphere. Meat Science 79, 648–655.

Zakrys, P.I., O'Sullivan, M.G., Allen, P., Kerry, J.P., 2009. Consumer acceptability and physiochemical characteristics of modified atmosphere packed beef steaks. Meat Science 81, 720–725.

Zakrys-Waliwander, P.I., O'Sullivan, M.G., Allen, P., O'Neill, E.E., Kerry, J.P., 2010. Investigation of the effects of commercial carcass suspension (24 and 48 h) on meat quality in modified atmosphere packed beef steaks during chill storage. Food Research International 43, 277–284.

Zakrys-Waliwander, P.I., O'Sullivan, M.G., Walshe, H., Allen, P., Kerry, J.P., 2011a. Sensory comparison of commercial low and high oxygen modified atmosphere packed sirloin beef steaks. Meat Science 88, 198—202.

Zakrys-Waliwander, P.I., O'Sullivan, M.G., O'Neill, E.E., Kerry, J.P., 2011b. The effects of high oxygen modified atmosphere packaging on protein oxidation of bovine *M. longissimus dorsi* muscle during chilled storage. Food Chemistry 2, 527—532.

Zapotoczny, P., Majewska, K., 2010. A comparative analysis of colour measurements of the seed coat and endosperm of wheat kernels performed by various techniques. International Journal of Food Properties 13, 75—89.

Zheng, C., Sun, D.W., Zheng, L.Y., 2006. Correlating colour to moisture content of large cooked beef joints by computer vision. Journal of Food Engineering 77 (4), 858—863.

Chapter 9

Nutritionally Optimised Low Fat Foods

INTRODUCTION

Fat has an important role in human nutrition and is part of a normal balanced diet. Fats are required for hormone production (prostaglandins) and are used in blood clotting, oxygen transport as well as the making of new cells and are required for energy production, without which muscles would cease to function. Animal fats play important functional, sensory and nutritional roles in many food products including processed meats. Animal fats have been also used for centuries in the manufacturing of meat products (sausages, hams, pies) and other foods (baked goods, dairy products) (Barbut, 2011). Additionally, the Food and Agriculture Organisation of the United Nations (FAO, 2010) publish the dietary guidelines from many countries around the world and these include dairy products which contain fat (Komorowski, 2011). Obesity is primarily caused by an over intake of food and a lack of exercise that results in an energy imbalance between calories consumed and calories expended. In particular, over consumption of fat has assumed epidemic proportions in the developed world due to the association between the type and level of dietary fat and the risk of coronary heart disease (CHD) and related health problems such as high blood pressure, osteoarthritis, sleep apnoea, cancer, etc. This has led to greater consumer awareness and thus the demand for healthier, fat-reduced foods (O'Sullivan, 2015). Consequently, food and regulatory bodies have targeted issues like fat reduction in processed products. Organisations like the World Health Organization (WHO) are driving measures to reduce saturated fat content in foods by raising the consumers' awareness and setting guidelines around ingredient usage for companies. Currently, the WHO recommends a daily intake of polyunsaturated fatty acids (PUFAs) between 6% and 11% based on daily energy intake (WHO, 2003). The World Health Organization (2003) has also recommended a daily intake of total fat comprising between 15% and 30% of dietary energy (Talbot, 2011). The prevalence of excess levels of adipose tissue build up in the human body is widely recognised as a leading world health problem and is referred to as an obesity epidemic (WHO, 2000).

A Handbook for Sensory and Consumer-Driven New Product Development.
http://dx.doi.org/10.1016/B978-0-08-100352-7.00009-9
177

The WHO found in 2008 that more than 1.4 billion adults, aged 20 and older, were overweight. Out of these overweight adults, 200 million men and nearly 300 million women were classified as obese (WHO, 2012). The WHO also found that more than 40 million children under the age of five were overweight in 2010 (WHO, 2012). Although the WHO and many governmental agencies across the world are all agreed that intake of saturated fat should be reduced to below 10−11% of dietary energy as a means of reducing the risks of cardiovascular disease, not all of the clinical (and even anecdotal) data would support this (Talbot, 2011). Mensink et al. (2003) were one group that established a link between intake of saturated fats and risk of cardiovascular disease. However, health benefits from reducing saturated fats in the diet are being increasingly challenged, with a recent meta-analysis concluding that there is no significant evidence that dietary saturated fat is associated with increased risk of heart disease (Siri-Tarino et al., 2010).

FAT REPLACERS

Fat replacers are ingredients used to mimic and replace fat in food products. The term 'fat replacer' is a generic term which covers a very wide variety of materials such as starches, proteins, emulsifiers, hydrocolloids and fibres (Atkinson, 2011). Fat levels are very much on the agenda for the consumer when purchasing food and food processors have endeavoured to tackle the issue by the production of fat-reduced foods. However, fat also has some important sensory properties and contributes greatly to texture, mouthfeel and satiety as well as assisting in the sensation of lubricity in foods and overall flavour. It is thus very challenging for food processors to compensate for these qualities in fat-reduced food variants (O'Sullivan, 2015). There are several ways in which reduced fat foods may be produced. The first and simplest method is the removal of fat by using alternative ingredients. This might include the incorporation of air (increase volume and change texture in ice cream) or water (low fat spread, sauce) in a product or by using baking instead of frying in the production of snack foods, skimmed milk in yoghurt production or leaner meat in a sausage formulation. The second method involves the replacement of fat with carbohydrate, protein or lipid-based fat replacers that can be classified into either fat substitutes, fat mimetics or bulking agents (O'Sullivan, 2015).

FAT SUBSTITUTES

Fat substitutes are materials that simulate the chemical and physical properties of fats and oils and can directly replace fat on a weight-to-weight basis. They quite often take on board the processing characteristics of conventional fats and can be used in baking and frying. These materials take the basic structure of a fat or triglyceride which is then modified to reduce the calorific level of the final material. The fat backbone, glycerol, has three linkages to fatty acids, which if increased decrease fat absorption by the human body (Mattson and Volpenhein, 1972). Fat substitutes can be produced by enzyme-modified oils and fats and can also be synthesised chemically. Examples of common fat substitutes include: fat-based sucrose polyesters such as olestra (Olean), esterified propoxylated glycerol, fatty acid esters of sorbitol and sorbitol anhydrides (Sorbestrin) (Table 9.1) (O'Sullivan, 2015). The famous and controversial fat substitute 'Olestra' developed by Proctor and Gamble, has a sucrose backbone with eight linkages and thus has essentially zero calorific value (Atkinson, 2011). One of the very significant drawbacks of using Olestra is that product labels must contain the warning 'This product contains Olestra. Olestra may cause abdominal cramping and loose stools. Olestra inhibits the absorption of some vitamins and other nutrients. Vitamins A, D, E, and K have been added' (Akoh, 1998). Due to this reason, Olestra is not used in the United Kingdom and Canada. Additionally, fat substitutes can be produced by modifying the fatty acids attached to the glycerol backbone of the fat molecule which may be long, medium or short chain saturated fats (Salatrim) or by cleaving fatty acids attached to glycerol (Enova) (Atkinson, 2011).

FAT MIMETICS

Fat mimetics may be protein (e.g. gelatin, egg, milk, whey, soy, gluten)- or carbohydrate (modified starches, dextrins, nondigestible complex carbohydrates)-based compounds which simulate the sensory properties of fat. As with substitutes, they a have a much lower calorific level compared to conventional fats and oils as they quite often bind relatively large volumes of water. For this reason, they can limit the processing characteristics of the reduced fat foods they are incorporated into and generally cannot be used in products that are fried and carry water-soluble flavours as opposed to the fat-soluble flavours of conventional fats (O'Sullivan, 2015). However, they generally are suitable for use as ingredients in foods that may undergo cooking, retorting and ultrahigh temperature processing. Protein-based fat mimetics are generally used in dairy products, salad dressings, frozen desserts and margarines (IFT, 2016). The heating characteristics can additionally be resolved by the incorporation of emulsifying agents and more nonpolar or lipophilic flavouring compounds. Examples of common fat mimetics with GRAS (Generally Recognised As Safe) status include: protein-based materials such as (Simplesse) or Dairy-Lo (Table 9.1). Simplesse is a spray-dried microparticulate protein powder used as a texturiser in many different foods such as dairy products (cheese, yoghurt, frozen desserts), margarine, salad dressings, sauces and soups. Simplesse is suitable for use in additional products that do not require frying, such as baked goods, dips, frostings, salad dressing, mayonnaise, margarine, sauces and soups. The caloric value of Simplesse, on a dry

TABLE 9.1 Fat Substitutes and Mimetics

| | Fat Mimetic | | | | |
| | | Carbohydrate | | | |
Fat Substitute	Protein	Starch	Cellulose	Gums	Polysaccharide
Salatrim Acyl triglyceride molecule Energy Density: 5 kcal/g **Functional properties:** Emulsify, stabilise, texturise, lubricate, flavour carrier **Application:** Cheese, dips, sauces, dairy desserts, confectionary	**Simplesse** Whey protein or white egg protein Energy Density: 4 kcal/g **Functional properties:** Stabilise, texturise, **Application:** Cheese, yoghurt, dips, mayo, dairy spreads, desserts	**Starches:** Corn, potato, tapioca, rice, maize, waxy maize Energy Density: 4 kcal/g **Functional properties:** Texturiser, water holding, thickening, gelling **Application:** Processed meats, margarines, sauces, dressings, baked foods	**Methylcellulose** Energy Density: 0 kcal/g **Functional properties:** Texturiser **Application:** Frozen desserts, powdered soups/sauces	**Guar, locust bean, xanthan** Energy Density: 0 kcal/g **Functional properties:** Retard staling in baked goods, retain moisture **Application:** Bakery goods	**Maltodextrin** Hydrolysed starch Energy Density: 4 kcal/g **Functional properties:** Water binding, imparts mouthfeel, body, viscosity **Application:** Processed meats, dairy spreads, salad dressings, baked foods, frozen desserts
Caprenin Caprocaprylobehenic Triglyceride molecule Energy Density: 5 kcal/g **Functional properties:** Emulsify, stabilise, texturise, lubricate, flavour carrier	**Trail Blazer** White egg protein and xanthan gum Energy Density: 4 kcal/g **Functional properties:** Stabilise, texturise,		**Microcrystalline cellulose** Energy Density: 0 kcal/g **Functional properties:** Stabiliser, texturiser, imparts mouthfeel	**Carrageenan** Energy Density: 0 kcal/g **Functional properties:** Texturiser, imparts mouthfeel, viscosity **Application:** Processed meats, yoghurts, salad	**Polydextrose** Energy Density: 1 kcal/g **Functional properties:** Texturiser, bulking agent **Application:** Confectionary,

Application: Candy, confectionary

Application: dairy, soups, sauces, baked foods

Olestra (Olean)
Sucrose polyester
Energy Density: 0 kcal/g
Functional properties: Texturise, lubricate, flavour carrier
Application: Reduced fat snacks, potato chips

Application: Dairy, salad dressing, sauces desserts

dressings, desserts, ice cream, chocolate

Pectin
Energy Density: 0 kcal/g
Functional properties: Texturiser, imparts mouthfeel, viscosity
Application: Sauces

Gum Arabic
Energy Density: 0 kcal/g
Functional properties: Texturiser, imparts mouthfeel, viscosity
Application: Salad dressings

spreads, salad dressings, baked foods, frozen desserts, sauces, toppings

β-Glucan
Energy Density: 1 kcal/g
Functional properties: Texturiser
Application: Baked foods

Reproduced from O'Sullivan, M.C., 2015. Chapter 430. Low-fat Foods: Types and Manufacture. Encyclopaedia of Food and Health.

basis, is 4 kcal/g (IFT, 2016). Dairy-Lo is a modified whey protein concentrate that can form a set matrix gel with heat, and can hold gel structure during shelf life and is quite often used as a fat replacer in cheese products (O'Sullivan, 2015). Carbohydrate-based compounds include cellulose (Avicel), starches (modified, dextrin), hydrocolloid gums (Slendid), pectins, β-glucan and inulin. Inulin is a prebiotic polydisperse carbohydrate dietary fibre that is easy to use, and contributes to the desired taste and texture of baked goods and cereals and can be used as a fat replacer (Niness, 1999; Rodríguez-García et al., 2014). Gel properties are important for flavour release from gelled foods, whereas the clarity and surface smoothness of gels mainly depend on the presence and structure of insoluble components. Hydrocolloids can form molecular gels through association of parts of the molecules forming 'junction zones' (Williams et al., 2004). Other common hydrocolloid food gelling agents include gelatine, starch, pectin, carrageenan and alginate, agar, celluloses, gellan gum, konjac, milk proteins and soy proteins. Gel-forming polysaccharides such as k-carrageenan that form gels in the presence of water can be used as fat replacers for emulsion-based semisolid foods such as mayonnaise and margarine, custards, gravies and pourable dressings (Sworn et al., 1995; Garrec and Norton, 2012). Blends of the hydrocolloid k-carrageenan and starch have been adopted in the food industry as a texture enhancer for several dairy products (Descamps et al., 1986; Williams et al., 2004; Matignon et al., 2014). These fluid gels offer a wide range of highly tuneable material properties, and as such, have potential use in numerous applications including reduced fat food formulations (Garrec and Norton, 2012).

MEAT PRODUCTS

Overconsumption of meat and meat products has been linked with obesity, cancer and cardiovascular diseases primarily due to the high amounts of sodium chloride and saturated fat present in processed products (Li et al., 2005; Cross et al., 2007; Halkjaer et al., 2009; Micha et al., 2010). However, fats and oils play vital functional and sensory roles in various food products and are important in processed meats for flavour and have a large impact in terms of eating quality. Fats and oils interact with other ingredients and help to develop texture, mouthfeel and provide a lubricating effect (Webb, 2006; Javidipour et al., 2005; Wood, 1990). Thus, many researchers have been working on strategies to reduce animal fat usage in meat products (Delgado-Pando et al., 2011; Özvural and Vural, 2008; Prosslow, 2016) and thus this area has received considerable recent research focus on products such as frankfurters, breakfast sausage, beef patties, pork patties, bologna, black and white puddings and corned beef (Ruusunen et al., 2005; Jeong et al., 2007; Ventanas et al., 2010; Tobin et al., 2012a,b, 2013; Fellendorf et al., 2015, 2016a,b,c,d; Prosslow, 2016). Processing extends shelf life, improves texture and enhances overall flavour. Additionally, irrespective of public opinion, the processing of underutilised meat is necessary on ethical grounds alone as it is responsible for converting inedible material to a more palatable form, thereby reducing food waste and generating more protein-based food products which also present product diversity in the marketplace (Fellendorf et al., 2015). However, meat processing inevitably leads to products having higher amounts of salt, saturated fatty acids and preservatives such as nitrates which have health implications (Barcellos et al., 2011). For this reason, processed meats have moved more into the public focus over the past 20 years, particularly with respect to health concerns. Meat product suppliers have already commenced reformulating their recipes, and now offer lower levels of nitrate, salt and fat, or even higher levels of PUFAs in processed meat products on the market (Verbeke et al., 2010). For meat, there are several methods to produce fat-reduced products such as using leaner meat, less fat and salt, more water, replacing parts of animal fat with plant oil, or the application of fat replacers (Fellendorf et al., 2015, 2016a,b,c,d; Weiss et al., 2010).

One of the first methods of fat reduction in processed meat systems is the reduction of fat in formulations without replacement using other ingredients. Tobin et al. (2012a,b, 2013) explored the sequential reduction of salt and fat in beef patties, frankfurters and breakfast sausage, respectively. Additionally, Fellendorf et al. (2015, 2016a) investigated salt and fat reduction in black and white pudding sausages and corned beef (Fellendorf et al., 2016d), by reduction without replacement. They used sensory methods to identify optimal formulations and found that different meat systems had varying optimisation criteria. Both salt and fat reduction were shown not to negatively impact on consumer acceptability in patties and sausages, while only fat reduction could be achieved in frankfurters without negative assessor feedback (Tobin et al., 2012a,b, 2013). Fellendorf et al. (2015) found that white pudding samples containing 15% fat and 0.6% sodium was highly accepted ($P < .05$), thereby satisfying the sodium target (0.6%) set by the Food Safety Authority of Ireland (FSAI, 2011). Additionally, black pudding samples (Fellendorf et al., 2016a) containing 0.6% sodium and 10% fat displayed a positive ($P < .05$) correlation to liking of flavour and overall acceptability. This also meets the sodium target level set by the FSAI (2011).

The use of unsaturated oils such as vegetable or fish oils as a partial substitute to animal fat in meat products has received a great deal of attention. By replacing animal fat in processed meats with vegetable oils, the fatty acid profile can be nutritionally optimised as well as producing lower fat and/or cholesterol containing products. Özvural & Vural (2008) reported that palm oil, palm stearin, cottonseed oil, hazelnut oil and mixes of these oils could be used to increase the ratio of unsaturated to saturated fatty acids and maintain product quality in processed meats. Also,

Delgado-Pando et al. (2011) reported similar findings for olive oil, linseed oil, fish oil and konjac oil. Both studies report that products scored as acceptable in consumer trials even with certain changes to physiochemical properties of the meat products.

As explained earlier in this chapter fat replacers are categorised into fat substitutes and fat mimetics. Fat substitutes display a chemical structure close to fats, have similar physicochemical properties and are either low in calorie or indigestible such as Olestra, Caprenin and Salatrim (Kosmark, 1996; Peters et al., 1997; Fellendorf et al., 2016b,c). With Salatrim, the fatty acids attached to the glycerol backbone of the fat molecule are substituted with a combination of long chain length (18−22 carbon chain), medium chain length (6−10 carbon chain) and short (2−4 carbon chain) saturated fats. This changes the way this material is metabolised by the body and thus has only 5 calories per gram compared to 9 for standard fat (Atkinson, 2011). By contrast, fat mimetics show distinctly different chemical structures from fat − usually protein and/or carbohydrate based, though some of the physico-chemical attributes and eating qualities of fat can be simulated like mouthfeel, appearance and viscosity (Harrigan and Breene, 1989; Duflot, 1996; Fellendorf et al., 2016b,c). The use of fat replacers can drastically change the sensory profile of a meat product. Over half the volatile com-pounds found in meat originate from the lipid fraction (Brewer, 2012). The use of fat mimetics is in general more common. Chevance et al. (2000) achieved a fat reduction of 46% in salami, 60−83% in frankfurters and 55% in beef patties with tapioca starch and oat fibre or maltodextrin and milk protein.

Numerous studies have lauded the benefits of human diets high in plant fibre on overall and digestive health through regular bowel movement. These include reduced risk of cancers, CHD, obesity and diabetes. Essentially, fibre is complex carbohydrate that can be incorporated into processed meat for-mulations as partial fat replacers. Fibre positively affects the physicochemical properties by improving water binding, cooking loss and texture quality of sausages, puddings, bologna and meat balls (Fellendorf et al., 2015, 2016a,b, Borderías et al., 2005; Fernandez-Ginés et al., 2004).

Hydrocolloids (a range of polysaccharides and proteins) can also be employed as fat replacers in the meat processing industry, as they have been used in processed meat products for many years to improve properties such as water binding and texture due to their ability to thicken, gel, bind, stabilise emulsions and pH (Andres et al., 2006). Hydrocolloids, based on animal proteins, include casein, whey, gelatin and blood-derived protein. Ingredients such as whey protein concentrate, sodium caseinate, lecithin and a variety of hydrocolloids like gelatine and carrageenan are commonly used in the meat industry for these reasons (Andrès et al., 2006). Several researchers (Murphy et al., 2004; Sampaio et al., 2004; Cengiz and Gokoglu, 2007) have reported the effectiveness of fat reduction using whey protein, soy protein and surimi, respectively. As with oil incorporation, textural differences were detected in the finished product but these changes did not have a negative impact on overall sensory acceptability.

Stabilisers, emulsifiers or thickeners are added to meat to help maintain a uniform texture (Fellendorf et al., 2016b,c). Additionally, a diverse range of polysaccharides are available on the market, such as starches (corn, wheat, maize, potato, tapioca, pea), celluloses (carboxymethylcellulose), gums (guar, alginate, pectin, locust bean), fibres (β-glucan),chitin/chitosan and xanthan derived from microorganisms (Cutter, 2006). García-García and Totosaus (2008) showed that starch added with either K-carrageenan or locust bean gum could produce similar results in terms of texture as full fat products. Ayo et al. (2008) developed consumer-acceptable low-fat products produced with walnut. Cold-set binders are of particular interest to the meat industry as they can affect the textural properties of a meat system without the application of heat, therefore keeping the meat product in its natural raw state. They also have applications in the replacement of salt and fat and the production of healthier processed meats. Examples of cold set binders include nonthermal gelatine, alginate and fibrin (Boles, 2011). Commercially available cold-set binding systems include Fibrimex, alginate and Activa.

Fellendorf et al. (2016b) investigated the effect of using ingredient replacers on the physicochemical properties and sensory quality of low-salt and low-fat white puddings. They produced 22 formulations comprised of 2 different fat (10%, 5%) and sodium (0.6%, 0.4%) levels and containing 11 different ingredient replacers. These authors found that adding replacers to low-sodium and low-fat white puddings showed a range of effects on sensory and physicochemical properties. Two formulations containing 10% fat and 0.6% sodium formulated with sodium citrate, as well as the combination of potassium chloride and glycine (KClG), were found to have high overall acceptance ($P < .05$) by assessors. These samples showed higher ($P < .05$) hardness values, scored lower ($P < .05$) in fatness perception and higher ($P < .05$) in spiciness perception. Hence, the recommended sodium target level of 0.6% set by the FSAI 2011 was achieved for white pudding products, in addition to a significant reduction in fat level from commercial levels, without causing negative sensory attributes. Fellendorf et al. (2016b) investigated the impact of ingredient replacers on the physicochemical properties and sensory quality of reduced salt and fat black puddings where 22 products possessing different fat (10%, 5%) and sodium (0.6%, 0.4%) levels were used as base formulations for 11 different salt and fat replacers. These authors found that samples made with 5% fat and 0.6% sodium containing pectin and a combination of potassium citrate, potassium phosphate and potassium chloride (KCPCl), as well as samples containing 10% fat and 0.4% sodium with waxy maize starch were liked ($P < .05$) for flavour and overall acceptance. The FSAI recommends a sodium target level of 0.6% and an even lower sodium level (0.4%) was achieved. Fellendorf et al. (2016d) investigated the impact of varying sodium levels (0.2−1.0%) and salt replacers in corned beef on physicochemical, sensory and microbiological properties. Samples formulated with CaCl2, MgCl2 and KCl scored higher ($P < .01$) in saltiness perceptions, but correlated negatively ($P > .05$) to liking of flavour and overall acceptability. However, a sodium reduction of 60% in corned beef was determined to be achievable as assessors liked ($P < .05$) the flavour of the sodium-reduced corned beef containing 0.4% sodium and formulated with potassium lactate and glycine (KLG), even with the noticeable lower salty taste.

DAIRY PRODUCTS

In Europe, a reduction of 30% of the fat content relative to the reference full-fat cheese is required to use the term 'reduced fat' in the description of the product, according to European Council regulation No.1924/2006 (Deegan et al., 2014). Whereas the FDA regulation mandates that food products claiming to be low fat must not contain more than 3 g of fat per reference amount (50 g). Reduced fat labelling can be used for food that contains 25% less fat than the regular version (FDA-DHHS, 2002; Amelie et al., 2013). Low fat dairy products are in demand by consumers for exactly the same reasons as outlined for processed meats. Consumers wish to reduce their calorie, particularly saturated fat, intake and thus have driven the market to develop healthier reduced fat variants of cheese, yoghurt and milk products. Fat has an important sensory role in dairy products contributing to both flavour and texture and thus technological difficulties can be encountered when levels are reduced. Cheeses like cheddar and mozzarella become firmer and more rubbery on removal of fat, thus increasing in corresponding protein content. Flavour is particularly altered as the fat/protein matrix changes which can affect flavour perception deleteriously. The greater the fat reduction, the more negative the sensory flavour and texture quality. Some of the flavour defects mentioned included meaty, brothy, burnt, bitterness, low flavour intensity and milk fat flavour (Amelie et al., 2013; Drake and Swanson, 1995; Mistry, 2001; Banks, 2004; Drake et al., 2010).

Henneberry et al. (2016) investigated the sensory quality of mozzarella-style cheeses with different fat, salt and calcium levels in an effort to develop a more healthy version of this cheese which is consumed in significant volumes, particularly as an ingredient of pizza. These authors investigated the reduction of fat from 23 w/w to 11 w/w and salt from 1.8 to 1.0 w/w on the texture profile attributes, volatile compounds and sensory characteristics of unheated and heated (95°C) mozzarella-style cheese after 15 and 35 days of storage at 4°C. Reducing fat content significantly impaired sensory acceptability, mainly because of associated increases in firmness, rubberiness and chewiness, and reductions in fat flavour and cheese flavour in the unheated and heated cheeses. Reducing salt content had relatively minor effects. Reducing calcium content counteracted some of the negative sensory attributes associated with reduced fat cheese (Henneberry et al., 2016).

Several studies have investigated the sensory optimisation of reduced fat cheese through the use of curd washing (Johnson et al., 1998) and adjunct cultures (Drake et al., 1997) in cheddar cheese. Fenelon et al. (2002) used the adjunct *Lactobacillus helveticus* in combination with *Leuconostoc cremoris*, *Lactococcus lactis* var. *diacetylactis* and Streptococcus thermophiles to produce 50% fat-reduced cheddar cheese with a higher preference score than a similar cheese made without the adjunct culture although the flavour profile was clearly different from regular cheddar and burnt off-flavour notes were produced. Fat mimetics (McMahon et al., 1996) were investigated in mozzarella cheese with two protein-based fat replacers (Simplesse D100 and Dairy-Lo) and two carbohydrate-based fat replacers (Stellar 100X and Novagel RCN-15) although sensory investigations were not included.

Nelson and Barbano (2004) investigated a novel approach to flavour optimisation of reduced fat cheddar cheese by only removing the fat from regular aged cheddar once the flavour had fully developed and according to sensory studies managed to retain the same flavour intensity although the texture was quite different. Deegan et al. (2014) investigated the chemical and textural changes of reduced fat Emmental cheese made using a preprocessing routine involving homogenisation of milk and comparison with full-fat Emmental cheese made with the same process. They found that the changes due to homogenisation of milk in cheese making improved sensory aspects and market position of these reduced fat cheeses.

Kavas et al. (2004) investigated the effect the protein-based fat replacers, Dairy-Lo and Simplesse D-100 and carbohydrate-based fat replacers Perfectamyl gel MB and Satiagel ME4 on the chemical, physical and sensory attributes of low-fat white pickled cheese. They found that low-fat cheeses, produced by adding Dairy-Lo, Perfectamyl gel MB and Satiagel ME4 were highly acceptable compared to the low-fat cheese without fat replacers. Also differences between cheeses made with Simplesse D-100 and full-fat cheese or low-fat cheese without fat replacer, as found by the taste panel were not detectable.

It is a difficult challenge for food processors to reduce the fat content of ice cream and maintain the fundamental qualities that are integral for the formation of its microstructure. Fat presence is crucial for aeration, as it stabilises the air phase during freezing and is important as a carrier of lipophilic flavours and their subsequent release during consumption. Fat also increases the buttery and creamy notes as well as mouth coating in full-fat ice creams compared with low-fat ice cream (Koeferli et al., 1996) and generally low-fat ice creams have lower flavour and texture ratings when measured by sensory analysis (Guinard and Marty, 1995) and display textural defects like coarseness, iciness, crumbliness, shrinkage on storage and reduced flavour release (Marshall and Arbuckle, 1996). Nonetheless, ice cream manufacturers have attempted to simulate the texture and flavour of fat-containing ice cream by adding bulking agents and concentrated milk components for fat-free formulations (Roland et al., 1999). Carbohydrate bulking agents, such as maltodextrin and poly-dextrose, are currently used in low-fat formulations because they produce minimal negative effects on ice cream production, shelf life, and price (Schmidt et al., 1993)

Roland et al. (1999) investigated the carbohydrate-based replacers, maltodextrin, milk protein concentrate and polydextrose to maintain the same sweetness intensity and freezing characteristics as 10% fat ice cream. Lactose-reduced, freeze-concentrated skim milk was used to formulate a 1.6% fat ice cream mix. The sample containing only maltodextrin had the greatest cream flavour, the best textural characteristics and the sensory analysis scored it as best overall as a single fat replacer in ice cream. Although the lactose-reduced, freeze-concentrated skim milk sample had the lowest level of detrimental flavours (i.e. corn syrup, milk powder and aftertaste) and, thereby, provided the best flavour of the fat replacers tested, however, its textural characteristics were scored as the least desirable (Roland et al., 1999). Schaller-Povolny and Smith (1999, 2001) investigated inulin as a fat replacer in ice cream and found that it increased the viscosity, decreased the freezing point and produced a product with acceptable sensory qualities. Prindiville et al. (2000) investigated the protein-based fat replacers, Dairy-Lo or Simplesse in low-fat and nonfat chocolate ice creams made with 2.5% fat. Consumer acceptance was the same for fresh ice creams at Wk 0. Simplesse was more similar to milk fat than Dairy-Lo on textural stability but was less similar in terms of thickness and mouth-coating.

Yoghurt is generally a healthy consumer product high in protein and calcium contents and is produced using skimmed milk or full fat milk. It is no surprise that consumers have also driven a demand for low or even zero fat variants. As with cheese and processed food the physicochemical and sensory properties of low-fat or fat-free yoghurt can be negative in comparison to higher fat variants. Fat replacement in yoghurt has been explored by several researchers (Barrantes et al., 1994a,b; Tamime et al., 1994, 1996). The prebiotic polysaccharide inulin, most often extracted from chicory root, is a fructan or dietary fibre and has applications as a fat replacer in yoghurt. Staffolo et al. (2004) investigated inulin as a fat replacer in yoghurt and found that it produced a product with a stable colour and water activity and syneresis did not prevail during storage. Srisuvor et al. (2013) compared the effects of inulin and polydextrose on physico-chemical and sensory properties of low-fat set yoghurt.

CONFECTIONARY PRODUCTS

Reduced fat cake products are available on the market but their scarcity could be an indication of either the difficulty of producing them or of a lack of acceptance of the product by the consumer (Talbot, 2011). Cookies are soft-type biscuits whose textural characteristics are mostly provided by their high fat content which provides flavour and mouthfeel and also contributes to appearance, palatability, texture and lubricity. Fat and sugar cannot be easily replaced, especially in a complex food system such as biscuits (Zoulias et al., 2002). Fat is responsible for soft and crisp texture of biscuits and imparts flavour, lubricity, mouthfeel, aeration and taste to the product (Aggarwal et al., 2016). The sensory properties of pastry products result from the use of saturated fats, particularly during their preparation and any reduction can reduce sensory quality (de Cindio and Lupi, 2011). As a general principle the reduction of saturate levels in a fat is brought about through an increase in the level of liquid oil in the product, as it is these liquid oils, such as rapeseed oil or sunflower oil, that have the lowest levels of saturated fats (Atkinson, 2011). Zoulias et al. (2002) investigated the fat mimetics, Oatrim and Z-Trim, both derived from oats, in cookies which produced acceptable sensory characteristics at 50% fat replacement. However, lower quality resulted at higher levels of replacement. Conforti et al. (1996) investigated three fat substitutes derived from pectin, gums and oat bran (Slendid, Kel-Lire BK and Trimchoice-5) for baking biscuits which were used to replace the fat at a level of 33%, 66% and 100%. They concluded that tenderness decreased as substitute use increased when biscuits were rated by the sensory panel. Consumer location testing rated the biscuits as moderately acceptable and indicated a desire for a product of this type to become available in the marketplace. Zoulias et al. (2002) explored the use of the fat replacers inulin (Raftiline, Orafti Active Food Ingredients, Oreye, France), Simplesse, C*deLight (Cerestar, Brussels, Belgium) and polydextrose (Litesse (Pfizer, Inc., New York, NY, USA)) for partial fat replacement and lactitol, sorbitol and maltitol, for sugar replacement in low-fat, sugar-free cookies. Polydextrose is a complex carbohydrate made from glucose, citric acid and sorbitol, which forms a highly viscous gel-like matrix contributing to creaminess and mouthfeel (Mitchell, 1996). Cookies prepared with Simplesse had the least acceptable flavour, while cookies prepared with C*deLight were rated as the most acceptable by a sensory panel. Bosman et al. (2000) used Simplesse as an oil replacer in the investigation of the sensory and physical characteristics and microbiological quality of high-fibre muffins. They concluded that oil replacement with Simplesse produced products with better stability towards batter refrigeration than the full-fat control muffins. The results also indicated that although the reduced fat muffins were different from the freshly baked full-fat control muffins, they were still of comparable sensory quality. Pszczola (1994) reported that Simplesse bakery blends did not affect the full-fat taste when they replaced fat up to 93% in cakes, muffins and brownies. Armbrister and Setser (1994) showed that the replacement of 50 or 75% of fat by six protein- or carbohydrate-based fat mimetics including polydextrose (Litesse), acid-treated cornstarch (Stellar, A.E Staley, Decatur, IL),

Slendid, mono- and diglycerides, Simplesse and potato maltodextrin produced products with different flavour characteristics compared to a full-fat control. Laguna et al. (2014) studied the effects of partial fat replacement with inulin and hydroxypropyl methylcellulose (HPMC) in biscuits. Consumer studies revealed that fat replacement up to 15/100 g with inulin or HPMC provided acceptable biscuits, but higher replacement decreased the overall acceptability. Campbell et al. (1994) managed to replace up to 25% of shortening in crisp oatmeal cookies using polydextrose (Litesse) while maintaining physicochemical and sensory properties. Hydrolysed oat flour (containing 5% β-glucan) has been used as a replacement for butter in baked desserts and for coconut cream in Thai desserts (Inglett et al., 2000).

The reduction of fat in sweet baked products such as cake, muffins and cookies is currently being studied in the SWEETLOW project funded by the Irish Department of Agriculture, Food and the Marine (Sweetlow, 2016). This project funded as part of the Food Industry Research Measure (FIRM) is investigating the use of fruit and vegetable fibres as fat replacers, as well as alternative fat ratios in formulation matrices, emulsifiers and bulking agents.

SALAD DRESSING AND SAUCES

One of the simplest methods for reducing the total fat level of an emulsion is to replace some of the fat with an ingredient that has similar properties. In such a way, it may be possible to avoid complex changes to the manufacturing process (Talbot, 2011). Liquid salad dressings are oil-in-water emulsions, typically acidified with vinegar or lemon juice and stabilised by emulsifiers such as egg yolk or milk proteins. Due to their relatively high oil content, much attention has been directed to the development of low fat variants, but because of this typically high oil content, any decrease negatively affects emulsion properties and thus physicochemical and sensory characteristics. Consumers do not wish to sacrifice flavour quality for nutritional properties in such products and thus the challenge for food technologists to produce acceptable products is great. Generally, thickeners are used to stabilise reduced fat salad dressings. Xanthan gum, because it can form weak gels, is also often used in salad dressings. Polysaccharides (carrageenan, xanthan, pectins, etc.) are usually incorporated in commercial salad dressing products, especially those of low oil content, to stabilise the oil droplets, prevent creaming and to compensate for the loss of thickening properties originating from their reduced oil content (Parker et al., 1995; Ma and Barbosa-Canovas, 1995). Fat mimetics such as Simplesse can be used as fat replacers and texturisers in margarines, salad dressings, sauces and soups. Trailblazer can be used as a texturiser in sauces and soups. Cooked starch may also be incorporated in these products since starch gelatinisation results in the release of polysaccharides which may then act as emulsion stabilisers and thickeners (Mantzouridou et al., 2013). Corn, waxy maize, wheat, potato, tapioca, rice and waxy rice starches can be used as fat mimetics in dressings, sauces margarines and spreads. Microcrystalline cellulose is used in reduced fat salad dressings and maltodextrin has found applications in low-fat margarines, salad dressings and sauces.

REFERENCES

Aggarwal, D., Sabikhi, L., Sathish Kumar, M.H., 2016. Formulation of reduced-calorie biscuits using artificial sweeteners and fat replacer with dairy—multigrain approach. NFS Journal 2, 1—7.

Akoh, C., 1998. Fat replacers. Food Technology 52, 47—53.

Amelia, I., Drake, M.A., Nelson, B., Barbano, D.M., 2013. A new method for the production of low-fat Cheddar cheese. Journal of Dairy Science 96, 4870—4884.

Andrès, S., Zaritzky, N., Califano, A., 2006. The effect of whey protein concentrates and hydrocolloids on the texture and colour characteristics of chicken sausages. International Journal of Food Science & Technology 41 (8), 954—961.

Armbrister, W.L., Setser, C.S., 1994. Sensory and physical properties of chocolate chip cookies made with vegetable shortening or fat replacers at 50% and 75% levels. Cereal Chemistry 71, 344—351.

Atkinson, G., 2011. CH 14. Saturated fat reduction in biscuits. In: Talbot, G. (Ed.), Reducing Saturated Fats in Foods. Woodhead Publishing Limited, pp. 301—317.

Ayo, J., Carballo, J., Solas, M.T., Jiménez-Colmenero, F., 2008. Physicochemical and sensory properties of healthier frankfurters as affected by walnut and fat content. Food Chemistry 107, 1547—1552.

Boles, J.A., 2011. Use of cold-set binders in meat systems. In: Kerry, J.P., Kerry, J.F. (Eds.), Processed Meats Improving Safety, Nutrition and Quality. Woodhead Publishing, UK, pp. 270—295.

Banks, J.M., 2004. The technology of low-fat cheese manufacture. International Journal of Dairy Technology 57, 199—207.

Barbut, S., 2011. Saturated fat reduction in processed meat products. In: Talbot, G. (Ed.), Reducing Saturated Fats in Foods. Woodhead Publishing Limited, pp. 210—233.

de Barcellos, M.D., Grunert, K.G., Scholderer, J., 2011. Processed meat products: consumer trends and emerging markets. In: Kerry, J.P., Kerry, J.F. (Eds.), Processed Meats. Improving Safety, Nutrition and Quality. Woodhead Publishing Ltd, Oxford, Cambridge, Philadelphia, New Dehli, pp. 30—53.

Barrantes, E., Tamime, A.Y., Davies, G., Barclay, M.N.I., 1994a. Production of low-calorie yoghurt using skim milk powder and fat substitutes. 2. Compositional quality. Milchwissenschaft 49, 135—139.

Barrantes, E., Tamime, A.Y., Sword, A.M., 1994b. Production of low-calorie yoghurt using skim milk powder and fat substitutes. 3. Microbiological and organoleptic qualities. Milchwissenschaft 49, 205—208.

Borderías, A.J., Sánchez-Alonso, I., Pérez-Mateos, A., 2005. New applications of fibres in foods: addition of fishery products. Trends in Food Science & Technology 16, 458—465.

Bosman, M.J.C., Vorster, H.H., Setser, C., Steyn, H.S., 2000. The effect of batter refrigeration on the characteristics of high-fibre muffins with oil replaced by a protein-based fat substitute. Journal of Family Ecology and Consumer Sciences 28, 1—15.

Brewer, M.S., 2012. Reducing the fat content in ground beef without sacrificing quality — a review. Meat Science 91, 385—395.

Campbell, L.A., Ketelsen, S.M., Antenucci, R.N., 1994. Formulating oatmeal cookies with calorie-sparing ingredients. Food Technology 48, 98—105.

Cengiz, E., Gokoglu, N., 2007. Effects of fat reduction and fat replacer addition on some quality characteristics of frankfurter-type sausages. International Journal of Food Science and Technology 42, 366–372.

Chevance, F.F., Farmer, L.J., Desmond, E.M., Novelli, E., Troy, D.J., Chizzolini, R., 2000. Effect of some fat replacers on the release of volatile aroma compounds from low-fat meat products. Journal of Agricultural and Food Chemistry 48 (8), 3476–3484.

Conforti, F., Charles, S.A., Duncan, S.E., 1996. Sensory evaluation and consumer acceptance of carbohydrate-based fat replacers in biscuits. Journal of Consumer Study and Home Economics 20, 285–296.

Cross, A.J., Leitzmann, M.F., Gail, M.H., Hollenbeck, A.R., Schatzkin, A., Sinha, R., 2007. A prospective study of red and processed meat intake in relation to Cancer risk. PLOS Medicine 4 (12), 345.

Cutter, C.N., 2006. Opportunities for bio-based packaging technologies to improve the quality and safety of fresh and further processed muscle foods. Meat Science 74, 131–142.

de Cindio, B., Lupi, F.R., 2011. CH 15 Saturated fat reduction in pastry. In: Talbot, G. (Ed.), Reducing Saturated Fats in Foods. Woodhead Publishing Limited, pp. 301–317.

Deegan, K.C., Holopainen, U., McSweeney, P.L.H., Alatossava, T., Tuorila, H., 2014. Characterisation of the sensory properties and market positioning of novel reduced-fat cheese. Innovative Food Science and Emerging Technologies 21, 169–178.

Delgado-Pando, G., Cofrades, S., Ruiz-Capillas, C., Solas, M.T., Triki, M., Jiménez-Colmenero, F., 2011. Low-fat frankfurters formulated with a healthier lipid combination as functional ingredient: microstructure, lipid oxidation, nitrite content, microbiological changes and biogenic amine formation. Meat Science 89, 65–71.

Duflot, P., 1996. Starches and sugars glucose polymers as sugar/fat substitutes. Trends in Food Science & Technology 7, 206.

Drake, M.A., Swanson, B.G., 1995. Reduced and low-fat cheese technology. A review. Trends in Food Science and Technology 6, 366–369.

Drake, M., Boylston, T., Spence, K., Swanson, B., 1997. Improvement of sensory quality of reduced fat Cheddar cheese by a Lactobacillus adjunct. Food Research International 30 (1), 35–40.

Drake, M.A., Miracle, R.E., McMahon, D.J., 2010. Impact of fat reduction on flavour and flavour chemistry of Cheddar cheeses. Journal of Dairy Science 93, 5069–5081.

Descamps, O., Langevin, P., Combs, D.H., 1986. Physical effect of starch carrageenaninteractions in water and milk. Food Technology 40, 81.

FAO, 2010. Food and Agriculture Organization of the United Nations, Food-Based Dietary Guidelines. Available from: http://www.fao.org/ag/humannutrition/nutritioneducation/fbdg/en/.

FDA-DHHS, 2002. 21CFR101.62b: Nutrient Content Claims for Fat, Fatty Acid, and Cholesterol Content of Foods. Food and Drug Administration Department of Health and Human Services, Washington.

Fellendorf, S., O'Sullivan, M.G., Kerry, J.P., 2015. Impact of varying salt and fat levels on the physiochemical properties and sensory quality of white pudding sausages. Meat Science 103, 75–82.

Fellendorf, S., O'Sullivan, M.G., Kerry, J.P., 2016a. The reduction of salt and fat levels in black pudding and the effects on physiochemical and sensory properties. International Journal of Food Science and Technology (Online).

Fellendorf, S., O'Sullivan, M.G., Kerry, J.P., 2016b. Effect of using replacers on the physicochemical properties and sensory quality of low salt and low fat white puddings. European Food Research and Technology (Online).

Fellendorf, S., O'Sullivan, M.G., Kerry, J.P., 2016c. Impact of using replacers on the physico-chemical properties and sensory quality of reduced salt and fat black pudding. Meat Science 113, 17−25.

Fellendorf, S., O'Sullivan, M.G., Kerry, J.P., 2016d. Impact on the physicochemical and sensory properties of salt reduced corned beef formulated with and without the use of salt replacers. Meat Science (submitted for publication).

Fenelon, M.A., Beresford, T.P., Guinee, T.P., 2002. Comparison of different bacterial culture systems for the production of reduced-fat Cheddar cheese. International Journal Dairy Technology 55, 194−203.

Fernandez-Ginés, J.M., Fernández-López, J., Sayas-Barberá, E., Sendra, E., Pérez-Álvarez, J.A., 2004. Lemon albedo as a new source of dietary fiber: application to bologna sausages. Meat Science 67, 7−13.

FSAI, 2011. Salt Reduction Programme (SRP)−2011 to 2012, p. 85.

García-García, E., Totosaus, A., 2008. Low-fat sodium-reduced sausages: effect of the interaction between locust bean gum, potato starch and K-carrageenan by a mixture design approach. Meat Science 78, 406−413.

Garrec, D.A., Norton, I.T., 2012. Understanding fluid gel formation and properties. Journal of Food Engineering 112, 175−182.

Guinard, J.X., Marty, C., 1995. Time-intensity measurement of flavour release from a model gel system: effect of gelling agent type and concentration. Journal of Food Science 60, 727−730.

Halkjaer, J., Tjønneland, A., Overvad, K., Sørensen, T., 2009. Dietary predictors of 5-year changes in waist circumference. Journal of the American Dietetic Association 109 (8), 1356−1366.

Harrigan, K.A., Breene, W.M., 1989. Fat substitutes: sucrose esters and Simplesse®. Cereal Foods World 34, 261−267.

Henneberry, S., O'Sullivan, M.G., Kilcawley, K.N., Kelly, P.M., Wilkinson, M.G., Guinee, T.P., 2016. Sensory quality of unheated and heated Mozzarella-style cheeses with different fat, salt and calcium levels. International Journal of Dairy Science 69, 38−50.

IFT, 2016. Fat Replacers. http://www.ift.org/knowledge-center/read-ift-publications/science-reports/scientific-status-summaries/fat-replacers.aspx.

Inglett, G.E., Maneepun, S., Vatanasuchart, N., 2000. Evaluation of hydrolysed oat flour as a replacement for butter and coconut cream in bakery products. Food Science and Technology International 6, 457−462.

Javidipour, I., Vural, H., Ozbas, O.O., Tekin, A., 2005. Effects of interesterified vegetable oils and sugar beet fibre on the quality of Turkish-type salami. International Journal of Food Science and Technology 5 (40), 177−185.

Jeong, J.Y., Lee, E.S., Choi, J.H., Lee, J.Y., Kim, J.M., Min, S.G., Kim, C.J., 2007. Variability in temperature distribution and cooking properties of ground pork patties containing different fat level and with/without salt cooked by microwave energy. Meat Science 75 (3), 415−422.

Johnson, M.E., Steele, J.L., Broadbent, J., Weimer, B.C., 1998. Manufacture of Gouda and flavour development in reduced-fat Cheddar cheese. Australian Journal of Dairy Technology 53 (2), 67−69.

Kavas, G., Oysun, G., Kinik, O., Uysal, H., 2004. Effect of some fat replacers on chemical, physical and sensory attributes of low-fat white pickled cheese. Food Chemistry 88, 381−388.

Koeferli, C.R.S., Piccinali, P., Sigrist, S., 1996. The influence of fat, sugar and non-fat milk solids on selected taste, flavour and texture parameters of a vanilla ice-cream. Food Quality and Preference 7, 69−79.

Komorowski, E.S., 2011. Saturated fat reduction in milk and dairy products. In: Talbot, G. (Ed.), Reducing Saturated Fats in Foods. Woodhead Publishing Limited, pp. 180−194.

Kosmark, R., 1996. Salatrim: properties and applications. Food Technology 50, 98−101.

Laguna, L., Primo-Martín, C., Varela, P., Salvador, A., Sanz, T., 2014. HPMC and inulin as fat replacers in biscuits: sensory and instrumental Evaluation. LWT — Food Science and Technology 56, 494–501.

Li, D., Siriamornpun, S., Wahlqvist, M.L., Mann, N.J., Sinclair, A.J., 2005. Lean meat and heart health. Asia Pacific Journal of Clinical Nutrition 14 (2), 113–119.

Ma, L., Barbosa-Canovas, G.V., 1995. Rheological characterization of mayonnaise. Part II: Flow and viscoelastic properties at different oil and xanthan gum concentrations. Journal of Food Engineering 25, 409–425.

Mantzouridou, F., Karousioti, A., Kiosseoglou, V., 2013. Formulation optimization of a potentially prebiotic low-in-oil oat-based salad dressing to improve *Lactobacillus paracasei* subsp. *paracasei* survival and physicochemical characteristics. LWT — Food Science and Technology 53, 560–568.

Marshall, R.T., Arbuckle, W.S., 1996. Ice Cream, fifth ed. Chapman & Hall, New York, p. 349.

Matignon, A., Barey, P., Desprairies, M., Mauduit, S., Sieffermann, J.M., Michon, C., 2014. Starch/carrageenan mixed systems: penetration in, adsorption on or exclusion of carrageenan chains by granules? Food Hydrocolloids 35, 597–605.

Mattson, F.H., Volpenhein, R.A., 1972. Rate and extent of absorption of the fatty acids of fully esterified glycerol, erythritol and sucrose as measured in thoracic duct cannulated rats. Journal of Nutrition 102, 1177–1180.

McMahon, D., Alleyne, M., Fife, R., Oberg, C., 1996. Use of fat replacers in low fat Mozzarella cheese. Journal of Dairy Science 79 (11), 1911–1921.

Mensink, R.P., Zock, P.L., Kester, A.D., Katan, M.B., 2003. Effects of dietary fatty acids and carbohydrates on the ratio of total to HDL cholesterol and on serum lipids an apolipoproteins: a meta-analysis of 60 controlled trials. American Journal of Clinical Nutrition 77, 1146–1155.

Micha, R., Wallace, S., Mozaffarian, D., 2010. Red and processed meat consumption and risk of incident coronary heart disease, stroke, and diabetes mellitus. A systematic review and meta-analysis. Circulation 121, 2271–2283.

Mistry, V.V., 2001. Low fat cheese technology. International Dairy Journal 11, 413–422.

Mitchell, H.L., 1996. The role of bulking agent polydextrose in fat replacement. In: Roller, S., Jones, S.A. (Eds.), Handbook of Fat Replacers. CRC Press, Boca Raton, Florida, pp. 235–248.

Murphy, S.C., Gilroy, D., Kerry, J.F., Buckley, D.J., Kerry, J.P., 2004. Evaluation of surimi, fat and water content in a low/no added pork sausage formulation using response surface methodology. Meat Science 66, 689–701.

Nelson, B.K., Barbano, D.M., 2004. Reduced-fat Cheddar cheese manufactured using a novel fat removal process. Journal of Dairy Science 87, 841–853.

Niness, K.R., 1999. Inulin and oligofructose: what are they? American Society for Nutritional Sciences 129, 1402S–1406S.

O'Sullivan, M.G., 2015. Chapter 430. Low-Fat Foods: Types and Manufacture. Encyclopedia of Food and Health.

Özvural, E.B., Vural, H., 2008. Utilization of interesterified oil blends in the production of frankfurters. Meat Science 78, 211–216.

Parker, A., Gunning, P.A., Ng, K., Robins, M.N., 1995. How does xanthan stabilise salad dressing. Food Hydrocolloids 9, 333–342.

Peters, J.C., Lawson, K.D., Middleton, S.J., Triebwasser, K.C., 1997. Assessment of the nutritional effects of olestra, a nonabsorbed fat replacement: introduction and overview. Journal of Nutrition 127, 1539–1546.

Prindiville, E.A., Marshall, R.T., Heymann, H., 2000. Effect of milk fat, cocoa butter, and whey protein fat replacers on the sensory properties of lowfat and nonfat chocolate ice cream. Journal of Dairy Science 83, 2216–2223.

Prosslow, 2016. Development of Consumer Accepted Low Salt and Low Fat Irish Traditional Processed Meats. Project Coordinator: Dr Maurice O' Sullivan. Ref:11F026. https://www. agriculture.gov.ie/media/migration/research/firmreports/CALL2011ProjectAbstracts240216.pdf.

Pszczola, D.E., 1994. Blends reduce fat in bakery products. Food Technology 46, 168–170.

Rodríguez-García, J., Sahi, S.S., Hernando, I., 2014. Functionality of lipase and emulsifiers in low-fat cakes with inulin. LWT – Food Science and Technology 58, 173–182.

Roland, A.M., Phillips, L.G., Boor, K.J., 1999. Effects of fat replacers on the sensory properties, colour, melting, and hardness of ice cream. Journal of Dairy Science 82, 2094–2100.

Ruusunen, M., Vainionpää, J., Lyly, M., Lähteenmäki, L., Niemistö, M., Ahvenainen, R., Puolanne, E., 2005. Reducing the sodium content in meat products: the effect of the formulation in low-sodium ground meat patties. Meat Science 69 (1), 53–60.

Sampaio, G.R., Castellucci, C.M.N., Pinto e Silva, M.E.M., Torres, E.A.F.S., 2004. Effect of fat replacers on the nutritive value and acceptability of beef frankfurters. Journal of Food Composition and Analysis 17, 469–474.

Schaller-Povolny, L.A., Smith, D.E., 1999. Sensory attributes and storage life of reduced fat ice cream as related to inulin content. Journal of Food Science 64, 555–559.

Schaller-Povolny, L.A., Smith, D.E., 2001. Viscosity and freezing point of a reduced fat ice cream mix as related to inulin content. Milchwissenschaft 56, 25–29.

Schmidt, K., Lundy, A., Reynolds, J., Yee, L., 1993. Carbohydrate or protein based fat mimicker effects on ice milk properties. Journal of Food Science 58, 761–763.

Siri-Tarino, P.W., Sun, Q., Hu, F.B., Krauss, R.M., 2010. Meta-analysis of prospective cohort studies evaluating the association of saturated fat with cardiovascular disease. American Journal of Clinical Nutrition 91, 535–546.

Srisuvor, N., Chinprahast, N., Prakitchaiwattana, C., Subhimaros, S., 2013. Effects of inulin and polydextrose on physicochemical and sensory properties of low-fat set yoghurt with probiotic-cultured banana purée. LWT – Food Science and Technology 51, 30–36.

Staffolo, M.D., Bertola, N., Martino, M., Bevilacqua, y. A., 2004. Influence of dietary fiber addition on sensory and rheological properties of yogurt. International Dairy Journal 14, 263–268.

Sweetlow, 2016. Development of Consumer Optimised Low Carbohydrate Irish Confectionary Products. Project Coordinator: Dr Maurice O' Sullivan. Ref:14/F/812. https://www.agriculture. gov.ie/media/migration/research/firmreports/CALL2014ProjectAbstracts240216.pdf.

Sworn, G., Sanderson, G.R., Gibson, W., 1995. Gellan gum fluid gels. Food Hydrocolloids 9, 265–271.

Talbot, G., 2011. CH1 Saturated fats in foods and strategies for their replacement: an introduction. In: Talbot, G. (Ed.), Reducing Saturated Fats in Foods. Woodhead Publishing Limited, pp. 1–28.

Tamime, A.Y., Barclay, M.N.I., Davies, G., Barrantes, E., 1994. Production of low-calorie yoghurt using skim milk powder and fat substitutes. 1. A review. Milchwissenschaft 49, 85–87.

Tamime, A.Y., Barrantes, E., Sword, A.M., 1996. The effect of starch based fat substitutes on the microstructure of set-style yogurt made from reconstituted skimmed milk powder. Journal of the Society of Dairy Technology 49, 1–10.

Tobin, B.D., O'Sullivan, M.G., Hamill, R.M., Kerry, J.P., 2012a. Effect of varying salt and fat levels on the sensory quality of beef patties. Meat Science 4, 460–465.

Tobin, B.D., O'Sullivan, M.G., Hamill, R.M., Kerry, J.P., 2012b. Effect of varying salt and fat levels on the sensory and physiochemical quality of frankfurters. Meat Science 92, 659–666.

Tobin, B.D., O'Sullivan, M.G., Hamill, R.M., Kerry, J.P., 2013. The impact of salt and fat level variation on the physiochemical properties and sensory quality of pork breakfast sausages. Meat Science 93, 145–152.

Ventanas, S., Puolanne, E., Tuorila, H., 2010. Temporal changes of flavour and texture in cooked bologna type sausages as affected by fat and salt content. Meat Science 85 (3), 410–419.

Verbeke, W., Pérez-Cueto, F.J.A., De Barcellos, M.D., Krystallis, A., Grunert, K.G., 2010. European citizen and consumer attitudes and preferences regarding beef and pork. Meat Science 84 (2), 284–292.

Webb, E.C., 2006. Manipulating beef quality through feeding. South African Animal Science 7, 5–15.

Weiss, J., Gibis, M., Schuh, V., Salminen, H., 2010. Advances in ingredient and processing systems for meat and meat products. Meat Science 86 (1), 196–213.

Williams, P.A., Phillips, G.O., de Vries, J., 2004. Hydrocolloid gelling agents and their applications. Gums and Stabilizers for the Food Industry 12, 23–31.

WHO, 2003. Diet, Nutrition and the Prevention of Chronic Disease. World Health Organization.

World Health Organization, 2012. Obesity and Overweight Fact Sheet. http://www.who.int/mediacentre/factsheets/fs311/en/.

World Health Organization, 2000. Obesity: Preventing and Managing the Global Epidemic. WHO Obesity Technical Report Series 894. World Health Organization, Geneva, Switzerland.

Wood, J.D., 1990. Consequences for meat quality of reducing carcass fatness. In: Wood, J.D., Fisher, A.V. (Eds.), Reducing Fat in Meat Animals. Elsevier Applied Science, London, pp. 344–397.

Zoulias, E.I., Oreopoulou, V., Kounalaki, E., 2002. Effect of fat and sugar replacement on cookie properties. Journal of the Science of Food and Agriculture 82, 1637–1644.

Chapter 10

Sensory and Consumer-Led Innovative Product Development — From Inception to the Shelf (Current and Future Methodologies)

INTRODUCTION

Speed to market is one of the primary considerations of product developers. If a product has taken too long to get in to the marketplace, the demographic consumer segments identified as the optimal users and potential purchaser's initially, from marketing sources, might have changed their minds and no longer desire it to the same degree. For this reason, all due speed should be utilised, but without cutting corners. However, there are risks and obstacles along the product development route, and after undertaking costly and resource consuming research and developments projects, there is no guarantee of success. Developing products is easy, developing products that appeal to consumers is less so and developing products that appeal to sufficient numbers of consumers and also achieve commercial success is very difficult (Stone and Sidel, 2007). What exactly are consumers buying when purchasing the products we manufacture? They may be buying nutrition, convenience and image, but most importantly, consumers are buying sensory properties, sensory performance and product consistency (O'Sullivan et al., 2011). Therefore, it is clear that sensory techniques must be an integral part in defining and controlling product quality. However, the most important feature of product quality in the marketplace is its direct relationship to consumer perception, satisfaction with and ultimate acceptance of a product's sensory attributes (O'Sullivan et al., 2011). Today's new product project teams and leaders seem to fall into the same traps that their predecessors did back in the 1970s; moreover, there is little evidence that success rates or R&D productivity have increased very much (Cooper, 1999). It is well documented that more than 90% of all new product development (NPD) in the food and beverage

A Handbook for Sensory and Consumer-Driven New Product Development.
http://dx.doi.org/10.1016/B978-0-08-100352-7.00010-5

industries fails, some claim the figure is in fact closer to 98%. However, the 2% that is successful accounts for billions of pounds, dollars, yen and euros worth of business every day, which begs the question: is the risk a worthy one? (Business Insights, 2004). A low rate of innovation, coupled with the high failure rate of food products following market launch implied that the methodology for new food product development was long overdue for a systematic rethink and clearly needed vast improvement (Stewart–Knox and Mitchell, 2003). It is clear that most companies are utilising sensory analysis, but quite often, the wrong methods are being utilised for the stated objectives of the studies (Stone and Sidel, 1993).

Thus, many new products fail in the marketplace because product production and development does not focus systematically on consumer preferences and perceptions of sensory properties (O'Sullivan et al., 2011). However, despite the very low success rate, about 10% or less, companies continue to make the investment because of the enormous profit and reputation opportunities expected by the winners (Stone and Sidel, 2007). Therefore, the objective of producers is to develop and manufacture products for those that will ultimately consume them, through the integration of the end user (Grunert et al., 2008). Thus, NPD strategies should be consumer driven and it is essential as product quality directly relates to customer satisfaction through sensory properties and ultimately to repeat sales (O'Sullivan et al., 2011). Additionally, hedonic measurements should use consumers which are users of the product in question, where possible, which will give better insight in to this subjective modality of affective testing instead of using consumers in an ad hoc fashion. However, the definitive correlation to hedonic (liking) or affective sensory assessment and actual consumer behaviour is still an unknown entity. The greater the number of consumers used, the more reliable the data. Sensory acceptance tests are very useful for screening experimental prototypes but are not a substitute for the more costly large-scale affective tests, using targeted consumers. Ultimately the final variants products through product development and optimisation should be validated by appropriately conducted larger scale consumer testing.

NEW PRODUCT DEVELOPMENT TEAMS – MANAGED AND RESOURCED FOR SUCCESS

In food and beverage organisations, who actually drives product development and optimisation? It could be the owner of the company, which is quite often the case for start-ups or small companies, or another employee with other dual duties in production or quality. It is only when companies start to grow and get bigger do research and development resources become more focussed. Initially this might involve the low cost solution of hiring a graduate with a dedicated research and development brief. Also, perhaps small- and medium-sized companies can liaise with a knowledge provider, such as a university and assisted by state aid, to sponsor a postgraduate student to undertake research within an academic institution in an innovation partnership-type scenario. As an operation gets bigger an efficient method of product development is the provision and maintenance of a dedicated product and development team with logistical support from the organisation as a whole. The team should be led by a strong accountable project manager and populated by motivated cross-functional individuals with the necessary experience to successfully navigate through the R&D landscape which may require individuals with skills such as: sensory and consumer science, flavour chemistry, statistics (multivariate is most useful), marketing, production, quality, regulatory affairs, procurement, and perhaps also microbiological and physicochemical analysis and validation. The R&D team might be composed principally of a core team of sensory and consumer scientist(s), flavour chemists and production personnel with strong cross-functioning support from individuals within the food or beverage business with experience in the other elements. Strong support from senior management is also fundamental to success through the allocation of adequate resources although day-to-day meddling and the pushing of projects by top management is not conducive to success (Cooper, 1999). Good production personnel are crucial for pilot plant production who must also possess an inherent and in-depth understanding of the commercial manufacturing processes as ultimately this is where successful prototypes will be made. The R&D team will thus have access to laboratory and pilot plant facilities for making experimental prototypes and will liaise extensively with suppliers of ingredients in order to optimise products as well as procurement in order that prototypes can be manufactured within the least cost formulation (LCF) limits. The R&D team should interact closely with the regulatory affairs section in order to ensure complete legal compliance with the product in the market(s) where ultimately the product will be sold. In many companies a dedicated regulatory affairs service may not be available in which case supplier companies can sometimes provide intelligence and up-to-date information regarding the compliance of ingredients. Also, the R&D team may be required to undertake their own regulatory 'due diligence' which might involve the obtaining of relevant information from the regulators directly (websites, databases, archives, etc.) or appropriate government department such as the Ministry of Agriculture in the United Kingdom, Department of Food, Agriculture and the Marine in Ireland, the US Food and Drug Administration (FDA) or ANSES, The French Agency for Food, Environmental and Occupational Health & Safety. Third party agencies, such as 'Leatherhead Food Research' in the United Kingdom also provide a toll provision regulatory service relating to compliance of ingredients and products in global markets.

It is essential that an effective record keeping procedure is implemented so that prototypes can be accurately made as well as documenting successes and failures so that the former can be built on and the latter reduced.

The team may undertake in-house sensory evaluation on a routine basis but use external panels or sensory agencies to validate commercial prototypes for both descriptive and affective evaluation. An important part of the brief is product validation which might involve safety and shelf life testing of experimental prototypes.

One system of facilitating and overseeing the NPD process is the implementation of a stage gate system, where defined criteria must be met in order to progress along the chain to the next gate finally culminating in an ultimate gate which is product approval to commercialise and upscale. Also, the sensory methodologies that are employed in product development become sequentially more complicated until ultimately the product goes live and is released on the market. Initial product concepts may be assessed by focus groups and early prototypes by affective analysis such as preference or acceptability testing. For reverse engineering or product nutritional optimisation projects (e.g., salt, fat, sugar reduction or replacement), difference testing will prove useful. As prototypes advance and pass successive stages in the product development process, descriptive analysis using quantitative descriptive analysis (QDA) might be required to better understand differences between variants which might also be analysed in parallel with instrumental measurements such as GC analysis or sensory affective analysis and consumer testing. Multivariate data analysis (MVA) could then be employed to correlate sensory descriptive terms to consumer drivers (positive or negative) for preference mapping and additionally to flavour volatiles which may be tweaked in order to optimise the product further (i.e., the modification of top-notes). This is a valuable but also expensive option for the product developer requiring significant resources. The case study section of this chapter describes product validation through the use of sensory acceptance testing combining a rapid descriptive analysis where the data are analysed using MVA for prototype optimisation. However, the ultimate prototype should be validated in the marketplace using appropriately selected consumers, in sufficient numbers who fit the demographic of the product to be sold.

R&D PROCESS STAGES

In order to operate efficiently, the product development project should be overseen by a process that has defined milestones which must be passed in order to proceed to the next successive step. Products fail because consumers do not want them, they may be too expensive to make, the marketing effort was insufficient or the whole process took too long and the demographic changed. It is thus crucial to have an optimally designed process whereby winning products can be assisted on their journey to launch with all due speed and less successful variants are weeded out, thus conserving resources for the next project. New product development systems can be implemented similar to the 'The Stage-Gate idea-to-launch model' developed by Robert Cooper are a result of comprehensive research on reasons why products succeed and why they fail (Brands, 2013). This robust system integrates numerous performance-driving practices from scoping ideas and building a business case to development, testing, validation and finally culmination in product launch. Close scrutiny of performance at each stage gate ensures resources are used efficiently with performing projects proceeding and subperforming projects killed decisively. This go/no-go or 'Go/Kill' ethos allows the efficient optimisation of R&D resources (Fig. 10.1). Proceeding to each stage, a project passes through a gate where a decision is made by the stakeholders whether progress meets defined criteria for success, or not, in order to continue investing in the project (a Go/Kill decision). These criteria include: strategic fit, product and competitive advantage, market attractiveness, technical feasibility, synergies/core competencies, financial reward/risk (Stage-Gate, 2016).

IDEATION

Ideation is the first port of call in the product development and optimisation process. According to Graham and Bachmann (2004), ideation can originate from a number of potential sources. One of the first is 'Problem Solution' where an individual(s) solves a particular problem. A 'Derivative idea' is where an existing product is modified. 'Symbiotic ideas' result when multiple ideas are brought together. 'Revolutionary ideas' are mould braking ideas. 'Serendipitous ideas' occur by chance with beneficial but accidental positive outcomes. 'Targeted ideas' result from extensive research paths of discovery. 'Artistic Ideas' and innovation generally do not have constraints and can disregard practicality. Less tangible ideas include 'Philosophical ideas' which are hypothetical and may never come in to reality and 'Computer-assisted discovery' uses algorithms to calculate simulated outcomes (Graham and Bachmann, 2004).

Where does ideation come from and what constitutes the commercial success of new ideas in the marketplace. As discussed above the failure rate of new products is very high and the risks are great, but the rewards are massive for those companies that succeed. An idea might come from a moment of inspiration or a brainstorming session with a selected group of individuals or from an individual who sees a gap in the market for a new product. Companies are increasingly relying on external partners as collaborators in product development projects such as suppliers or external marketing agencies. Suppliers have become an integral part of the development team and are also often required to contribute financially to development efforts (Stone and Sidel, 2007). Many ideas in the food manufacturing industry come from the ingredient suppliers, the food processors, who not only supply the ingredients but increasingly supply the formulation and also the relevant consumer and market research (Earle and Earle, 1998). This idea might be inspired by a new ingredient or process or packaging technology. An individual might have seen

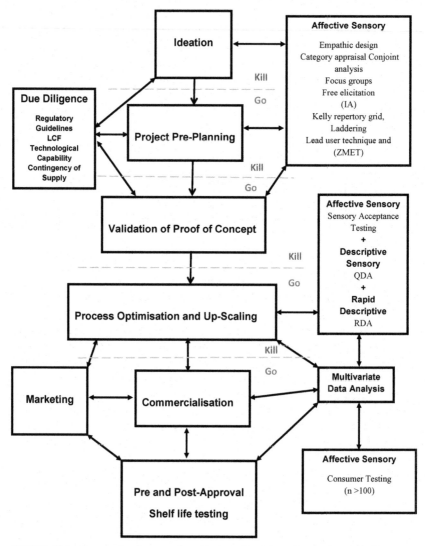

FIGURE 10.1 Process development flow diagram. From inception to the shelf. Go = meet criteria to proceed to next stage (as determined by management committee), Kill = Terminate project, progression criteria not met.

and be impressed by a certain product on their travels and feel a market exists also for that unknown product in their own market or country.

PROJECT PREPLANNING

Once ideas are generated, they must then undergo their own quality control or screening processes. The idea/concept must be thoroughly scrutinised, researched and defined before progression. The pros and cons of each must be investigated and carefully scrutinised within and outside the R&D team with a predevelopment business or financial analysis as well as SWOT (strengths, weaknesses, opportunities and threats) analysis implemented and in comparison with competitor products to determine and establish potential for success. A go/no-go stage gate should be employed here with ideas that do not make the grade directed to the archives and ideas that pass progressing to the next 'Proof of Concept' phase. NPD teams need to plan before developing and thus 'do their homework' in order to clearly define the product/concept prior to initiation of a stage gate protocol. Upfront homework drives up new product success rates significantly and is strongly correlated with financial performance (Cooper, 1999; Montoya-Weiss and Calantone, 1994). Too many projects move from the idea stage right into development with little or no assessment or upfront homework. The results of this 'ready, fire, aim' approach are usually disastrous (Cooper, 1999). Resources and effort are thus required to complete this 'homework' and NPD teams need to present evidence that their research is not just speculative in order to progress through the go/no-go stage gate process. Cooper and Kleinschmidt (1995) state that often in too many companies, projects move far into development without serious scrutiny: once a project begins, there is very little chance that it will ever be killed. This results in the approval of marginal projects and a misallocation of scarce resources. Tough Go/Kill decision points or gates are thus strongly correlated with the profitabilities of businesses' new product efforts (Cooper and Kleinschmidt, 1995). New products aimed at international markets (as opposed to domestic) also tend to fair better than products designed specifically for a domestic market. According to Cooper (1999), this is because such project generally utilise cross-functional teams with members from different counties who have to resolve a broader range of consumer problems and have to gather market information from multiple international markets as an input to the new product's design thus amplifying the challenge of understanding the different human behaviours (Cooper, 1999). Lean, mean and scalable are the key points to keep in mind. During the NPD process, keep the system nimble and use flexible discretion over which activities are executed (Brands, 2013).

Therefore, NPD teams must create a great product with a very clearly defined and large consumer market where the product resolves a real consumer issue and offers superb differentiation over its competitors. Secondly, the product must be well-defined across consumer, technology and business prior to full execution commencing (Cooper and Kleinschmidt, 1990).

VALIDATION OF PROOF OF CONCEPT

Consumer-driven innovation somehow infers that the consumer is involved in the R&D process. New product projects that feature high-quality marketing, preliminary and detailed market studies, customer tests, field trials and test markets, as well as launch are likely to have more than double the success rates and 70% higher market shares than those projects with poor marketing actions (Cooper et al., 1998).

Focus group interviews, as discussed in detail in Chapter 3, are a rapid method for collecting data through group interaction on defined topics in order for the researcher to try and understand the consumer and their environment. Focus groups are critical in order to generate concepts for an area when previous knowledge about the area to be investigated quantitatively is limited (Powell and Single, 1996). Targeted individuals, who fit the demographic brief for the product concept, and in the absence of 'cheaters' and 'repeaters' (see Chapter 3) can provide useful information about product concepts and even suggest potential modifications for uses and improvements. A carefully constructed recruitment 'screening' questionnaire can assist in weeding out these unsuitable individuals as well as informal conversations with the moderator to determine their inclusion. Different targeted demographics can be used to select the focus groups. It must be remembered that sensory perception is not standard response but is effected by age and gender (Michon et al., 2010a,b,c) as well as cultural influences (Yusop et al., 2009a,b) and many other factors must be taken in to account during recruitment where appropriate. Two or multiple focus groups can be undertaken as a means of validation where similar or trending responses will be indicative of consensus and the moderator can have confidence that the data obtained are correct. Focus groups can be used to build an understanding of a product category from the consumer viewpoint, inspiring idea generation, but can also be used to evaluate new product concepts and even samples (Edmunds, 1999). A product concept could be evaluated initially with actual prototypes tasted in later sessions. Additionally, focus groups can help in understanding packaging and even marketing concepts. Products which deliver real and unique benefits to customers are far more likely to succeed in the marketplace. Products in the top 20% by this criterion have a success rate of 98%. Those in the bottom 20% have a success rate of just 18.4%. Does the product meet the needs of consumers to a greater extent, is it perhaps the first of its kind, a category creating product. Does the product have higher quality, solve a problem, lower cost or possess unique features that are absent in competitor products (Brands, 2013).

REGULATORY GUIDELINES

Regulatory requirements must be carefully considered early on as part of the NPD approval process. This is also a critical stage in the go/no-go decision to advance. All ingredients, packaging and labelling requirements must be legally permitted in the jurisdiction(s) in which the product is to be sold. What is legal for one country may not be legal for another. Thus, the ingredients, packaging format and labelling requirements must be carefully researched in order not to be in contravention of the law. A good example of this would be the production of a flavoured alcoholic beverage for a European and US market. Flavouring ingredients in the EU may be nature identical, whereas those required to satisfy the regulator, the TTB (Alcohol and Tobacco Tax and Trade Bureau), in the US must be natural flavours. This can also have dramatic cost implications as natural flavours can be far more expensive than nature identical flavourings. Thus a legally compliant US variant could be sold in Europe, but the European variant manufactured with nature identical flavours cannot be sold in the US. Thus, it is common practice for manufacturers of such beverages to have two different variants, a US version and a version that can be sold in the rest of the world (ROW). This presents a challenge to an NPD team whose goal would be to make cost-effective products for the US and European or ROW markets separately but with the same flavour profile and using different flavour sources, either natural or nature identical.

LEAST COST FORMULATION

Product development programs should always be focussed on risk and as highlighted earlier the risks are high considering the huge failure rate of new product launches in the marketplace. The initial obvious risks of development that need to be analysed include: the costs that will be incurred in making the product, the equipment and personnel required, packaging, approval and testing, shipping, storage and marketing. An LCF calculation should be one of the very first activities a product development scientist performs once ideation is complete and an initial product prototype formulation has been realised. The cost of production of the concept product has to be considered as part of the approval process and can be relatively straightforward to determine. Weights or approximate values must be calculated for each component ingredient and multiplied by the optimal price at which the company can purchase that ingredient. This is simply performed as a paper exercise but more usually using spreadsheet software such as Excel (Microsoft). Purchasing agreements and strategic alliances with suppliers can optimise value of ingredients with economy of scale also being factored in. The greater the quantity purchased, generally the cheaper the price the ingredient will be supplied at. Developers can also factor in these projected costs of production once certain volume thresholds are achieved.

TECHNOLOGICAL CAPABILITY

Technological capability is a crucial consideration when developing an NPD project. Essentially, it is an evaluation to determine if the manufacturing capability for the product, if it ever gets to the commercialisation stage, exists within an organisation. Cooper and Kleinschmidt (1990) in their article 'New Products, Key Factors in Success' define technological synergy as a measure of how far from their current technology a firm must stray to build the product.

The further away they stray, the less likely the project will be successful. As well as considering LCF calculations, processing costs must also be factored into the LCF protocol. Questions must be considered like: can the product be made on existing equipment and using the available personnel and can these resources sustain production. There are three potential scenarios to this situation. (1) The existing process line resources can accommodate the potential manufacture of the new product. (2) The existing process line can technically produce the product but is currently operating at capacity or near capacity. If a process line is unable to accommodate the increased capacity requirements, then increasing the pressure further by introducing a new product for scheduling in the production rota will only lead to inefficiencies and downtime. Thus, it might be necessary to (3) invest capital in developing a new line which adds further to the risk to the product development program. Option 3 also comes into play if the new product requires a bespoke process. Obviously, this is the riskiest scenario and potentially could be foolhardy, especially for an NPD project. In this latter case, the best solution might also be to take the lesser of two evils by muddling through with option 2 with existing resources and to invest in option 3 when the new product has demonstrated a sufficient degree of commercial success.

When faced with initiating a new production process, there are further options available. Perhaps the necessary equipment is available secondhand at a much lower cost than a brand new setup. Producers, particularly SMEs or start-ups, can sometimes make the mistake of not considering the secondhand approach which could be the difference between success and failure. One other factor to consider with the scenario presented by option 3 is the potential to trial a process line, gratis or on a lease/rental basis, from an equipment supplier with the option of purchasing at a later date. If the equipment supplier can provide this service, usually with a company that it has established a positive track record of synergy, then this is potentially the best option. Another scenario where an equipment supply company might provide this service is in the demonstration of new technology. They might agree to supply the equipment, again either free or on a rental basis, on the proviso that they can demonstrate the line, in operation, to new clients which facilitates future sales without having potentially very expensive equipment lying idle for extended periods. An added benefit to this is the availability of the best technical expertise for product and process optimisation and efficiency.

If none of these options are available, then the commercialisation team, under management approval, could potentially source a third party manufacturer to make the new product for them under agreement. This is quite a common scenario with SMEs and considerably dilutes the risk involved with an NPD project. The origin company still completely owns the product, but on payment and potential completion of an NDA (nondisclosure agreement), also known as a confidential disclosure agreement, the third party can manufacture on their behalf. These agreements can identify sensitive information regarding ingredients and processes that fundamentally remain confidential to both parties and are legally binding, the breach of which can lead to litigation.

Another factor which must be considered in the production of the new product is the new products stock keeping unit (SKU). An SKU refers to the unique identifier that is allocated to all unique products. It can identify a specific volume or weight of product in a specific package with specific labels for the market it will be sold in. SKUs are essential for production but can be challenging for any production scheduler. Anything that requires reconfiguring a process line, retooling, swapping out webs and packaging materials, recalibration, personnel changes, etc., increase the equipment downtime and effect direct costs. A company with too many SKUs allocated to a specific process line can reduce efficiency and increase costs dramatically. In such cases, option 3 or third party manufacture may be the only option available.

CONTINGENCY OF SUPPLY

Once a formulation has been developed, it is prudent for manufacturers to continue with the optimisation process, but using ingredients from different suppliers, even if there is also a price differential on the negative end of an LCF strategy. This is important as it is never good practice to have all of ones eggs in one basket. A sole supplier can take comfort in their premier status and be inflexible to pricing negotiations, supply demands, economies of scale, etc. It is also a dangerous situation to be in to have a sole supplier for an ingredient because any catastrophic event that this company encounters will also be passed on to the origin company. A limited number of suppliers could have a negative effect on production costs. It is thus important to consider contingency of supply pretty early on in the NPD process and in conjunction with LCF calculations and projections. Anything that could terminate or retard supply can result in reduced levels of inventory which will also interrupt production. Extreme weather can result in crop failures in certain parts of the world which could have a dramatic effect on supply. Natural vanilla extracts are important flavouring for food and beverage products, but are particularly used in some liqueur-type alcoholic beverage products supplied into the US market. In 2000 Cyclone 'Hudah' in the Indian Ocean destroyed 20% of the vanilla crop which resulted in very dramatic price increases, twice those of the previous year, which left those exposed to increased production costs. The extract producers thus passed on this price increase to their customers. There is little that could have been undertaken as a contingency of supply in this instance as natural vanilla from other parts of the world also increased in price due to demand. The only solution would have been a hedge agreement between customer and supplier with fixed price option agreements over a mutually agreed timescale. This is the strategy used with airlines to stay competitive in a volatile oil market.

COPYCATTING AND REVERSE ENGINEERING

Reverse engineering is a topic seldom discussed openly outside of the confidential research and development team environment but can take several different forms. The first scenario is the blatant copying of a competitor's product. This might be performed with the hope of selling the developed product to a new, or common, customer with a specific advantage in order to leverage sales, perhaps at a lower price or other such concession. A product may also be initially reverse engineered, or at least partly reverse engineered before being used as a base formulation for a nutritionally or ingredient optimised product. This can then be supplied to a customer with the leverage of being a nutritionally or ingredient optimised formulation. The above cases may also come into play where a company wants to launch a product similar to a successful competitor product in order to steal market share and actively compete with them in that market space.

LINE EXTENSION, BRAND EXTENSION AND CANNIBALISATION!

Line extensions are products that diversify an origin product. Typically, this involves the flavour or ingredient modification of an established product. Examples of this include new flavour varieties of soft drinks like cola (cherry, vanilla, lemon, lime, grape, etc.) or chocolate bars with a standard and popular base chocolate with added fruit, nuts, mint honeycomb, Turkish delight, toffee, coconut, etc. Essentially, consumer choice is broadened with the anticipated effect of also broadening the consumer base and their interest and attention span in the extended product range. Line extensions may initially be developed to cater for niche markets, but ultimately become significant brands in themselves. For example, diet drinks which have definitely established themselves as separate to their full sugar counterparts. This is also seen in nutrient optimisation of established products, new recipe, lower; fat, salt, sugar, caffeine, etc.

Brand extension is where an established brand licenses their brand name and essential characteristics of that brand for inclusion in a completely different brand or product. Examples include the collaboration of alcoholic liqueur producers and a chocolate manufacturer or branded chocolates extended into an equivalently flavour ice cream.

Cannibalisation is a marketing term used to describe reduction in sales of a product as a consequence of the introduction of a new product by the same producer, i.e., a line extension. The intention is to grow market share by providing more choice to the consumer but results in initial negative impact on sales of the origin product with the hope of eventually increasing profitability over all.

PROCESS OPTIMISATION AND UPSCALING

Once a product is produced on a pilot scale and has succeeded in the necessary stage gate approval process, then the next step is to upscale production. This can sometimes be easier said than done as multiplying recipe or formulation weights and volumes to fit with a commercial batch production run can lead to final product differences. This is because unknown factors come in to play that cannot be predicted for such as physical interaction of the ingredients. For example, a pilot scale product when mixed with the component ingredients may produce an optimal texture in a final product, but when this process is upscaled, the texture may be unsatisfactory. This is because the actual physics of mixing have changed, the commercial batch weight of ingredients in the blender as well as blender design allows mixing to be completed more efficiently. In this case, an optimisation trial on a commercial batch scale needs to be undertaken to tweak the process in order to obtain optimal rheological properties.

MARKETING

Marketing is the method of communication that is used to promote products between a company and the consumer audience by multimodal means of communication from quite cheap (free ads, press release) to very expensive (TV advert). There are a number of factors the marketing team must determine with respect to the product in development which includes: If the product is a first of its kind, will consumers even get the concept. How well is the product matched with the existing marketing machinery of the business? How big is the market for the new product and what is trending in that market? As well as how important is the product (developed) to that market? (Cooper and Kleinschmidt, 1990; Brands, 2013). Prelaunch the marketing team can additionally arrange further private tests groups to validate the product, potential launching beta versions, and then forming test panels after the product or products have been tested which will provide valuable information allowing last minute improvements and tweaks to be made (Brands, 2013).

PRE- AND POSTAPPROVAL – SHELF LIFE TESTING

During the product development process, shelf life testing should be initiated at the earliest opportunity once products have passed through a defined stage and are allowed to progress through the development route. Retention samples can be stored under defined storage conditions and tested periodically for perhaps, microbiological, sensory and physicochemical quality. As described in Chapter 6, real-time shelf life studies have the potential of taking a very long time to complete and because of this other methods, such as accelerated ageing assist in estimating shelf life (O'Sullivan, 2011, 2016). Accelerated storage tests are designed to reduce shelf life testing time by speeding up the deteriorative changes that occur in the product by exposure to factors such as high temperatures or humidity (O'Sullivan, 2011, 2016). In this fashion a predicted shelf life may be determined or problems such as off-flavour development identified. After product commercialisation and launch the product needs to be monitored, as with any commercial product, to ensure that it is performing satisfactorily in the market and achieving the appropriate quality and shelf life criteria.

TECHNOLOGICAL DEVELOPMENTS, INTERNET TESTING, IMMERSIVE TECHNOLOGY, MOBILE APPLICATIONS AND EYE TRACKING

Recent developments in technology have created new opportunities for obtaining consumer data. Immersive virtual reality (IVR) and smart phone (mobile) applications can enhance and optimise qualitative and quantitative sensory data collection. Additionally, eye tracking technologies allow product developers to practically 'see through the eyes' of consumers and now to an even greater extent when augmented with IVR technology, potentially offering very exciting prospects for the future.

The survey method of data collection has served consumer science well for many years. However, overly long unengaging surveys as well as spam surveys, unsolicited telephone surveys and pop up surveys, etc., have caused over exposure and fatigue amongst the consumer population with the result of far fewer response rates than obtained in the past. Frankovic (2015) states that in 1997 a rigorous effort by a public poll, in the context of an election survey, could achieve a contact rate of over 90%, and a response rate of over 60%. By 2003, rates had dropped by 10 points or more, and by 2012, the overall response rate was as low as 9% (Frankovic, 2015). In some circles the response rate is predicted to fall below 1%. Thus, the traditional approach no longer seems to work and the data collected seem to represent the opinions of the minority. However, many developers are moving their research away from traditional surveys, towards innovative technology-based options. Internet-based testing has several advantages over traditional methods, in that large numbers of participants can be interacted with and data are collected automatically. However, the researcher has no control over the assessment environment or who has access to the online survey, and can have difficulty in verification of respondents. Incentives, like prizes or payment, can also encourage multiple responses from individuals. Additionally, people who respond to a request to complete an online survey are likely to be more interested in or enthusiastic about the topic or product and therefore more willing to complete the survey, which biases the results (Duda, 2016).

Mobile phone use, particularly the adoption of smartphones, has permeated society globally and presents an opportunity to get closer to current consumer reality. Additionally, the mobile phone allows researchers to reach consumers 'in the moment' as well as, observe behaviour, capturing emotive responses in real time and obtain behavioural as well as situational consumer feedback. Data collection is also spontaneous and negates having to input collected data manually. Questionnaires applied directly in the situation of interest also have the advantage that memory effects are prevented, which are a common issue with retrospective measures and interrupt the ongoing interactive experience (Hektner et al., 2007; Capatu et al., 2014). In the near future, geo-positioning system (GPS) and mobile device sensor technology can provide background information to user location and context thus allowing the consumer to be contacted and requested to take a short survey based on their location. This could be as they enter or leave the cinema, supermarket, leisure centre, etc., and thus are requested to complete a short, easy, interactive survey in real time or soon after a specific context. The process should be enjoyable and the survey kept concise or else fatigue and over exposure will also affect completion rates as in the case of classical surveys.

Traditionally, sensory consumer testing is completed in isolated sensory booths where the influence of nonproduct (e.g., environmental) attributes is controlled. These highly controlled environments strip away meaningful contextual (visual, auditory and olfactory) information important in forming consumer perceptions, liking and behaviours. Additionally, boredom and a lack of engagement can result in misleading consumer data (Bangcuyo et al., 2015) with some respondents failing to discriminate between samples through lack of attention to the test (Köster, 2009). In effect by controlling this internal validity, we introduce the banal and thus boredom. External validity, the context in which we experience the world can influence sensory perception. Immersive technologies allow the consumer to be transported to a virtual reality in the form of an immersive environment with controlled external validity and provide context which impacts positively to hedonic and food-related behaviours in the sensory component (Bell et al., 1994; Delarue and Boutrolle, 2010; Bangcuyo et al., 2015). Thus, immersive technologies may allow for the collection of more meaningful consumer data. Testing in a real environment such as a supermarket is very difficult as it requires the co-operation of the often reluctant retailer and conditions can be very difficult to control such as unpredictable consumers and shelf stocking characteristics changing on a minute by minute basis. The virtual store can be an appropriate solution to overcome these constraints by enhancing everyday experiences in a controlled, realistic and engaging context and creating a sense of actual presence in real or imagined worlds. VR headsets such as the Oculus rift combined with the appropriate simulation software allow the consumer to experience this controlled but realistic virtual world.

Eye tracking technology measures where an individual is looking or what they are looking at and has been used for several years as a tool in marketing and product design. By presenting an image to the respondent, the device will track their gaze in real time and allows developers to understand the consumer as an individual and also how their product colour shape, package, location on a supermarket shelf, etc., affects consumer perception of that product. In this fashion the semiotics of a product can be optimised to a specific consumer group. For example, a certain livery of colours, product package size and shape may cause the observer to be drawn more to that product than others. A further development of eye tracking is the augmentation of this technology with virtual reality.

SENSORY METHODOLOGY – CONSUMER EVALUATION, FOCUS GROUPS, VALIDATION, IDEAL PROFILING

Once a prototype has passed through the initial stages of approval from the stage gate process, it will be necessary to obtain some consumer hedonic feedback on both the concept and the product itself. This is achieved using a number of methods. It is often difficult to select a consumer research method in the early stages of product development, especially for developers from a technological (food) disciplinary background. However, a lack of consumer relevance and poor application of consumer research at the early stages of NPD are the key determinants of failure (van Kleef et al., 2005; van Kleef and van Trijp, 2007). There are a number of techniques that are used most frequently to uncover unmet consumer needs and wants and include empathic design, category appraisal (including preference analysis), conjoint analysis, focus groups, free elicitation, information acceleration (IA), Kelly repertory grid, laddering, lead user technique and Zaltman metaphor elicitation technique (ZMET) (van Kleef and van Trijp, 2007). Preference analysis, focus groups and conjoint analysis have been covered in detail in Chapter 3 of this book. Conjoint analysis (see Chapter 3) is a statistical technique used to identify the value placed by individuals on different product attributes and can also assist in the NPD process. The process measures a mapping from more detailed descriptors of a product or service onto an overall measure of the customer's evaluation of that product (Hauser and Rao, 2004). This method determines what combination of attributes influences consumer choice and decision-making and is used frequently in market research to study the effects of controlled stimuli or information on a particular consumer response. Statistical experimental design, analysis of variance and cluster analysis enable the response of each consumer to be analysed for the relative importance of each factor, and, similarly performing subjects can be clustered. It is a method used for understanding how consumers trade off product features (Green and Rao, 1971). Specific products or concepts are presented to consumers and the manner in which they make preferences can be analysed by modelling to determine information such as profitability and market share. In empathic design, consumers are observed by researchers in their natural environment or surroundings and their actions recorded (product use, frustration, latent needs, etc.). Free elicitation is where consumers respond spontaneously to questions regarding a product(s). IA is a method where preference and purchase intentions of consumers are recorded using virtual stimuli. Kelly repertory grid is a personal interview method as is laddering which uses layered questions. With laddering product attributes form the basis of an in-depth interview where the respondent is asked 'why is that important to you' continually, up the ladder of abstractness, until they can no longer respond (Krystallis, 2007). Consumers are better at articulating opinion on familiar categories of products than new category or mould breaking ones (van Kleef and van Trijp, 2007). The lead user technique, as it uses more advanced users, may thus be a good methodology for brand new concepts as these consumers are better able to articulate their viewpoints (von Hippel and Katz, 2002). Also, the ZMET uses collages of pictures and metaphors to stimulate emotive responses to products (van Kleef et al., 2005) and allow researchers to tap in to the consumer's deep thoughts at the subconscious level.

The concept, and even an early prototype, with also potentially including packaging (mock up) can be evaluated by an appropriately selected focus group. As discussed in Chapter 3 on qualitative methods. However, quantitative modification from hedonic cues can follow the traditional approach to variant optimisation with the utilisation of preference mapping, and the multivariate exploration of both hedonic (consumer study data) and descriptive data acquired from the traditional consumer test and QDA. In this fashion, sensory drivers (positive or negative) can be identified which correlate the hedonic and descriptive data. Additionally, quantitative instrumental analysis like GCMS volatile profile data could also be modelled to identify compounds which correlate to these sensory drivers. Thus, it is possible to tweak the profile by turning down negative drivers and modifiable associated top notes or increase slightly positive drivers and their corresponding volatile compound(s). Chapter 2 of this book demonstrates a sensory term reduction process in QDA, while Chapter 5 presents a case study which demonstrates the MVA of multimodal data from descriptive analysis (QDA) and instrumental analysis from GCMS and electronic nose analysis.

Alternatively, once the process and product development stage has progressed further Ranking Descriptive Analysis and sensory acceptance testing are another approach in optimising the prototypes. As discussed in Chapters 2 and 11. It must be remembered though that ultimately it will be required to validate the product using a conventional consumer test (Chapter 2). Fig. 10.1 presents a flow diagram depicting the product development steps and sensory inputs.

As discussed previously 'The Ideal Profile Method' (IPM) and just about right (JAR) scales (Chapter 3) can be used to determine the optimum level of an attribute intensity of a food or beverage product. IPM is a descriptive analysis technique in which consumers are asked to rate products on both their 'perceived' and 'ideal' intensities from a list of attributes. In addition, overall liking is asked. Each consumer provides a sensory profile of the products, their hedonic ratings and their ideal profile (Worch et al., 2014; Worch and Ennis, 2013). In theory, the method could be used to create ideal products for consumers, but the reliability of the data can be fragile. With this method, each consumer evaluates several products following standard randomisation and monadic presentation and rates products on both perceived and ideal intensity for a list of attributes. So, if the first question is: 'Please rate the sweetness of this product', the second question will be: 'Please rate your ideal sweetness for this product'. This methodology has been adopted with the aim to mimic the JAR scale, but using the perceived and ideal intensities instead of the difference with an imagined ideal. Additional hedonic questions are also asked for each tested product. The ideal profiles are directly actionable to guide for products improvement. However, this particular information should be carefully managed since it is obtained from consumers and it describes virtual products (Worch and Ennis, 2013).

CASE STUDIES

O'Sullivan et al. (2011) in their book chapter on sensory science as a practical commercial tool in the development of consumer-led processed meat products describe a simple case study where a Small Food Business Operator (FBO) and manufacturer of healthy, additive-free, meat-based ready meals wanted to optimise their product range (five products). They demonstrated that five simple steps were required for the optimisation process. (1) Identifying all competitor products, which was achieved by conducting an extensive supermarket survey of all the competitor products available along with relevant information such as stocking densities. (2) Development of the consumer questionnaire. All products within the FBO's product range, as well as all the relevant competitor products were sensory evaluated by experts with product knowledge to compile a descriptor list (O'Sullivan et al., 2011). A modified Ranking Descriptive Analysis-type protocol was utilised similar to that described by Richter et al. (2010) except the method used both affective and descriptive attributes, with the former presented first to reduce bias (overall acceptability only). Descriptors included those for appearance, aroma, taste, flavour, texture, off-flavour, aftertaste, etc. Additional questions were asked of these naïve assessors regarding product observations and packaging quality. (3) Assessor evaluation. A group of 25 assessors (naïve), who were consumers of similar convenience-type ready meals and who also fitted the consumer demographic (males and females (50/50) 20−30 years of age), were recruited. These naïve assessors were asked to evaluate all products (test products and competitor products) and to rank their scores on a continuous 10-cm line scale ranging from 0 (none) on the left to 10 (extreme) on the right for the hedonic attribute followed after by each descriptive attribute. The presentation order for all samples presented over two sessions was randomised to prevent first order and carryover effects (MacFie et al., 1989), but all samples were presented simultaneously. (4) Data mining was then performed using ANOVA-partial least squares regression to process the raw data accumulated from the 25 test subjects during the consumer sensory evaluation. From these data, consumer product variation and ranking could be determined for the test product range as well as all competitor products. In effect, the test products position in the consumer landscape was determined along with positive and negative sensory drivers identified for each product (O'Sullivan et al., 2011). (5) Product optimisation. Positive sensory drivers were increased and negative drivers were decreased by reformulation. Reformulated products were then evaluated using the same consumer panel as before. A clear increase in product ranking was observed for each test product. When the optimised products went into commercial production and subsequent retail sale, over a short period of time, the FBO observed increased sales and an increase in market share (O'Sullivan et al., 2011). Furthermore, some of the products went on to win primary food awards for their food category in internationally recognised food award competitions. The optimisation protocol was thus effective. It must also be noted that chemical (i.e., lipid oxidation or compositional data) or instrumental data (i.e., GC−MS) may also be included with sensory or consumer data in the MVA models. This may provide additional information for identification of sensory drivers for optimisation (O'Sullivan et al., 2011).

O'Sullivan et al. (2011) state that one may question using such a small group of consumers (25 naïve assessors, consumers) in making such important commercial decisions. The reality is that some small food businesses cannot afford large-scale consumer evaluations using perhaps 200, 300 or even 1000 consumers (O'Sullivan et al., 2011). Additionally, larger scale consumer evaluations take a considerably longer time to organise, conduct and data mine. From start to finish, months can elapse and in such an extended period of time, the consumer marketplace or demographic can change, which questions the accuracy of the information accumulated (O'Sullivan et al., 2011).

Sensory measurements are clearly integral to user-driven innovation adding much fundamental and applied insight as to why consumers form preferences for certain foods and not others. When it comes to making choices about food, the underlying reasons for our likes and dislikes are not easily accessible to our reasoning (Dijksterhuis and Byrne, 2005), but still, sensory objective descriptive methods rely largely on panellists' conscious action (Frandsen et al., 2003; Köster, 2003). As stated earlier in Chapters 2 and 4, descriptive profiling is one of the most powerful, sophisticated and extensively used tools in sensory science, which provides a complete description of the sensory characteristics of food products (Varela and Ares, 2012). However, these methods can be expensive and time-consuming because of the necessity to train and profile individual panellists over extended periods of time; days or even weeks. It is also not a method that can be readily used for routine analysis (O'Sullivan et al., 2011).

On the opposite end of the sensory spectrum the affective methods described in Chapter 3 are restricted to only measure hedonics, which includes acceptability or preference, but do not describe the product. As described in Chapter 3. Traditionally this issue was solved by combining descriptive data and hedonic data using data analytical tools involving predominantly chemometrics in a method called preference mapping (Chapter 5). Objective sensory measurements combined with affective sensory analyses are particularly suitable for testing the effectiveness of product improvements/optimisation and NPD potential. Furthermore, they allow a targeted adjustment of sensory properties with the purpose of obtaining a higher degree of consumer satisfaction (Moskowitz et al., 2006; Muñoz, 2002).

Due to government and consumer pressure, many sectors of the food and beverage industries are undertaking a very active approach to reformulation in order to provide healthier products. Overconsumption of fat and sugar is associated with many diseases, such as obesity, type 2 diabetes, high blood cholesterol and coronary heart diseases. Obesity is increasing around the world and it is a significant public health problem in many countries (International Obesity Task Force, 2002; Melanson et al., 2009; World Health Organisation, 2003). High sugar consumption is associated with obesity, insulin resistance, and diabetes mellitus type 2, as well as caries and fatty liver. Therefore, low sugar intake is strongly recommended (FAO/WHO, 2003). This recommendation is addressed not only to the consumer, but also requires the food industry to reduce sugar content in processed food (Barclay et al., 2008). The processed meat industry have already commenced reformulating their recipes, and now offer lower levels of salt and fat, or even higher levels of polyunsaturated fatty acids in processed meat products on the market (Verbeke et al., 2010). As for sugar, the food industry knows that it has a problem. Sugar is low cost, sugar tastes good and sugar sells, so companies have little incentive to change societal intervention to reduce the supply and demand for sugar faces an uphill political battle against a powerful sugar lobby, and will require active engagement from all stakeholders (Lustig et al., 2012). Duffey and Gordon-Larsen (2010) evidenced that food consumption is linked with food prices, which has a direct influence on human health. Furthermore, Niebylski et al. (2015) revealed consistent evidence that taxation (of unhealthy food/beverage products) and subsidy (of healthy food/beverage products) intervention positively influenced the dietary behaviour of consumers. Many countries are thus considering the introduction of fat and sugar taxes, the latter particularly on soft beverage products, in an effort to help fight the war on obesity. Nutritional improvement through product reformulation for producers will be very much on the product optimisation agenda for many years to come.

As a consequence of this, some producers are taking a scattergun approach to reformulation. They start with their current recipe and modify the ingredient of interest (sugar, fructose, fat, salt, etc.) to reflect current levels within the industry as well as those used by their competitors. They also, typically, include those stipulated by current and predicted regulatory guidelines as well as levels beyond these ranges, above and below, to fully explore the limits of modification in a full factorial design. In one recent example of this, myself and my team were asked to assist in modifying a cheese product with varying levels of salt and fat which produced 48 different product variants, not including replications. Faced with these vast numbers of products to be analysed affectively and descriptively, it is obvious the traditional QDA, consumer study, preference map approach would not be a useful tool; however, it certainly would be a most expensive one. It could be suggested that the company undertake some initial optimisation studies to reduce the volume of samples to produce a more manageable number but they were clear that they wanted to investigate how variations in the matrix effected sensory properties from first principles and that earlier assumptions would be disregarded. This approach was also undertaken within the realms of food safety, LCF and product functionality.

It is important to remember that effective, robust consumer-driven innovation strategies can assist companies in getting their products into the marketplace in a speedy fashion and also ensure the optimised products are not obsolete from a consumer relevance perspective, but are specifically designed to meet consumer requirements (O'Sullivan et al., 2011). The success of the consumer-driven strategy described above can be measured by the increased profits and market share of the companies that have been engaged with. Additionally, consumer sensory work undertaken recently (Zakrys-Walliwander et al., 2010, 2011) has produced similar findings to parallel studies using thousands of consumers which further validates this methodology (O'Sullivan et al., 2011).

Rapid sensory methods such as flash profiling have been demonstrated for a diverse selection of products including: pear/apple puree and fresh cheese (Loescher et al., 2001), yoghurt (Delarue and Sieffermann, 2004), jams (Dairou and Sieffermann, 2002), beer (Hempel et al., 2013a), bakery products (Lassoued et al., 2008; Hempel et al., 2013c), Gouda cheese (Cavanagh et al., 2014; Yarlagadda et al., 2014b), Cheddar cheese (Yarlagadda et al., 2014a), French dry sausages (Rason et al., 2003), beef patties (Tobin et al., 2012a), Frankfurters (Tobin et al., 2012b), breakfast sausages (Tobin et al., 2013), ready-to-eat mixed salad products (Hempel et al., 2013a,b). Chapter 13 presents a case study demonstrating this technique on a beer product.

Additionally, an evolution of this approach, sensory acceptance testing combined with Ranking Descriptive Analysis with the resulting data analysed by MVA is an approach that has been successfully demonstrated for a number of other products including: chocolate pudding (Richter et al., 2010), white pudding (Fellendorf et al., 2015, 2016b), black Pudding (Fellendorf et al., 2016a,c), corned beef (Fellendorf et al., 2016d), and butter (O'Callaghan et al., 2016). Chapter 12 presents a case study on this technique for Mozzarella cheese (Henneberry et al., 2016) and Chapter 2 a case study for salt and fat reduction in processed meats.

The above methods are very useful tools for product development and optimisation until the last stages of development are reached; however, it is prudent that validation using traditional consumer testing is still implemented to confirm acceptance in the marketplace on a large scale ($n > 100$) for the ultimate prototype(s).

REFERENCES

Bangcuyo, R.G., Smith, K.J., Zumach, J.L., Pierce, A.M., Guttman, G.A., Simons, C.T., 2015. The use of immersive technologies to improve consumer testing: the role of ecological validity, context and engagement in evaluating coffee. Food Quality and Preference 41, 84–95.

Barclay, A.W., Petocz, P., McMillan-Price, J., Flood, V.M., Prvan, T., Mitchell, P., Brand-Miller, J.C., 2008. Glycemic index, glycemic load, and chronic disease risk—a meta-analysis of observational studies. American Journal of Clinical Nutrition 87, 627–637.

Bell, R., Meiselman, H.L., Pierson, B.J., Reeve, W.G., 1994. Effects of adding an Italian theme to a restaurant on the perceived ethnicity, acceptability, and selection of foods. Appetite 22, 11–24.

Brands, R.F., 2013. Step Process Perfects New Product Development, p. 8. In: http://www.innovationexcellence.com/blog/2013/05/27/8-step-process-perfects-new-product-development/.

Business Insights, 2004. Future Innovations in Food and Drinks to 2006: Forward-Focused NPD and Consumer Trends.

Capatu, M., Regal, G., Schrammel, J., Mattheiss, E., Kramer, M., Batalas, N., Tscheligi, M., 2014. Capturing mobile experiences: context- and time-triggered in-situ questionnaires on a smartphone. In: Spink, A.J., van den Broek, E.L., Loijens, L.W.S., Woloszynowska-Fraser, M., Noldus, L.P.J.J. (Eds.), Proceedings of Measuring Behaviour, Wageningen, The Netherlands, August 27–29, 2014.

Cavanagh, D., Kilcawley, K.N., O'Sullivan, M.G., Fitzgerald, G.F., McAuliffe, O., 2014. Assessment of wild non-dairy lactococcal strains for flavour diversification in a mini Gouda type cheese model. Food Research International 62, 432–440.

Cooper, R.G., Kleinschmidt, E.J., 1990. New Products, Key Factors in Success. The American Marketing Association, Chicago, Illinois.

Cooper, R.G., Kleinschmidt, E.J., 1995. Benchmarking the firm's critical success factors in new product development. Journal of Product Innovation Management 12 (5), 374–391.

Cooper, R.G., 1999. From experience: the invisible success factors in product innovation. Journal of Product Innovation Management 16 (2), 115–133.

Cooper, R.G., Edgett, S.J., Kleinschmidt, E.J., 1998. Portfolio Management for New Products. Addison-Wesley Publishing, Reading, Mass.

Dairou, V., Sieffermann, J.M., 2002. A comparison of 14 jams characterized by conventional profile and a quick original method, the Flash profile. Journal of Food Science 67 (2), 826–834.

Delarue, J., Boutrolle, I., 2010. The effects of context on liking: implications for hedonic measurements in new product development. In: Jaeger, S.R., MacFie, H. (Eds.), Consumer-Driven Innovation in Food and Personal Care Products. Woodhead Publishing Ltd, Cambridge, pp. 175–218.

Delarue, J., Sieffermann, J.M., 2004. Sensory mapping using Flash profile. Comparison with a conventional descriptive method for the evaluation of the flavour of fruit dairy products. Food Quality and Preference 15, 383—392.

Dijksterhuis, G., Byrne, D.V., 2005. Does the mind reflect the mouth? Sensory profiling and the future. Critical Reviews in Food Science and Nutrition 45, 527—534.

Duda, M.D., 2016. The Fallacy of Online Surveys: No Data Are Better Than Bad Data. Responsive Management. Issue, May 2010.

Duffey, K., Gordon-Larsen, P., 2010. Food price and diet and health outcomes: 20 years of the CARDIA Study. Archives of Internal Medicine 170, 420—427.

Earle, M.D., Earle, R.L., 1998. Creating New Foods the Product Developer's Guide. CH3. Product Strategy Development: Idea Generation and Screening. Chandos Publishing. http://www.nzifst.org.nz/creatingnewfoods/idea_generation3.htm.

Edmunds, H., 1999. The Focus Group Handbook. NTC Contemporary Publishing Group, Chicago, USA.

FAO/WHO, 2003. Diet, Nutrition and the Prevention of Chronic Diseases. Pages 1—149 in WHO Technical Report Series 916. Food and Agriculture Organization of the United Nations (FAO), Rome, Italy, and World Health Organization (WHO), Geneva, Switzerland.

Fellendorf, S., O'Sullivan, M.G., Kerry, J.P., 2015. Impact of varying salt and fat levels on the physiochemical properties and sensory quality of white pudding sausages. Meat Science 103, 75—82.

Fellendorf, S., O'Sullivan, M.G., Kerry, J.P., 2016a. The reduction of salt and fat levels in black pudding and the effects on physiochemical and sensory properties. International Journal of Food Science and Technology (Online).

Fellendorf, S., O'Sullivan, M.G., Kerry, J.P., 2016b. Effect of using replacers on the physico-chemical properties and sensory quality of low salt and low fat white puddings. European Food Research and Technology (Online).

Fellendorf, S., O'Sullivan, M.G., Kerry, J.P., 2016c. Impact of using replacers on the physico-chemical properties and sensory quality of reduced salt and fat black pudding. Meat Science 113, 17—25.

Fellendorf, S., O'Sullivan, M.G., Kerry, J.P., 2016d. Impact on the physicochemical and sensory properties of salt reduced corned beef formulated with and without the use of salt replacers. Meat Science (submitted for publication).

Frandsen, L.W., Dijksterhuis, G., Brockhoff, P., Nielsen, J.H., Martens, M., 2003. Subtle differences in milk: comparison of an analytical and an affective test. Food Quality and Preference 14, 515—526.

Frankovic, K., 2015. Is the only poll that counts the one on Election Day? Research World 55, 16—19.

Graham, D., Bachmann, T.T., 2004. Ideation: The Birth and Death of Ideas. John Wiley & Sons, Ltd, New Jersey.

Green, P.E., Rao, V.R., August 1971. Conjoint measurement for quantifying judgmental data. Journal of Marketing Research 8, 355—363.

Grunert, K.G., Jensen, B.B., Sonne, A.M., Brunsø, K., Byrne, D.V., Clausen, C., Friis, A., Holm, L., Hyldig, G., Kristensen, N.H., Lettl, C., Scholderer, J., 2008. User-oriented innovation in the food sector: relevant streams of research and an agenda for future work. Trends in Food Science and Technology 19, 590—602.

Hauser, J.R., Rao, V., 2004. Conjoint analysis, related modeling, and applications. In: Green, P.E., Wind, Y. (Eds.), Advances in Marketing Research: Progress and Prospects, 2004. Springer Science & Business Media, pp. 141—168.

von Hippel, E., Katz, R., 2002. Shifting innovation to users via toolkits. Management Science 48, 821–833.

Hempel, A., O'Sullivan, M.G., Papkovsky, D., Kerry, J.P., 2013a. Use of optical oxygen sensors to monitor residual oxygen in pre- and post-pasteurised bottled beer and its effect on sensory attributes and product acceptability during simulated commercial storage. LWT-Food Science and Technology 50, 226–231.

Hempel, A., O'Sullivan, M.G., Papkovsky, D., Kerry, J.P., 2013b. Non-destructive and continuous monitoring of oxygen levels in modified atmosphere packaged ready-to-eat mixed salad products using optical oxygen sensors. Journal of Food Science 78, S1057–S1062.

Hempel, A., O'Sullivan, M.G., Papkovsky, D., Kerry, J.P., 2013c. Use of smart packaging technologies for monitoring and extending the shelf-life quality of modified atmosphere packaged (MAP) bread: application of intelligent oxygen sensors and active ethanol emitters. European Food Research and Technology 237, 117–124.

Henneberry, S., O'Sullivan, M.G., Kilcawley, K.N., Kelly, P.M., Wilkinson, M.G., Guinee, T.P., 2016. Sensory quality of unheated and heated Mozzarella-style cheeses with different fat, salt and calcium levels. International Journal of Dairy Science 69, 38–50.

Hektner, J.M., Schmidt, J.A., Csikszentmihalyi, M., 2007. Experience Sampling Method: Measuring the Quality of Everyday Life. SAGE Publications, Inc.

International Obesity Task Force, 2002. About Obesity: Incidence, Prevalence & Co-morbidity.

van Kleef, E., van Trijp, H.C.M., Luning, P., 2005. Consumer research in the early stages of new product development: a critical review of methods and techniques. Food Quality and Preference 16, 181–201.

Köster, E.P., 2003. The psychology of food choice: some often encountered fallacies. Food Quality and Preference 14, 359–373.

Köster, E.P., 2009. Diversity in the determinants of food choice: a psychological perspective. Food Quality and Preference 20, 70–82.

Krystallis, A., 2007. CH8. Using Means-End Chains to Understand Consumer's Knowledge Structures. In: Macfie, H. (Ed.). Woodhead, Cambridge, UK, pp. 158–196.

Lassoued, N., Delarue, J., Launay, B., Michon, C., 2008. Baked product texture: correlations between instrumental and sensory characterization using flash profile. Journal of Cereal Science 48, 133–143.

Loescher, E., Sieffermann, J.M., Pinguet, C., Kesteloot, R., Cuvlier, G., 2001. Development of a List of Textural Attributes on Pear/apple Puree and Fresh Cheese: Adaptation of the Quantitative Descriptive Analysis Method and Use of Flash Profiling, 4th Pangborn, Dijon, France.

Lustig, R.H., Schmidt, L.L., Brondis, C.D., 2012. Public health: the toxic truth about sugar. Nature 482, 27–29.

MacFie, H.J., Bratchell, N., Greenhoff, K., Vallis, L.V., 1989. Designs to balance the effect of order of presentation and first-order carry-over effects in hall tests. Journal of Sensory Studies 4, 129–148.

Melanson, E.L., Astrup, A., Donahoo, W.T., 2009. The relationship between dietary fat and fatty acid intake and body weight, diabetes, and the metabolic syndrome. Annals of Nutrition and Metabolism 55 (1–3), 229–243.

Michon, C., O'Sullivan, M.G., Delahunty, C.M., Kerry, J.P., 2010a. Study on the influence of age, gender and familiarity with the product on the acceptance of vegetable soups. Food Quality and Preference 21, 478–488.

Michon, C., O'Sullivan, M.G., Delahunty, C.M., Kerry, J.P., 2010b. The investigation of gender related sensitivity differences in food perception. Journal of Sensory Studies 24, 922–937.

Michon, C., O'Sullivan, M.G., Sheehan, E., Delahunty, C.M., Kerry, J.P., 2010c. Investigation of the influence of age, gender and consumption habits on the liking for jam-filled cakes. Food Quality and Preference 21, 553−561.

Montoya-Weiss, M.M., Calantone, R.J., 1994. Determinants of new product performance: a review and meta analysis. Journal of Product Innovation Management 11 (5), 397−417.

Moskowitz, H.R., Beckley, J.H., Resurreccion, A.V.A., 2006. Sensory and Consumer Research in Product Development and Design. Blackwell Publishing, Iowa, USA.

Muñoz, A.M., 2002. Sensory evaluation in quality control: an overview, new developments and future opportunities. Food Quality and Preference 13, 329−339.

Niebylski, M.L., Redburn, K.A., Duhaney, T., Campbell, N.R., 2015. Healthy food subsidies and unhealthy food taxation: a systematic review of the evidence. Nutrition 31 (6), 787−795.

O'Callaghan, T., O'Sullivan, M.G., Kerry, J.P., Kilcawley, K.N., Stanton, C., 2016. Characteristics, composition and sensory properties of butter from cows on pasture versus indoor feeding systems. Journal of Dairy Science (submitted for publication).

O'Sullivan, M.G., Kerry, J.P., Byrne, D.V., 2011. Use of sensory science as a practical commercial tool in the development of consumer-led processed meat products. In: Kerry, J.P., Kerry, J.F. (Eds.), Processed Meats. Woodhead Publishing Ltd, United Kingdom.

O'Sullivan, M.G., 2011. Ch 25, case studies: meat and poultry. In: Kilcast, D., Subramaniam, P. (Eds.), Food and Beverage Shelf-Life and Stability. Woodhead Publishing Limited, Cambridge, UK.

O'Sullivan, M.G., 2016. CH18. The Stability and Shelf Life of Meat and Poultry. The Stability and Shelf Life of Food. In: Subramaniam, P. (Ed.). Elsevier Academic Press, Ltd, Oxford, UK.

Powell, R.A., Single, H.M., 1996. Focus groups. International Journal for Quality in Health Care 8, 499−504.

Rason, J., Lebecque, A., leger, L., Dufour, E., 2003. Delineation of the sensory characteristics of traditional dry sausage. I − Typology of the traditional workshops in Massif Central. In: The 5th Pangborn Sensory Science Symposium, July 21−24. Boston, USA.

Richter, V., Almeida, T., Prudencio, S., Benassi, M., 2010. Proposing a ranking descriptive sensory method. Food Quality and Preference 21 (6), 611−620.

Stage-Gate, 2016. Innovation Process. Stage-Gate® Idea-to-Launch Model. http://www.stage-gate. com/resources_stage-gate_full.php.

Stewart-Knox, B., Mitchell, P., 2003. What separates the winners from the losers in new food product development? Trends in Food Science and Technology 14, 58−64.

Stone, H., Sidel, J.S., 1993. Sensory Evaluation Practices, second ed. Academic Press, Orlando, FL, p. 327.

Stone, H., Sidel, J.L., 2007. CH13. Sensory research and consumer-led food product development. In: Macfie, H. (Ed.), Consumer-Led Food Product Development. Woodhead, Cambridge, UK, pp. 307−320.

Tobin, B.D., O'Sullivan, M.G., Hamill, R.M., Kerry, J.P., 2012a. Effect of varying salt and fat levels on the sensory quality of beef patties. Meat Science 4, 460−465.

Tobin, B.D., O'Sullivan, M.G., Hamill, R.M., Kerry, J.P., 2012b. Effect of varying salt and fat levels on the sensory and physiochemical quality of frankfurters. Meat Science 92, 659−666.

Tobin, B.D., O'Sullivan, M.G., Hamill, R.M., Kerry, J.P., 2013. The impact of salt and fat level variation on the physiochemical properties and sensory quality of pork breakfast sausages. Meat Science 93, 145−152.

Van Kleef, E., Van Trijp, H.C.M., 2007. CH14. Opportunity identification in new product development and innovation in food product development. In: Macfie, H. (Ed.), Consumer-Led Food Product Development. Woodhead, Cambridge, UK, pp. 321–338.

Varela, P., Ares, G., 2012. Sensory profiling, the blurred line between sensory and consumer science. A review of novel methods for product characterization. Food Research International 48, 893–908.

Verbeke, W., Pérez-Cueto, F.J.A., De Barcellos, M.D., Krystallis, A., Grunert, K.G., 2010. European citizen and consumer attitudes and preferences regarding beef and pork. Meat Science 84, 284–292.

WHO, 2003. Diet, Nutrition and the Prevention of Chronic Diseases. Report of a Joint WHO/FAO Expert Consultation. In: WHO Technical Report Series, vol. 919, 148 pp. http://www.who.int/dietphysicalactivity/publications/trs916/summary/en/print.html.

Worch, T., Ennis, J.M., 2013. Investigating the single ideal assumption using Ideal Profile Method. Food Quality and Preference 29, 40–47.

Worch, T., Crine, A., Gruel, A., Lê, S., 2014. Analysis and validation of the Ideal Profile Method: application to a skin cream study. Food Quality and Preference 32, 132–144.

Yarlagadda, A., Wilkinson, M.G., Ryan, S., Doolan, A.I., O'Sullivan, M.G., Kilcawley, K.N., 2014. Utilisation of a cell free extract of lactic acid bacteria entrapped in yeast to enhance flavour development in Cheddar cheese. International Journal of Dairy Science Technology 67, 21–30.

Yarlagadda, A., Wilkinson, M.G., O'Sullivan, M.G., Kilcawley, K.N., 2014. Utilisation of microfluidisation to enhance enzymatic and metabolic potential of lactococcal strains as adjuncts in Gouda type cheese. International Dairy Journal 38, 124–132.

Yusop, S.M., O'Sullivan, M.G., Kerry, J.F., Kerry, J.P., 2009a. Sensory evaluation of Indian-style marinated chicken by Malaysian and European naïve assessors. Journal of Sensory Studies 24, 269–289.

Yusop, S.M., O'Sullivan, M.G., Kerry, J.F., Kerry, J.P., 2009b. Sensory evaluation of Chinese-style marinated chicken by Chinese and European naïve assessors. Journal of Sensory Studies 24, 512–533.

Zakrys-Walliwander, P.I., O'Sullivan, M.G., Allen, P., O'Neill, E.E., Kerry, J.P., 2010. Investigation of the effects of commercial carcass suspension (24 and 48 h) on meat quality in modified atmosphere packed beef steaks during chill storage. Food Research International 43, 277–284.

Zakrys-Waliwander, P.I., O'Sullivan, M.G., Walshe, H., Allen, P., Kerry, J.P., 2011. Sensory comparison of commercial low and high oxygen modified atmosphere packed sirloin beef steaks. Meat Science 88, 198–202.

Part III

Case Studies: Sensory and Consumer Driven NPD in Action

Chapter 11

Sensory Properties Affecting Meat and Poultry Quality

INTRODUCTION

The microbiological safety and stability of meat products (also poultry) is the primary parameter that must be determined before sensory quality factors can be even considered. The first default criteria for any producer must be that products must be fit for human consumption and this is governed by the regulatory authorities, which in Europe is the European Commission and in the United States is the FDA. The shelf life of products is thus determined by microbial loadings, the limits of which determine the length of time a product can legally be sold and or consumed by, with cut-off levels set by the above-mentioned or similar regulators. However, meat is not only affected by microbial deterioration, but also negative physicochemical reactions principally through oxidation processes. The first sensory parameter that can be affected by oxidation is appearance which is the initial queue that the consumer has in determining quality, of in particular, red meat products. The cherry red pigment oxymyoglobin can become oxidised to the brown pigment metmyoglobin with negative sensory consequences. Similarly, polyunsaturated fatty acids present in meat products including poultry, pork, lamb and beef can become oxidised with the generation of off-flavours which are negatively perceived by the consumer. Finally, meat proteins can be oxidised resulting in an increase in sensory perceived toughness. These deleterious reactions to sensory quality can be countered using antioxidants and packaging solutions and sometimes a compromise must be made as the packaging solution itself (i.e. high oxygen modified atmosphere packaging (MAP)) can ultimately accelerate oxidation processes in meat (pigment, flavour and protein). Thus packaging effects on sensory properties are presented as well as future developments in packaging technologies such as active and intelligent packaging (IP). Additionally, salt and nitrate are important functional and preservative ingredients in processed meats but have negative health implications. Reduction strategies in processed meats for these ingredients are thus of interest and in the context of nutritional and sensory optimisation.

A Handbook for Sensory and Consumer-Driven New Product Development.
http://dx.doi.org/10.1016/B978-0-08-100352-7.00011-7
225

MICROBIOLOGICAL STABILITY

The microbial loading of meat and poultry products will determine food safety but also affect shelf life. European Commission regulation No 2073/2005 outlines the microbial testing criteria for meat and poultry products. The regulation specifies the maximum counts for pathogenic microorganisms in meat and poultry products, e.g. minced meat products and meat preparations made from poultry meat placed on the market during their shelf life and intended to be eaten cooked should be completely absent for *Salmonella* in 25 g of product (EC No 2073/2005). The microorganisms that are principally found on the surface of animal carcasses are Gram-negative bacteria such as *Acinetobacter, Aeromonas, Pseudomonas, Moraxella, Enterobacter* and *Escherichia*. Gram-positive organisms such as Bronchotrix, other lactic acid bacteria (LAB) and Micrococcaceae can also be found. The microorganisms quantified, in order to set shelf life limits, will depend very much on the product. Each specific meat product will have its own defined set of microbiological tests that must be used to quantify shelf life. Comprehensive specific tolerance limits are set for microbial loading by which meat and poultry production must comply in the European regulatory guidelines, EC 2073/2005 and have been explained in detail in Chapter 6. Essentially the species and population of microorganisms on meat are influenced by animal species; animal health, handling of live animals, slaughter practices; plant and personnel sanitation, and carcass chilling; fabrication sanitation, type of packaging, storage time and storage temperature (Nottingham, 1982; Grau, 1986; McMillin, 2008). Challenge testing can be used to determine the likelihood of the growth of particular microorganisms, pathogens or resistant spoilage organisms, by inoculating selected microorganisms into products and the growth monitored through a storage test (Kilcast and Subramaniam, 2004). Samples are analysed at specific regular time points for counts of spoilage bacteria, at specific incubation temperatures, and when a maximum is reached, a shelf life period may be determined. A safe margin of time should also be built in to the shelf life to ensure that the microbial count limits are not reached during the normal shelf life of the product (O'Sullivan, 2016, 2011).

For European legislation, shelf life is essentially the 'date of minimum durability' or the date until which a foodstuff retains its specific properties when properly stored, and is defined under Council Directive 2000/13/EC on the labelling, presentation and advertising of foodstuffs. The date mark in European countries itself can be a 'best before' or 'use-by date'. The 'use-by' date will indicate the date up until which the product can be safely consumed. Therefore, unlike the 'best-before' date, the accurate determination of the 'use-by' date to ensure product safety is critical (EC No, 2073/2005). The 'best-before' date will reflect the quality, e.g. taste, aroma, appearance rather than safety of a food product. A food which has passed its 'best-before' date may not necessarily be unsafe to consume but it may no longer be of optimum quality. Typically, a 'best-before' date is required on products such as canned, dried and frozen foods (FSAI, 2005). Food products which, from a microbiological point of view, are highly perishable and are therefore likely, after a short period of time, to constitute a danger to human health must have a 'use-by' date (EC No, 2000/13/EC).

The way in which meat is packed directly affects the microorganisms that grow over the course of product shelf life. High O_2 MAP is used to maintain the cherry red bloom of red meats in order to make them appealing to consumers. However, oxygen also stimulates the growth of aerobic bacteria and inhibits the growth of anaerobes (McMillin, 2008). On the other hand, CO_2 inhibits microbial growth of aerobic spoilage organisms by penetrating membranes and lowering intracellular pH in refrigerated storage, and is used typically at 20–40% CO_2 in MAP (Clark and Lentz, 1969; Smith et al., 1990). One major concern in MAP containing CO_2 is the inhibition of normal aerobic spoilage bacteria and the possible growth of psychrotrophic food pathogens, which may result in the food becoming unsafe for consumption before it appears to be unacceptable regarding sensory quality (Devlieghere et al., 2003). From a microbial loading perspective, Zakrys-Walliwander et al. (2010) showed that microbiological growth of LAB was the highest for commercially packaged sirloin steaks (75% O_2, 25% CO_2, 5% N_2) in comparison to non-commercially packaged samples ($O_2$80, $O_2$70 and $O_2$50). Thus, LAB bacteria were dominant in MAP meats, and due to their metabolic activity, the spoilage appeared as off-flavours and off-odours. Consequently, commercially packaged beef steaks were the least acceptable as determined by sensory naive assessors in comparison to other MAP samples. This is an important finding which suggests that slightly better plant hygiene in this case, especially with respect to LAB bacteria, could have a beneficial effect on the subsequent consumer quality of the meat packaged under MAP conditions (Zakrys-Walliwander et al., 2010; O'Sullivan, 2016, 2011).

Processed meat shelf life can be extended significantly by curing as the nitrate/nitrite has a strong antibacterial effect, most importantly against *Clostridium botulinum* and contributes to control of other microorganisms such as *Listeria monocytogenes* (Sebranek and Bacus, 2007). Processed meat products that contain nitrites include bacon, bologna, corned beef, frankfurters, luncheon meats, ham, fermented sausages, shelf-stable canned, cured meats, perishable canned, cured meat (e.g. ham) and a variety of fish and poultry products (Pennington, 1998). As well as contributing to sensory properties, particularly appearance and flavour, the addition of nitrates/nitrites can retard lipid oxidation, as they are potent antioxidants and thus virtually eliminate oxidative off-flavour development.

FACTORS AFFECTING MEAT COLOUR

Consumers consider colour the primary purchase criterions of fresh meat and thus relate the bright red colour to freshness, while discriminating against meat that has turned brown in colour (Hood and Riordan, 1973; Morrissey et al., 1994). Carpenter et al. (2001) noted a strong association between colour preference and purchasing intent with consumers discriminating against beef that is not red (i.e. beef that is purple or brown). This is more an issue with red meat due to the greater concentration of myoglobin present in the muscle compared to chicken, but poultry such as turkey, duck and ostrich have higher concentrations of myoglobin which can also make them susceptible to discolouration during retail display. Discolouration in retail meats during display conditions may occur as a combined function of muscle pigment oxidation (oxymyoglobin to metmyoglobin) and lipid oxidation in membrane phospholipids (Sherbeck et al., 1995). Myoglobin pigment is responsible for meat colour and can exist in three forms. Deoxymyoglobin (purple), is rapidly oxygenated to oxymyoglobin (cherry red) on exposure to air, but as meat ages, the myoglobin oxidises to metmyoglobin (brown) which denotes a lack of product freshness (Kropf, 1993). Beef is high in myoglobin and thus is more red than lamb or pork colour due to the proportionate lower levels of the myoglobin pigment in these respective muscles. The degree of meat redness depends on the state of myoglobin oxygenation or oxidation (Fig. 11.1.) High-oxygen atmospheres (80% O_2), which are typically used to package beef cuts, promote pigment oxygenation, which prolongs the period of time before metmyoglobin is visible on the muscle surface (O'Sullivan, 2016, 2011). Thus, the principal method of improving the sensory colour quality of meat is through the use of MAP, because consumers use meat colour as an indicator of freshness and wholesomeness (Mancini and Hunt, 2005). MAP usually contain mixtures of two or three gases: O_2 (to enhance colour stability), CO_2 (to inhibit microbiological growth) and N_2 (to maintain pack shape) (Sorheim et al., 1999; Jakobsen and Bertelsen, 2000). In European countries such as Ireland,

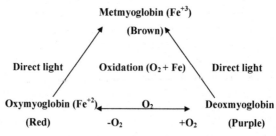

FIGURE 11.1 Oxidation of oxymyoglobin and deoxymyoglobin to metmyoglobin (O'Sullivan and Kerry, 2008).

the United Kingdom and France, beef steaks are commonly displayed in 70 mL O_2 and 30 mL CO_2 per 100 mL pack gas in MAP, whereas the concentrations used in the United States are 80 mL O_2 and 20 mL CO_2 per 100 mL pack gas (O'Sullivan et al., 2010). The drawback to high O_2 MAP is that although it maintains redness of the meat during storage, rancidity often develops while colour is still desirable (Jayasingh et al., 2002; Jackson et al., 1992).

Meat in vacuum skin packaging (VSP) is being used increasingly by processors and anecdotal evidence suggests that consumers have widely accepted this format due to its convenience. VSP minimises drip formation and air pockets and wrinkles by heating the upper cover film and making it shrink tightly around the meat. The product appears very differently to MAP-displayed meat in that the meat is purple in colour, similar to vacuum-packaged meat and also with a considerably longer shelf life. Consumers can freeze the product if they decide not to consume it immediately without having to repackage (O'Sullivan, 2016, 2011). Since the product is displayed in the myoglobin (purple) state, there is no loss of colour in the display case and oxidation issues are minimised using this packaging format (Belcher, 2006). A recent Scandinavian study found that consumers were willing to pay more for VSP meat than for various types of gas-packaged meat (Aaslyng et al., 2010).

Red muscles contain higher proportions of myoglobin than white muscles and are thus more susceptible to pigment oxidation. Thus, over time, the cherry red colour of oxymyoglobin is oxidised to the grey-brown pigment of met-myoglobin (Fig. 11.1). Oxymyoglobin is a haeme protein in which iron exists in the ferrous form (Fe^{+2}), unlike metmyoglobin that possesses the ferric form (Fe^{+3}). The conversion of the ferrous to the ferric form is a result of oxidation (Liu et al., 1995; Yin et al., 1993). Free iron and copper accelerate the autooxidation (Snyder and Skrdlant, 1966) and the photooxidation of oxy-myoglobin (Assef et al., 1971). Oxymyoglobin may be maintained in meat by also delaying oxidation to metmyoglobin (Lynch et al., 1999; O'Sullivan et al., 2003d). This can be achieved by antioxidants retarding this autocatalytic chain reaction. These antioxidants can be incorporated into meat through the animal feed. α-Tocopherol acetate may be added with feed concentrate, but is also present in quite high concentrations in pasture grass. The colour, but also the lipid stability of meat can be thus improved by vitamin E supplementation in animal diets (Faustman and Cassens, 1990; Morrissey et al., 1994; Lanari et al., 1995; O'Sullivan et al., 2002b). However, Carpenter et al. (2001) showed that once a decision to purchase beef is made in the market by the consumer, whether the beef is cherry red fresh-bloomed beef, the brown of discounted beef, or the purple of vacuum-packaged beef, eating satisfaction at home will depend only on the beef quality attributes of tenderness, juiciness and flavour (Carpenter et al., 2001).

FACTORS AFFECTING MEAT FLAVOUR

Over 1000 volatile compounds have been identified in cooked meat (Elmore and Mottram, 2009). Meat flavour is a combination of taste and odour; however, mouthfeel and juiciness of meat also affect the individual flavour perception (Farmer, 1992; Robbins et al., 2003). Meat flavour is characteristic of volatiles produced as a result of reactions of nonvolatile components that are induced thermally (Khan et al., 2015). Uncooked meat has little or no aroma and only a blood-like taste. Only after cooking and a series of thermally induced complex reactions that occur between many different nonvolatile compounds of the lean and fatty tissues does meat become flavoursome (Mottram, 1998; Calkins and Hodgen, 2007). Fats and additionally low-molecular-weight water-soluble compounds constitute the most important precursors of cooked meat flavour (Resconi et al., 2013). The broad array of flavour compounds found in meat include hydrocarbons, aldehydes, ketones, alcohols, furans, thriphenes, pyrrols, pyridines, pyrazines, oxazols, thiazols, sulphurous compounds and numerous others (MacLeod, 1994). Many of these sulphur compounds contribute sulphurous, onion-like and, sometimes, meaty aromas (Fors, 1983). Roast flavours in foods are usually associated with the presence of heterocyclic compounds such as pyrazines, thiazoles and oxazoles (Mottram, 1998).

Additionally, lipid oxidation, Maillard's reaction, interaction of lipid oxidation products with Maillard's reaction products and vitamin degradation are thermally induced reactions producing volatile flavour components responsible for the characteristic cooked meat aroma (MacLeod, 1994).

Typically, fresh red meats are stored in MAP containing 80% O_2:20% CO_2 (Georgala and Davidson, 1970) and cooked meats are stored in 70% N_2:30% CO_2 (Smiddy et al., 2002). Zakrys et al. (2008) reported that sensory panellists expressed a preference for cooked beef steaks stored in packs containing $O_2$50% and $O_2$80%, despite detecting oxidised flavours under these conditions. In general, muscle foods are susceptible to oxidative activity of their lipid, protein, pigment, vitamin and carbohydrate composition (Kanner, 1994). Additionally, there is evidence that off-flavours or taints may develop in MAP meat due to CO_2 dissipation into tissue and the formation of carbonic acid (Nattress and Jeremiah, 2000; O'Sullivan et al., 2010).

Polyunsaturated fat levels vary with species and are higher in poultry meat followed by pork, lamb and beef. Thus meats like chicken are particularly susceptible to lipid oxidation. Duck meat has higher lipid content than chicken and turkey meat and is more susceptible to oxidation as it contains high levels of unsaturated fatty acids (around 60% of total fatty acids) and also high levels of haemoglobin and myoglobin (Baéza et al., 2002). The oxidation of polyunsaturated fatty acids in meat causes the rapid development of meat rancidity and also affects colour, nutritional quality and meat texture (Kanner, 1994; Zakrys et al., 2009). The products of fatty acid oxidation produce off-flavours and odours usually described as rancid (Gray and Pearson, 1994). As rancid flavours develop, there is also a loss of desirable flavour notes; however, it is difficult to determine the limiting point at which beef can be rejected due to lipid oxidation, based on sensory perceptions (Campo et al., 2006).

The proportion of polyunsaturated fat in muscle can be also altered by feed intervention. Feeding a more unsaturated diet will result in greater concentrations of polyunsaturated fat within the muscle tissue which in turn affects meat flavour and reduces oxidative stability. Antioxidants are compounds that inhibit or retard the free radical generating chain mechanism of lipid oxidation (O'Sullivan et al., 1998). Many studies have been conducted on the basis that incorporation of α-tocopherol into the cell membrane will stabilise the membrane lipids and consequently enhance the quality of meat during storage. Oxidation studies with chicken (Jensen et al., 1995), turkeys (Marusich et al., 1975), pigs (O'Sullivan et al., 1997, 1998), cattle (Faustman et al., 1989), veal (Shorland et al., 1981) and fish (Frigg et al., 1990) have all demonstrated reduced lipid oxidation in muscles and adipose tissue from animals supplemented with dietary α-tocopherol compared to the same muscles from non-supplemented animals.

The oxidation of fatty acids occurs due to the exposure to O_2 and is accelerated in the presence of light and catalysts, such as free iron, and similar to the mechanism described previously for pigment oxidation (O'Sullivan and Kerry, 2011). The oxidation of unsaturated fatty acids is a free-radical chain reaction with three stages. (1) initiation, the formation of free-radicals; (2) propagation, the free-radical chain reactions; (3) termination, the formation of nonradical products (Tappel, 1962; O'Sullivan and Kerry, 2008). Preformed fatty acid hydroperoxides react with haeme compounds and undergo homolytic decomposition (Fig. 11.2). The alkoxy (LO•) radical formed in turn propagates the peroxidation reaction. Lipid hydroperoxides may also be decomposed by ferrous iron (Fig. 11.3), which form very reactive alkoxy radicals. However, ferric iron produces the less reactive peroxy (LOO•) radicals from fatty acid hydroperoxides (Ingold, 1962). Transition metals, notably iron, are believed to be pivotal in the generation of species capable of abstracting a proton from an unsaturated fatty acid (Gutteridge and Halliwell, 1990; Kanner, 1994). Seman et al. (1991) suggested that ferritin may be responsible for catalysing lipid peroxidation in muscle foods. Model systems with water-extracted muscle residues implied that myoglobin was not the principal prooxidant in meat and that nonhaeme iron was the main catalyst (Sato and Hegarty, 1971; Tichivangana and Morrissey, 1985).

1. Initiation:

$$\text{LH (unsaturated fatty acid)} + O_2 \rightarrow L\cdot + \cdot OOH$$

2. Propagation:

$$L\cdot + O_2 \rightarrow LOO\cdot$$

$$LH + LOO\cdot \rightarrow LOOH + L\cdot$$

$$LOOH \rightarrow LO\cdot + \cdot OH$$

3. Termination:

$$L\cdot + L\cdot \rightarrow L\text{-}L$$

$$L\cdot + LOO\cdot \rightarrow LOOL$$

$$LOO\cdot + LOO\cdot \rightarrow LOOL + O_2$$

FIGURE 11.2   Diagram displaying the mechanism and stages of lipid oxidation (O'Sullivan and Kerry, 2008).

$$LOOH + Fe^{2+} \rightarrow LO\cdot + OH^- + Fe^{3+}$$

$$LOOH + Fe^{3+} \rightarrow LOO\cdot + OH^+ + Fe^{2+}$$

FIGURE 11.3   Diagram displaying the decomposition of lipid hydroperoxides by ferrous iron and the resultant formation of the very reactive alkoxy radicals (O'Sullivan and Kerry, 2008).

FACTORS AFFECTING MEAT TENDERNESS

It has been shown that a certain level of tenderness is crucial in order that meat quality can be acceptable (Huffman et al., 1996) and that tenderness of beef is such an important quality attribute to consumers that they are willing to pay more for this tenderness (Boleman et al., 1997). Tenderness has often been described as the most important factor in terms of high eating quality, especially in beef and is principally an issue with red meats compared to poultry. Tenderness of meat is also affected by ageing, type of rigor (Pearson, 1987), chilling, freezing and storage. It is well established that stretching of the muscle by certain hanging methods (aitch bone suspension) improves tenderness of meat. Stretched muscle has greater sarcomere lengths resulting in increased tenderness (Fisher et al., 2000).

Protein oxidation (PO) is a closely associated deteriorative process occurring in meat, although relatively little is known about the repercussions of the latter on the quality of meat products (Rhee and Ziprin, 1987; Estevez and Cava, 2004). PO may lead to decreased eating quality (Xiong, 2000). Rowe et al. (2004) reported that increased PO during the first 24-h postmortem can substantially decrease beef tenderness even in steaks aged 14 days. High O_2 MAP has been shown to be detrimental to beef tenderness (Decker et al., 1993; Seyfert et al., 2005; Torngren, 2003; Zakrys et al., 2010, 2011, 2008; Zakrys-Waliwander et al., 2012) and pork due to protein cross-linking caused by oxidative processes (Lund et al., 2007; McMillin, 2008). Zakrys et al. (2010, 2011) showed that PO increased cooked meat toughness due to storage in higher MAP oxygen atmospheres. Lund et al., (2007) investigated the effect of MAP (70% O_2/30% CO_2) and skin packaging (no oxygen) on PO and texture of pork *M. longissimus dorsi* muscle (stored for 14 days at 4°C). SDS−PAGE data revealed cross-linking of myosin heavy chain through disulphide bonding, and the content of protein thiols was reduced indicating PO. The high oxygen atmosphere stored meat thus had lower tenderness and juiciness.

The consumer is willing to pay a higher price in the marketplace for beef as long as it is guaranteed to be tender (Miller et al., 2001). For many scientists, tenderness is considered the most important qualitative characteristic of meat (Savell et al. 1987, 1989; Smith et al., 1987; Destefanis et al., 2008). As such, the requirements for colour stability of high O_2 MAP meat must be balanced against the deteriorative action of lipid oxidation (Torngren, 2003; Zakrys et al., 2008) and any reduction in meat tenderness (O'Sullivan and Kerry, 2011).

SALT AND NITRATE REDUCTION STRATEGIES IN PROCESSED MEATS

Consumers are demanding variations of meat products that are low in salt, fat, cholesterol, nitrites and calories in general and contain in addition health-promoting bioactive components such as, for example, carotenoids, unsaturated fatty acids, sterols and fibres (Weiss et al., 2010). The use of salt, nitrates and nitrites in preserving processed food products was vital in the past, but the advent of modern packaging and refrigeration reduced its primary role and necessity.

SALT IN PROCESSED MEATS

Salt is basic to all meat curing mixtures and is the primary ingredient necessary for curing. It acts by dehydration and alters the osmotic pressure, inhibiting bacterial growth and subsequent spoilage (Pearson and Tauber, 1984). Processed meat products comprise one of the major sources of sodium in the diet in the form of sodium chloride (salt) (Desmond, 2006). Fresh meat is low in sodium but processed meats contain 2% added salt, a value that may increase to 6% in dried products. Processed meats contribute 20−30% to the daily salt (NaCl) intake in industrialised countries amounting to between 9 and 12 g/day, a much larger value than the recommended value of <5 g/d (Jiménez-Colmenero et al., 2001; WHO, 2003). Intake of dietary sodium has been linked to hypertension in about 20% of the population and consequently increased risk of cardiovascular disease (CVD). The estimated cost of CVD to both the EU and US economies is €169B and $403B, respectively (Desmond, 2006). The clear association between consumption of processed meats and the incidence of hypertension (Paik et al., 2005) confirms the importance of meat technology in relation to salt intake. Apart from the recommendation to 'limit consumption of salty foods and foods processed with salt (sodium)' (WCRF, 2007), a possible association of the processed meat—colorectal cancer relationship with the 'salt problem' should not be discarded. In Ireland and the United Kingdom, the daily sodium adult intake is approximately three times the recommended daily allowance and therefore public health and regulatory authorities are recommending reducing dietary intake of sodium to 2.4 g (6 g salt) per day (Desmond, 2006).

Meat product suppliers have already commenced reformulating their recipes, and now offer lower levels of nitrate, salt and fat, or even higher levels of polyunsaturated fatty acids in processed meat products on the market (Verbeke et al., 2010). Many studies have looked at sensory focused salt and fat reduction, without utilising replacement ingredients, in processed meats including beef patties (Tobin et al., 2012a), breakfast sausage (Tobin et al., 2013), frankfurters (Tobin et al., 2012b) white pudding (Fellendorf et al., 2015; PROSSLOW, 2016) and black pudding (Fellendorf et al., 2016a; PROSSLOW, 2016) to mention but a few. These studies sequentially reduced salt and fat, without using alternative ingredients, in order to determine sensory optima but maintaining safety, functionality and adequate shelf life. Reductions in salt content could also be optimised by using packaging technologies to compensate for loss of safety or shelf life (Fellendorf et al., 2016d; PROSSLOW, 2016).

Salt and fat replacers can offer even further possibilities with respect to reduction of salt and fat in processed meats. The basis of using salt replacers is to reduce sodium cations with, for instance, potassium, magnesium, calcium or to reduce the chloride anions with ingredients such as glutamates, phosphates, etc., as a means of providing salty tastes or flavours (Wheelock and Hobbiss, 1999). The most commonly used salt replacer is potassium chloride, although it is self-limiting due to its bitterness and metallic flavour when used above certain concentrations (Dzendolet and Meiselman, 1967). However, the bitterness perception of potassium chloride can be suppressed by using it in combination with other salt replacers. Zanardi et al. (2010) reduced successfully the sodium content in Cacciatore salami, a typical Italian dry fermented sausage, by using a mixture of KCl, $CaCl_2$ and $MgCl_2$.

Salt and fat reduced processed meats have even been proposed as delivery systems for bioactive ingredients (Tobin et al., 2014a,b). However, there is still a huge potential to produce even healthier and more sensory accepted products (Fellendorf et al., 2016c). The use of ingredient replacers such as hydrocolloids (a range of polysaccharides and proteins) could also be utilised in the meat processing industry, as they have been used in processed meat products for many years to improve properties such as water binding and texture due to their ability to thicken, gel, bind, stabilise emulsions and pH (Fellendorf et al., 2015, 2016b,c; Andrès et al., 2006). Hydrocolloids, based on animal proteins, include casein, whey, gelatin and blood-derived protein. Additionally, an enormous range of polysaccharides are available on the market, such as starches (corn, wheat, maize, potato, tapioca, pea), celluloses (carboxymethylcellulose), gums (guar, alginate, pectin, locust bean), fibres (β-glucan), chitin/chitosan and xanthan derived from microorganisms (Cutter, 2006). Recently, published studies have also presented the use of different types of edible seaweed (Sea Spaghetti, Wakame and Nori) in meat products (Cofrades et al., 2008; Jiménez-Colmenero et al., 2010; López-López et al., 2010; Fellendorf et al., 2015). Seaweeds are a rich source of minerals, trace elements, proteinaceous compounds and flavour precursors (reducing sugars) which can act as flavour enhancers, or even as reactants in the flavour developing process (Maillard reaction, caramelisation), within processed meats. Additionally, seaweeds comprise a unique taste profile, which might replenish lost flavour in reduced salt and fat-processed meat products (Hotchkiss, 2012). Furthermore, salt enhancers such as amino acids (glycine, glutamate), lactates and yeast extracts have found applications in processed meat. These substances have no salty taste themselves, although with the combination of sodium chloride, they are able to enhance the salty flavour (Desmond, 2006). The authors Guàrdia et al. (2008) substituted successfully 50% of NaCl by a mixture of KCl and potassium lactate in small fermented sausages. Recently, dos Santos et al. (2014) reported that fermented cooked sausages containing monosodium glutamate combined with lysine, taurine, disodium inosinate and disodium guanylate masked the unpalatable sensory attributes linked with the replacement of 50% and 75% NaCl with KCl.

Phosphates are very effective water binders in processed meats and can dramatically improve cook yield by increasing ionic strength and freeing negative ion sites on the meat protein surface thus increasing the binding of more water. Sodium polyphosphate and sodium tripolyphosphate are very effective water binders and contain sodium salts which contribute partially to salt taste, but the addition rate to formulations is typically below 0.05%. Barbut et al. (1988) found that sodium tripolyphosphate and other phosphates (sodium hexametaphosphate, sodium acid pyrophosphate) used at a 0.4% addition improved the emulsion stability and sensory properties of 20% and 40% reduced salt turkey frankfurters. Ruusunen et al. (1999) fond that the NaCl content of cooked bologna sausages made with added phosphates may be reduced to 1.4% added NaCl without loss of flavour. Additionally, these authors (Ruusunen et al., 2001) reduced the salt content of cooked ham to 1.7%

NaCl. However, due to consumer demand for cleaner labelled meats, the concurrent demand for products made without added phosphate is increasing. Fellendorf et al. (2016c) explored the application of traditional and clean label ingredients as salt and fat replacers in black pudding. These authors produced 22 black puddings possessing different fat (10%, 5%) and sodium (0.6%, 0.4%) levels which were used as base formulations for 11 different salt and fat replacers. Compositional, physicochemical and sensory analyses (hedonic and descriptive) were conducted. Black pudding samples with 5% fat and 0.6% sodium containing potassium chloride (KCl), potassium chloride and glycine mixture (KClG) and seaweed, respectively, and 10% fat and 0.4% sodium containing carrageen were rated higher ($p < .05$) for spiciness and saltiness. Samples with 10% fat and 0.4% sodium containing KClG were rated positively ($p < .05$) to fatness. Samples with 5% fat and 0.6% sodium containing pectin and a combination of potassium citrate, potassium phosphate and potassium chloride (KCPCl), as well as samples containing 10% fat and 0.4% sodium with waxy maize starch were liked ($p < .05$) for flavour and overall acceptance. The Food Safety of Ireland (FSAI) recommends a sodium target level of 0.6% and an even lower sodium level (0.4%) was achieved. Additionally, Fellendorf et al. (2016b) produced 22 white pudding formulations which were comprised of 2 different fat (10%, 5%) and sodium (0.6%, 0.4%) levels and containing 11 different traditional and clean label ingredient replacers. Compositional, texture and sensory analyses were conducted. Adding replacers to low-sodium and low-fat white puddings showed a range of effects on sensory and physicochemical properties. Two formulations containing 10% fat and 0.6% sodium formulated with sodium citrate, as well as the combination of potassium chloride and glycine (KClG), were found to have overall acceptance ($p < .05$) by assessors. These samples showed higher ($p < .05$) hardness values, scored lower ($p < .05$) in fatness perception and higher ($p < .05$) in spiciness perception. Hence, the recommended sodium target level of 0.6% set by the Food Safety Authority of Ireland (FSAI, 2011) was achieved for white pudding products, in addition to a significant reduction in fat level from commercial levels, without causing negative sensory attributes.

Fellendorf et al. (2016d) investigated the impact of varying sodium levels (0.2–1.0%) and salt replacers in corned beef on physicochemical, sensory (affective and descriptive) and microbiological properties. Potassium nitrite levels were kept constant for all variants and conformed to EU guidelines (95/2/EC). Significant differences in colour, hardness and cooking loss were measured. Corned beef samples low in sodium (0.2%, 0.4%) showed reduced ($p < .05$) saltiness perception, but were positively correlated to liking of flavour and overall acceptability. Samples formulated with the salt replacers $CaCl_2$, $MgCl_2$ and KCl scored higher ($p < .01$) in saltiness perceptions, but correlated negatively ($p > .05$) to liking of flavour and overall acceptability. However, a sodium reduction of 60% in corned beef was determined to be achievable as assessors liked ($p < .05$) the flavour of the sodium-reduced corned beef containing 0.4% sodium and formulated with potassium lactate and glycine (KLG), even with the noticeable lower salty taste. Sodium reduction in corned beef (packaged under modified atmosphere) did not negatively impact on the microbiological shelf life.

NITRATE AND NITRITE IN PROCESSED MEATS

The addition of nitrate-based compounds to cured meats is thought to have arisen from the salting of meats contaminated with saltpetre (KNO_3). Saltpetre is a commonly occurring impurity in salt which enhanced its preserving action and produced a red colour in the product (Pearson and Gillett, 1995; Honikel, 2008). It was subsequently adopted for the colour and flavour properties it contributed. Nowadays, the usual process for the accelerated manufacture of cured meats is to incorporate salt, nitrite, a reducing agent (such as ascorbate) and other ingredients such as seasonings (Kramlich et al., 1973). Meat products that may contain nitrites include bacon, bologna, corned beef, frankfurters, luncheon meats, ham, fermented sausages, shelf-stable canned cured meats, perishable canned cured meat (e.g. ham) and a variety of fish and poultry products (Pennington, 1998). The use of nitrates and nitrites in food products must comply with the provisions set out in Annex III part C of Directive 95/2/EC on additives other than colours and sweeteners as amended (**95/2/EC**). The regulation stipulates that potassium nitrite (E249) in nonheat-treated, cured, dried meat products must have a residual level of 50 mg/kg and sodium nitrite (E 250) a residual level in cured bacon must be 175 mg/kg. Cured meat products containing sodium nitrite (E 251) must have 250 mg/kg as a residual level. It is thus vital when considering removing nitrite/nitrate from meat products to ensure that the reformulated meat product is safe during its whole shelf life under reasonable conditions of consumer misuse (FSAI, 2012). Nitrate used to be used as the primary source for nitrite, supplied as either sodium nitrate or potassium nitrate, but nitrite is now preferred (Sebranek and Bacus, 2007). Nitrate is converted to nitrite by microbiological processes, therefore acts more slowly than nitrite and is generally not extensively used today in the accelerated manufacture of cured meats; however, it is still used in slower curing processes such as that utilised in Wiltshire ham manufacture and may be used in modern cured meat manufacture as a reservoir source for nitrite replacement in meats as it decomposes or becomes more depleted within the cured product. Nitrate and nitrite have strong antibacterial effects, particularly in relation to the growth of *Clostridium botulinum*. Nitrite is strongly inhibitory to anaerobic bacteria, like that of *C. botulinum* and contributes to the limited control of other microorganisms such as *Listeria monocytogenes* (Sebranek and Bacus, 2007). Nitrite, more stable that nitrate, is also commonly added to cured meats dissolved in water where it converts myoglobin to metmyoglobin (brown pigment). On cooking, the typical pink cured meat heat-stable pigment nitrosylhemochrome is formed. However, this colour can oxidise and fade over time thus requiring products to be packed in vacuum or oxygen-free MAP. Nitrite has powerful antioxidant properties and prevents the development of off-flavour like warmed-over flavour (WOF) or rancidity.

HUMAN HEALTH IMPACT OF NITRATE AND NITRITE

Both nitrate and nitrite can be hazardous to humans if ingested in large amounts. Since the 1970s, there has been concern about a possible link between nitrite consumption and cancer. Nitrite can cause the formation of carcinogenic N-nitrosamines in cured products due to its reaction with secondary amines and amino acids in muscle proteins. Furthermore, residual nitrite in cured meats may form nitrosamines in the gastrointestinal tract (Shahidi and Pegg, 1991). There is no conclusive evidence that nitrite is directly carcinogenic (Cantor, 1997); however, in high doses it has been implicated as a cocarcinogen (Schweinsberg and Burkle, 1985). The lethal oral doses for human beings are established as 80–800 mg nitrate/kg body weight and 33–250 mg nitrite/kg body weight (Schuddeboom, 1993). This may increase the incidence of colorectal cancer, and we are recommended to 'Limit intake of red meat and avoid processed meat' as one of the 10 universal guidelines for healthy nutrition (World Cancer Research Fund (WCRF), 2007). The consumption of red meat and, in particular, processed meats, has been related to the incidence of colorectal cancer since 1975 in several epidemiological studies, mainly in the United States and the United Kingdom (Demeyer et al., 2008). In cured meats, nitrosamines occur only in small amounts and they are easily avoidable through proper frying, grilling and pizza baking (Honikel, 2008). Although negative reports and scientific studies have demonstrated health risks associated with the consumption of processed foods, results from these reports and studies are variable and not wholly conclusive. The WCRF report, among others, published in peer-reviewed journals is subject to some criticism because the large variability in composition and nature of meat and meat products is not sufficiently taken into account, nor are processed meats defined with sufficient precision (Demeyer et al., 2008). The WCRF report itself states that 'there is no generally agreed definition of processed meat. The term is used inconsistently in epidemiological studies. Judgments and recommendations are therefore less clear than they could be' (Demeyer et al., 2008).

In most countries, the use of potassium or sodium salts is limited. Either the ingoing or the residual amounts are regulated by laws (Honikel, 2008). The curing process has been regulated in the United States, by the United States Department of Agriculture (USDA), since the early 1900s. Sodium nitrite is allowed to be added at a maximum of 156 ppm (Cassens, 1997). In Europe, the curing process is regulated by Directive, 2006/52/EC. In this directive, the use of nitrates is limited to nonheated meat products with 150 mg sodium nitrite/

kg and nitrite up to 100 mg, and 150 mg nitrite/kg meat in all meat products (Directive, 2006; Honikel, 2008).

The meat industry continues to search for alternative methods to produce nitrite-free meats that maintain the colour characteristics of nitrite-cured meat products (Zhang et al., 2007). Acceptable alternatives for the use of nitrate and nitrite exist in relation to colour development, flavour and microbiological safety (Demeyer et al., 2008). Zhang et al. (2007) showed that nitro-sylmyoglobin could be generated in Harbin red sausage when *L. fermentum* (AS1.1880) was inoculated into the meat batter, and the formation of a characteristic pink colour with an intensity comparable to that in nitrite-cured sausage could be achieved by its use. This treatment did not seem to have a negative impact on the product texture and flavour, although the oxidative stability and microbial shelf life require further investigation to determine limits. Sindelar et al. (2007) compared uncured, no-nitrate/nitrite-added hams, frankfurters and bacons against nitrite-cured products considered to be in-dustry standards in their respective product category. Consumer sensory dif-ferences existed between all the Brands, nonetheless, the hams tested were considered acceptable by a majority of consumers. A greater amount of variation was identified between frankfurters than hams and consumer sensory results for the bacon products revealed that a majority of the nonnitrate products had similar sensory scores to the nitrite-added control. These studies have proved promising, but greater work is required before such products become widely adopted, primarily because of the greater effectiveness of ni-trite as an antimicrobial agent and its inhibitory effects on *Clostridium* and *Listeria* species and because of its positive effects on sensory flavour and colour qualities.

Natural and organic foods are not permitted to use chemical preservatives; therefore the traditional curing agents used for cured meats, nitrate and/or nitrite cannot be added to natural and organic processed meat products. However, alternative processes that utilise ingredients with high nitrate con-tent, such as vegetable-based ingredients, and a nitrate-reducing starter culture can produce processed meats with very typical cured meat properties (Sebranek and Bacus, 2007). However, when manufacturing 'natural' and 'organic' meat products using natural ingredients, the inherent variability of natural ingredients must be considered (Sebranek and Bacus, 2007).

INNOVATIVE PACKAGING (ACTIVE, INTELLIGENT) OF MEAT PRODUCTS

Active packaging is a recent technological development which has the po-tential of extending the shelf life of meat and poultry products. Active packaging has the advantage of maintaining the preservative effects of various compounds (antimicrobial, antifungal or antioxidant), but without being in direct contact with the food product (O'Sullivan and Kerry, 2012).

This is an important development, considering the consumer drive towards clean labelling of food products and the desire to limit the use of food additives (O'Sullivan and Kerry, 2009). These technologies can be employed to improve the food safety and shelf life of processed meat products and allow the subsequent reduction of the preservative ingredients such as salt and nitrates in these processed foods. By incorporation of the preservative effects directly into packaging, preservation may be maintained which will compensate for the lesser preservative effect of the optimised processed meats (O'Sullivan and Kerry, 2011; O'Sullivan, 2016). The antimicrobial agent is incorporated into the packaging material by either spraying, coating, physical mixing or chemical binding (Berry, 2000). Chemical preservatives can be employed in antimicrobial-releasing film systems, including organic acids and their salts (sorbates, benzoates and propionates), parabens, sulphites, nitrites, chlorides, phosphates, epoxides, alcohols, ozone, hydrogen peroxide, diethyl pyrocarbonate, antibiotics and bacteriocins (Ozdemir and Floros, 2004). In meat systems, the preservative effect of active packaging can substitute for the reduced preservative effects of component ingredients such of salt or nitrate. It is important to note that nitrite cannot be completely replaced as there is a legal minimum concentration of nitrate that can be used in cured meats (O'Sullivan, 2016). By reducing the growth and spread of spoilage and pathogenic microorganisms in meat products, antimicrobial packaging materials can inhibit or kill the microorganisms and thus extend the shelf life of perishable products and enhance the safety of packaged products (Han et al., 2005).

Packages integrated with a sensor or an indicator such as time−temperature indicators, gas indicators and biosensors are regarded as IP systems because they inform the consumer about the kinetic changes related to the quality of the food or the environment that is contained within the package (Yucel, 2016). An active and IP case study is presented in Chapter 14 for an MAP product. This case study demonstrates the use of smart packaging technologies for monitoring and extending the shelf life quality of MAP bread with the application of intelligent oxygen sensors and active ethanol emitters (Hempel et al., 2013). Additionally, the case study in Chapter 13 demonstrates an intelligent oxygen sensors system for beer freshness.

EVALUATING THE SHELF LIFE OF MEAT AND POULTRY − SENSORY ANALYSIS

Meat product must be fit for human consumption which are defined by regulation and law, but there are no defined sensory limits to negative sensory parameters by which meat and poultry should be sold, in any case these would be very difficult if almost impossible to set. However, it is in the commercial interest to meat and poultry producers to sell their products with optimum sensory quality. The appearance, aroma, taste, texture and flavour quality will deteriorate over time and off-flavours are likely to form beyond a certain point

FIGURE 11.4   Retail display cabinet used to simulate commercial retail display conditions. It is important when using such equipment that the light intensity (LUX) is recorded as well as operating temperature fluctuations using a data logger.

(O'Sullivan, 2016, 2011). A refrigerated display cabinet (Fig. 11.4) under a defined Lux strength light can be used to store the test product to simulate commercial shelf life storage conditions. The test samples should then be tested at regular intervals over the shelf life as determined by microbiological safety limits (O'Sullivan, 2016, 2011).

Rapid sensory methods such as flash profiling (Chapter 4) has been demonstrated for a diverse selection of meat products including French dry sausages (Rason et al., 2003), beef patties (Tobin et al., 2012a), frankfurters (Tobin et al., 2012b) and breakfast sausages (Tobin et al., 2013). Additionally, Ranking Descriptive Analysis (Chapter 4) and Sensory Acceptance Testing (Chapter 3) have been successfully demonstrated for a number of other processed meat products including white pudding (Fellendorf et al., 2015, 2016b), black Pudding (Fellendorf et al., 2016a,c) and corned beef (Fellendorf et al., 2016d).

For Descriptive analysis (Chapter 2), the starting list of sensory terms is compiled, by individuals with sufficient product knowledge, using defined lexicons as a starting point along with any additional terms that are deemed relevant or useful. This is usually undertaken with a subset of samples, such as samples at the beginning or end of the shelf life period along with an inter-mediate sample or alternative muscle or meat cut, depending on the design, that reflect the sensory variation in the full samples set to be profiled (O'Sullivan, 2016, 2011). Sensory terms can be initially determined from lexicons which have been compiled for a large number of meat and poultry products; Byrne et al. (1999a), O'Sullivan et al. (2003b) and Chu et al. (2015) for pork, Johnson and Civille (1986), Maughan et al. (2012) and Miller (2010) for beef and Lyon (1987); Byrne et al. (1999b) for chicken. These sensory terms should fit the purpose of the analysis, correspond to the samples, have relevant definitions (Civille and Lawless, 1986; Civille, 1987; Piggott, 1991), must allow differentiation between sensations, identification of the object it describes and recognition of the object by others seeing the term (Harper et al., 1974). As described in Chapter 2, this metalist of terms is then reduced during training (Byrne et al., 1999a,b, 2001; O'Sullivan et al., 2003a,b,c). In summary, the criteria by which a sensory term can be used in a sensory profile include (1) the sensory terms selected must be relevant to the samples, (2) discriminate between the samples, (3) have cognitive clarity and (4) be nonredundant (Byrne et al., 1999a,b, 2001; O'Sullivan et al., 2003a,b,c). The final sensory profile will display quantifiable sensory changes in meat or poultry products over the course of the shelf life, which is an important tool in defining sensory shelf life changes over time but this does not reflect consumer sentiment or acceptance of these products (O'Sullivan, 2011, 2016).

CASE STUDY – SENSORY PROFILING (QUANTITATIVE DESCRIPTIVE ANALYSIS) OF PORK MEAT SAMPLES OVER DIFFERENT TIME POINTS (DAYS, SHELF LIFE) AND CORRELATION WITH PHYSICOCHEMICAL DATA

The following case study investigated the sensory and chemical assessment of pork supplemented with iron and vitamin E (O'Sullivan et al., 2003b) and is a good example of quantitative descriptive analysis of a meat product undertaken over different storage times and analysed in conjunction with physicochemical data. This research was conducted at The Department of Dairy and Food Science, The Royal Veterinary and Agricultural University (now amalgamated with Copenhagen University), Frederiksberg, Copenhagen, Denmark, in collaboration with the Danish Institute of Agricultural Sciences. The main aims of this study were to determine the effects of iron and vitamin E supplementation on animal in vivo tissue levels. The 'meat avoiding part' of the female population (especially adolescents) is at risk from iron deficiency (Ryan, 1997). This study wanted to investigate if iron levels could be boosted in pork, a cheap readily available and the most popular meat product consumed in Denmark, so that a very low consumption levels would contribute sufficiently to iron intake in the above-mentioned group at risk. Moreover, a second aim was to determine the sensory consequences of supplementation of pig diets with iron and vitamin E and the resulting effects on WOF (warmed over flavour) development (O'Sullivan et al., 2003b).

Pork muscle samples (*M. longissimus dorsi* and *M. psoas major*) were obtained from pigs given one of four dietary treatments, (1) control diet, (2) supplemental iron (7 g iron (II) sulphate/kg feed), (3) supplemental vitamin E (200 mg dL-α-tocopheryl acetate/kg of feed) and (4) supplemental vitamin E + supplemental iron. Vitamin C was supplemented to all dietary treatments to facilitate iron uptake. Vitamin E and iron tissue levels were determined for each treatment. WOF was evaluated by a trained sensory panel ($n = 8$) for the four treatments which were cooked and refrigerated at 4°C for up to 5 days.

Thawing loss, drip loss and thiobarbituric acid reactive substances (to measure lipid oxidation) were also determined (O'Sullivan et al., 2003b).

Training and Profiling

An eight-member sensory panel (four males/four females, aged from 24 to 62 years) was recruited from the public and students of The Royal Veterinary and Agricultural University (KVL), Frederiksberg, Denmark. Selection criteria for panellists were availability and motivation to participate on all days of the experiment. Sensory training was carried out prior to sensory profiling as per O'Sullivan et al. (2002a). Sensory analysis was carried out in the panel booths at the university sensory laboratory that conforms to ISO (1988) international standard. See Chapter 3 case study for sensory term reduction process for the same data set (O'Sullivan et al., 2002a).

During both training and sensory profiling, analysis was performed with assessors not having any knowledge of sample history so as not to introduce bias into their sensory assessment. Unstructured 15-cm line scales anchored on the left by the term 'none' and on the right by the term 'extreme' were used for all the sensory descriptors (Meilgaard et al., 1999). The responses of the panellists were recorded by measuring the distance in millimetre (1−150) from the left side of the scale for the odour, flavour, taste and aftertaste sensory terms (Table 11.1) (O'Sullivan et al., 2003b).

Monte Carlo Power Estimation of Experimental Design

The statistical power of the experimental design in this study was tested by Monte Carlo simulation as described by Martens et al. (2000). This method involved testing many different experimental designs with respect to balancing the risk of committing type I errors, i.e. being fooled into believing that random errors in the data represent real effects and type II errors, overlooking interesting effects as if they were just random errors. Overall, Monte Carlo power estimation of a design involves initial generation of artificial data for a number of (e.g. 10,000) hypothetical experiments, then based on the experimental design and on the assumption that each artificial data set is analysed in the same way that the future, real data is intended to be analysed and from the distributions of the obtained parameter estimates the risks associated with a given experimental design can be studied (Marten and Martens, 2000; O'Sullivan et al., 2003b).

Data Analysis

Descriptive statistical analysis was performed to determine means, standard deviations, ranges, medians and percentile distributions for each of the sensory terms. The sensory data used was the raw data averaged over assessor and replicate.

TABLE 11.1 Significance of ANOVA-Partial Least Squares Regression (APLSR)-Derived Estimated Regression Coefficients[c] for the Relationships of Sensory Terms[a] and Chemical Measurements[b] to the Design Indicators[e] as Derived by Jack-Knife Uncertainty Testing (O'Sullivan et al., 2003b)

Sensory Term[f]	Vitamin E[e] (LD)		Control (LD)		Iron Vitamin E (LD)		Iron (LD)		Vitamin E (PS)		Control (PS)		Iron Vitamin E (PS)		Iron (PS)	
1. Cardboard-O[a]	-0.07[c]	ns[d]	-0.17	*	-0.10	ns	-0.19	*	0.12	ns	0.11	ns	0.19	ns	0.12	ns
2. Linseed oil-O	-0.10	ns	-0.12	ns	-0.10	ns	-0.13	ns	0.10	ns	0.05	ns	0.19	*	0.12	ns
3. Rubber/Sulphur-O	-0.03	ns	-0.13	ns	-0.07	ns	-0.15	*	0.10	ns	0.09	ns	0.12	ns	0.08	ns
4. Nut-O	0.03	ns	0.02	ns	0.02	ns	0.02	ns	-0.02	ns	-0.01	ns	-0.04	ns	-0.02	ns
5. Green-O	-0.01	ns	0.12	ns	0.04	ns	0.13	ns	-0.08	ns	-0.09	ns	-0.06	ns	-0.05	ns
6. Fatty-O	-0.05	ns	-0.13	*	-0.08	ns	-0.14	*	0.10	ns	0.08	ns	0.14	ns	0.10	ns
7. Sweet-T	-0.01	ns	-0.13	ns	-0.05	ns	-0.15	*	0.09	ns	0.10	ns	0.09	ns	0.06	ns
8. Sour-T	-0.04	ns	0.09	ns	-0.01	ns	0.11	ns	-0.06	ns	-0.09	ns	-0.01	ns	-0.02	ns
9. Salt-T	-0.07	ns	-0.19	*	-0.10	ns	-0.21	*	0.13	ns	0.12	ns	0.19	ns	0.13	ns
10. Bitter-T	-0.04	ns	-0.04	ns	-0.04	ns	-0.05	ns	0.03	ns	0.02	ns	0.07	ns	0.05	ns
11. MSG/Umami-T	-0.09	ns	-0.21	*	-0.12	ns	-0.24	*	0.15	ns	0.13	ns	0.23	ns	0.15	ns
12. Metallic-F/Bloody-F	0.01	ns	-0.02	ns	0.01	ns	-0.02	ns	0.01	ns	0.02	ns	-0.01	ns	0.01	ns
13. Fresh cooked pork-F	0.05	ns	0.04	ns	0.05	ns	0.04	*	-0.03	ns	-0.01	ns	-0.09	ns	-0.05	ns

	1		2		3		4		5		6		7		8	
14. Rancid-F	−0.10	ns	−0.10	ns	−0.09	ns	−0.11	ns	0.08	ns	0.04	ns	0.17	ns	0.10	ns
15. Lactic acid/fresh sour-F	0.07	ns	0.17	*	0.09	ns	0.19	ns	−0.12	ns	−0.10	ns	−0.18	*	−0.12	ns
16. Vegetable oil-F	−0.06	ns	−0.07	ns	−0.06	ns	−0.07	ns	0.05	ns	0.03	ns	0.11	ns	0.07	ns
17. Piggy/Animal-F	−0.05	ns	−0.16	*	−0.08	ns	−0.17	*	0.11	ns	0.10	ns	0.14	ns	0.10	ns
18. Fish-F	−0.07	ns	−0.06	ns	−0.07	ns	−0.07	ns	0.05	ns	0.02	ns	0.13	ns	0.08	ns
19. Tinny-F	0.01	ns	0.05	ns	−0.02	ns	0.05	ns	−0.03	ns	−0.03	ns	−0.04	ns	−0.03	ns
20. Livery-F	−0.07	ns	−0.19	*	−0.11	ns	−0.21	*	0.13	ns	0.12	ns	0.19	*	0.13	ns
21. Astringent-AT	−0.01	ns	0.09	ns	−0.03	ns	0.10	ns	−0.06	ns	−0.07	ns	−0.05	ns	−0.04	ns
22. Vitamin E[b,g]	0.51	***	Na	Na	0.01	ns	−0.01	***	−0.01	ns	0.23	**	Na	Na	−0.33	***
23. Iron	−0.12	**	Na	Na	−0.14	ns	−0.15	***	0.15	ns	0.14	*	Na	Na	0.11	ns
24. Thiobarbituric acid reactive substances	−0.14	***	Na	Na	0.15	ns	−0.14	***	0.04	ns	−0.27	***	Na	Na	0.35	***

AT, aftertaste; F, flavour; O, odour; T, taste.
[a]Sensory measurement.
[b]Chemical measurement of cooked sample.
[c]Estimated regression coefficients from APLSR.
[d]Significance of regression coefficients; ns = not significant; * = $p < .05$; ** = $p < .01$; *** = $p < .001$.
[e]Sensory design indicators, LD = M. longissimus dorsi, PS = M. psoas major.
[f]2 Principal Components.
[g]4 Principal Components (Model optima-derived principal components).

Comparison of mean vitamin E and iron levels were performed using The Tukey Honestly Significant Difference test at the 5% level (Statistical Package for the Social Sciences, SPSS, Chicago, USA).

ANOVA-Partial Least Squares Regression (APLSR) was used to investigate product sensory variation in the profiling data and was performed using full cross-validation. The X-matrix was designated as 0/1 design variables for assessor, replicate and product. The Y-matrix was designated as sensory, instrumental and physical variables. In this model, assessor and replicate level effects were removed (Martens and Martens, 2000). To derive significance indications for the relationships determined in the quantitative APLSR, regression coefficients were analysed by jack-knifing which is based on cross-validation and stability plots (Martens and Martens, 1999, 2000, 2001). This allowed determination of the regression coefficients $(\overline{b})$ with uncertainty limits that correspond to ± 2 standard uncertainties estimated by leave-one-replicate-out jack-knifing, i.e. $\overline{b} \pm 2\,\overline{s}\left(\overline{b}\right)$ (Martens and Martens, 2001). From these, the significances ($p < .05$) of the variable relationships in the X- and Y-matrices were determined, i.e. $\alpha \approx 0.05$, defined as the Type I probability that the observed effects could have been caused by random measurement errors. All analyses were performed using the Unscrambler Software, version 7.6 (CAMO ASA, Trondheim, Norway) (O'Sullivan et al., 2003b).

RESULTS AND DISCUSSION

Table 11.1 displays the regression coefficients and significances for the design indicators. The significances are derived from jack-knife uncertainty testing of the regression coefficients in the APLSR model (Martens and Martens, 2000). The regression coefficients were standardised and correspond to an average standardised response to each design indicator. Fig. 11.5 displays an APLSR plot of (a. *M. longissimus dorsi*; b. *M. psoas major*) for animals fed the Vitamin E, control, iron/vitamin E and iron diets for days 0, 1, 3 and 5 of sensory profiling (O'Sullivan et al., 2003b).

For Fig. 11.5a (APLSR plot of *M. longissimus dorsi*), the validated explained variance for this model was PC1: 58% and PC2: 18%. The vitamin E-treated group covaried with the Fresh Cooked-F and Sweet-T sensory descriptors on profile day 0 and progressed, almost in a linear fashion, to and covaried with the Liver-F and Astringent-AT descriptors by day 5 of the profile. The control group covaried with Rubber-Like−O on day 0 and Tinney-F and Bitter-T by day 5. The vitamin E-treated group and to a lesser extent the control group displayed a negative covariance to the oxidative descriptors Rancid-F, Linseed Oil-Like-O and Green-O on the day 5 of profiling. The iron/vitamin E group covaried with Animal-Like-F initially on day 0, Cardboard-Like-O on day 3 and with Rancid-F by day 5. The iron group covaried with the oxidative descriptors Rancid-F and Cardboard-Like-O by profile day 5, but had a negative correlation to Fresh Sour-Like-F and nonoxidised Fat-Like-O

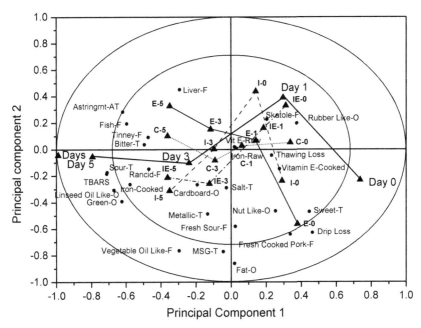

FIGURE 11.5A An overview of the variation found in the mean data from the ANOVA-Partial Least Squares Regression correlation loadings plot for each of the four dietary treatment groups for *M. longissimus dorsi*. Shown are the loadings of the X- and Y-variables for the first 2 PCs for mean data. ▲ = sensory descriptor and ● = sample. D = day in refrigerated display cabinet. ——— = Control, —— = Vitamin E, – – – = Iron/Vitamin E, — — = Iron. The concentric circles represent 100% and 50% explained variance, respectively (O'Sullivan et al., 2003b).

on profile day 1. In general, oxidation appears to be described along PC1 from right to left with the iron supplemented groups exhibiting the greatest degree of oxidation followed by the control and vitamin E-treated groups after 5 days of sensory profiling (O'Sullivan et al., 2003b).

Fig. 11.5b displays an APLSR plot of *M. psoas major* for animals fed the vitamin E, control, iron/vitamin E and iron diets for days 0, 1, 3 and 5 of sensory profiling. The validated explained variance for this model was PC1: 59% and PC2: 18%. Oxidation occurs along PC1 from left to right with samples covarying with the Sweet-T and Fresh Cooked Pork-F descriptors initially and the control group with Green-O and the iron supplemented groups with Astringent-AT, Linseed Oil-Like-O, Fish-F, Rancid-F and Fish-F by the fifth day of profiling. These results are in general agreement with those found by Byrne et al. (2001). The vitamin E-treated group in the present study had a negative covariance to the oxidative descriptors Linseed Oil-Like-O, Fish-F and Rancid-F on the fifth day of profiling. As with the *M. longissimus dorsi*, the iron supplemented groups displayed the greatest degree of oxidation by day 5 of sensory profiling with the iron group showing a slightly higher

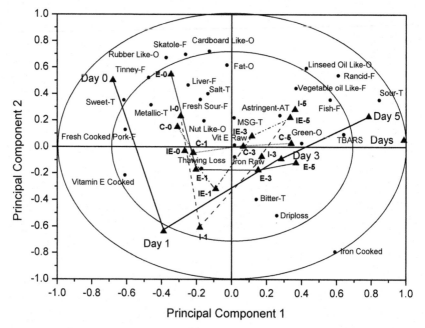

FIGURE 11.5B   An overview of the variation found in the mean data from the ANOVA-Partial Least Squares Regression correlation loadings plot for each of the four dietary treatment groups for *M. psoas major*. Shown are the loadings of the X- and Y-variables for the first 2 PCs for mean data. ▲ = sensory descriptor and • = sample. D = day in refrigerated display cabinet. —— = Control, — — = Vitamin E, - – – = Iron/Vitamin E, — — = Iron. The concentric circles represent 100% and 50% explained variance, respectively (O'Sullivan et al., 2003b).

covariance with the oxidative terms by profile day 5 followed by the control and vitamin E-treated groups. These results are in general agreement as for those found for *M. longissimus dorsi* in this study except vitamin E was marginally more effective at inhibiting the prooxidative nature of iron with regard to WOF development in the iron/vitamin E-treated group. For both Figs. 11.5a and 11.5b, oxidation appears to be described along PC 1 with the main effects of storage days, but not in a linear fashion. An initial increase in oxidation is observed on Day 1 for both muscles (O'Sullivan et al., 2003b).

Additionally, vitamin E muscle tissue levels were greatest in the iron/vitamin E-treated group followed by the vitamin E group, control and iron-treated groups, respectively, for *M. longissimus dorsi*. Whereas, for *M. psoas major* vitamin E tissue levels were in order of magnitude, vitamin E > iron/vitamin E > iron > control group. Iron tissue levels were in the order vitamin E > iron/vitamin E > control > iron for *M. longissimus dorsi* and iron > vitamin E > control > iron/vitamin E for *M. psoas major*. Thus, vitamin E and vitamin C promoted nonsupplemental iron absorption in the vitamin E-treated group for *M. longissimus dorsi* and to a lesser extent for

M. psoas major. M. psoas major was more susceptible to WOF development than *M. longissimus dorsi* for all treatments as determined by sensory profiling due to higher tissue iron levels. From sensory profiling, WOF development in *M. longissimus dorsi* and *M. psoas major* was highest in the iron-supplemented groups followed by the control and vitamin E-supplemented groups (O'Sullivan et al., 2003b).

REFERENCES

Aaslyng, M.D., Tørngren, M.A., Madsen, N.T., 2010. Scandinavian consumer preference for beef steaks packed with or without oxygen. Meat Science 85, 519–524.

Andrès, S., Zaritzky, N., Califano, A., 2006. The effect of whey protein concentrates and hydro-colloids on the texture and colour characteristics of chicken sausages. International Journal of Food Science & Technology 41 (8), 954–961.

Assef, S.A., Bratzler, L.J., Cameron, B.F., Yunis, A.A., 1971. Photo-oxidation of bovine oxy-myoglobin in frozen solutions. The effect of redox active inorganic elements in muscle ex-tracts. Comparative Biochemical Physiology 39B, 395–407.

Baéza, E., Dessay, C., Wacrenier, N., Marché, G., Listrat, A., 2002. Effect of selection for improved body weight and composition on muscle and meat characteristics in Muscovy duck. British Poultry Science 43, 560–568.

Barbut, S., Maurer, A.J., Lindsey, R.C., 1988. Effects of reduced sodium chloride and added phosphates on physical and sensory properties of turkey frankfurters. Journal of Food Science 35, 62–66.

Belcher, J.N., 2006. Review; industrial packaging developments for the global meat market. Meat Science 74, 143–148.

Berry, D., August 2000. Packagings Role. Brief Article, Dairy Foods.

Boleman, S.J., Boleman, S.L., Miller, R.K., Taylor, J.F., Cross, H.R., Wheeler, T.L., Koohmaraie, M., Shackelford, S.D., Miller, M.F., West, R.L., Johnson, D.D., Savell, J.W., 1997. Consumer evalu-ation of beef of known categories of tenderness. Journal of Animal Science 75, 1521–1524.

Byrne, D.V., Bak, L.S., Bredie, W.L.P., Bertelsen, G., Martens, M., 1999a. Development of a sensory vocabulary for warmed-over flavour 1: in porcine meat. Journal of Sensory Studies 14, 47–65.

Byrne, D.V., Bredie, W.L.P., Martens, M., 1999b. Development of a sensory vocabulary for warmed-over flavour: Part II in chicken meat. Journal of Sensory Studies 14, 67–78.

Byrne, D.V., O'Sullivan, M.G., Dijksterhuis, G.B., Bredie, W.L.P., Martens, M., 2001. Sensory panel consistency during development of a vocabulary for warmed-over flavour. Food Quality and Preference 12, 171–187.

Calkins, C.R., Hogden, J.K., 2007. A fresh look at meat flavour. Meat Science 77, 63–80.

Campo, M.M., Nute, G.R., Hughes, S.I., Enser, M., Wood, J.D., Richardson, R.I., 2006. Flavour perception of oxidation in beef. Meat Science 72, 303–311.

Cantor, K.P., 1997. Drinking water and cancer. Cancer Causes & Control 8, 292–308.

Carpenter, E., Cornforth, D.P., Whittier, D., 2001. Consumer preferences for beef colour and packaging did not affect eating satisfaction. Meat Science 57, 359–363.

Cassens, R.G., 1997. Composition and safety of cured meats in the USA. Food Chemistry 59 (No. 4), 561–566.

Chu, S., Miller, R., Kerth, C., 2015. Development of an intact muscle pork flavor lexicon. Meat Science 101, 148.

Civille, G.V., Lawless, H.T., 1986. The importance of language in describing perceptions. Journal of Sensory Studies 1, 203–212, 15.

Civille, G.V., 1987. Development of vocabulary for flavor descriptive analysis. In: Martens, M., Dalen, G.A., Russwurm Jr., H. (Eds.), Flavor Science and Technology. John Wiley & Sons Ltd, Chichester, pp. 357–368.

Clark, D.S., Lentz, C.P., 1969. The effect of carbon dioxide on the growth of slime producing bacteria on fresh beef. Canadian Institute of Food Science and Technology Journal 2 (2), 72–75.

Cofrades, S., López-López, I., Solas, M.T., Bravo, L., Jiménez-Colmenero, F., 2008. Influence of different types and proportions of added edible seaweeds on characteristics of low-salt gel/emulsion meat systems. Meat Science 79 (4), 767–776. http://doi.org/10.1016/j.meatsci.2007.11.010.

Cutter, C.N., 2006. Opportunities for bio-based packaging technologies to improve the quality and safety of fresh and further processed muscle foods. Meat Science 74 (1), 131–142. http://doi.org/10.1016/j.meatsci.2006.04.023.

Decker, E.A., Xiong, Y.L., Calvert, J.T., Crum, A.D., Blanchard, S.P., 1993. Chemical, physical, and functional properties of oxidized turkey white muscle myofibrillar proteins. Journal of Agricultural and Food Chemistry 41, 186–189.

Demeyer, D., Honikel, K., De Smet, S., 2008. The World Cancer Research Fund Report 2007: a challenge for the meat processing industry. Meat Science 80, 953–959.

Desmond, E., 2006. Reducing salt: a challenge for the meat industry. Meat Science 74, 188–196.

Destefanis, G., Brugiapaglia, A., Barge, M.T., Dal Molin, E., 2008. Relationship between beef consumer tenderness perception and Warner–Bratzler shear force. Meat Science 78, 153–156.

Devlieghere, F., Debever, J., Gil, M.I., 2003. MAP, product safety and nutritional quality. In: Ahvenainen, R. (Ed.), Novel Food Packaging Techniques. Woodhead Publishing Limited, Cambridge.

Directive, 2006. Directive 2006/52/EC of the European Parliament and of the Council of 5 July 2006 Amending Directive 95/2/EC on Food Additives Other than Colours and Sweeteners and Directive 95/35/EC on Sweeteners for Use in Foodstuffs, O.J. L204 of 26.7.2006.

dos Santos, B.A., Campagnol, P.C.B., Morgano, M.A., Pollonio, M.A.R., 2014. Monosodium glutamate, disodium inosinate, disodium guanylate, lysine and taurine improve the sensory quality of fermented cooked sausages with 50% and 75% replacement of NaCl with KCl. Meat Science 96 (1), 509–513.

Dzendolet, E., Meiselman, H.L., 1967. Gustatory quality changes as a function of solution concentration. Perception & Psychophysics 2, 29–33.

EC No 2000/13/EC, May 06, 2000. European Commission Regulation of 20 March 2000 on the approximation of laws of the Member States relating to the labelling, presentation and advertising of foodstuffs. Official Journal, L Series 109, p. 29, Brussels.

EC No 2073/2005, December 22, 2005. European Commission Regulation on Microbiological Criteria for Foodstuffs. Official Journal, L series 338, p. 1, Brussels.

Elmore, J.S., Mottram, D.S., 2009. Flavour Development in Meat in Improving the Sensory and Nutritional Quality of Fresh Meat. Woodhead Publishing Limited and CRC Press LLC, pp. 112–146 (Chapter 5).

Estevez, M., Cava, R., 2004. Lipid and protein oxidation, release of iron from heme molecule and colour deterioration during refrigerated storage of liver pate. Meat Science 68, 551–558.

Farmer, L.J., 1992. In: Johnston, D.E., Knight, M.K., Ledward, D.A. (Eds.), Meat Flavor in the Chemistry of Muscle-Based Foods. Royal Society of Chemistry, London, pp. 169–182.

Faustman, C., Cassens, R.G., 1990. Influence of aerobic metmyoglobin reducing capacity on colour stability of beef. Journal of Food Science 55, 1279−1283.

Faustman, C., Cassens, R.G., Schaefer, D.M., Buege, D.R., Williams, S.N., Scheller, K.K., 1989. Improvement of pigment and lipid stability in Holstein steer beef by dietary supplementation with vitamin E. Journal of Food Science 54, 858−862.

Fellendorf, S., O'Sullivan, M.G., Kerry, J.P., 2015. Impact of varying salt and fat levels on the physiochemical properties and sensory quality of white pudding sausages. Meat Science 103, 75−82.

Fellendorf, S., O'Sullivan, M.G., Kerry, J.P., 2016a. Effect of different salt and fat levels on the physicochemical properties and sensory quality of black pudding. Food Science and Nutrition (Online).

Fellendorf, S., O'Sullivan, M.G., Kerry, J.P., 2016b. Effect of using replacers on the physico-chemical properties and sensory quality of low salt and low fat white puddings. European Food Research and Technology (Online).

Fellendorf, S., O'Sullivan, M.G., Kerry, J.P., 2016c. Impact of using replacers on the physico-chemical properties and sensory quality of reduced salt and fat black pudding. Meat Science 113, 17−25.

Fellendorf, S., O'Sullivan, M.G., Kerry, J.P., 2016d. Impact on the physicochemical and sensory properties of salt reduced corned beef formulated with and without the use of salt replacers. Meat Science (submitted for publication).

Fisher, A.V., Pouros, A., Wood, J.D., Young-Boong, K., Sheard, P.R., 2000. Effect of pelvic suspension on three major leg muscles in the pig carcass and implications for ham manufacture. Meat Science 56, 127−132.

Fors, S., 1983. Sensory properties of volatile Maillard reaction products and related compounds. In: Waller, G.R., Feather, M.S. (Eds.), The Maillard Reaction in Foods and Nutrition. American Chemical Society, Washington, DC, pp. 185−286.

Frigg, M., Prabucki, A.L., Ruhdel, E.U., 1990. Effect of dietary vitamin E levels on oxidative stability of trout fillets. Aquaculture 84, 145−158.

FSAI, 2005. Guidance Note No. 18 Determination of Product Shelf-Life. www.fsai.ie.

FSAI, 2012. Use and Removal of Nitrite in Meat Products. https://www.fsai.ie/faq/use_and_removal_of_nitrite.html#use_of.

FSAI, 2011. Currently Agreed FSAI/Industry Guidelines for Sodium From the Salt Reduction Programme (SRP) − 2011 to 2012.

Georgala, D.L., Davidson, C.M., 1970. Food Package. British Patent 1 199 998.

Grau, F.H., 1986. Microbial ecology of meat and poultry. In: Pearson, A.M., Dutson, T.R. (Eds.), Advances in Meat Research, vol. 2. AVI Publishing Company, Westport, Connecticut, pp. 1−47.

Gray, J.L., Pearson, A.M., 1994. Lipid-derived off-flavours in meat formation and inhibition. In: Shahidi, F. (Ed.), Flavour of Meat and Meat Products. Blackie Academic, London, pp. 116−143.

Guàrdia, M.D., Guerrero, L., Gelabert, J., Gou, P., Arnau, J., 2008. Sensory characterisation and consumer acceptability of small calibre fermented sausages with 50% substitution of NaCl by mixtures of KCl and potassium lactate. Meat Science 80 (4), 1225−1230.

Gutteridge, J.M.C., Halliwell, B., 1990. The measurement and mechanism of lipid peroxidation in biological systems. Trends in Biochemical Sciences 15, 129−135.

Han, J.H., Zhang, Y., Buffo, R., 2005. Chapter 4. Surface chemistry of food, packaging and biopolymer materials. In: Han, J.H. (Ed.), Innovations in Food Packaging. Elsevier Academic Press, London.

Harper, R., Bate Smith, E.C., Land, D.G., 1974. Odour Description and Odour Classification: A Multidisciplinary Examination. American Elsevier Publishing Co., New York.

Hempel, A., O'Sullivan, M.G., Papkovsky, D., Kerry, J.P., 2013. Use of smart packaging technologies for monitoring and extending the shelf-life quality of modified atmosphere packaged (MAP) bread: application of intelligent oxygen sensors and active ethanol emitters. European Food Research and Technology 237, 117–124.

Honikel, K., 2008. The use and control of nitrate and nitrite for the processing of meat products. Meat Science 78, 68–76.

Hood, D.E., Riordan, E.B., 1973. Discoloration in pre-packed beef. Journal of Food Technology 8, 333–348.

Hotchkiss, S., 2012. Edible Seaweeds. A rich source of flavor components for sodium replacement. Agro Food Industry Hi Tech 23 (6), 30–32.

Huffman, K.L., Miller, M.F., Hoover, L.C., Wu, C.K., Brittin, H.C., Ramsey, C.B., 1996. Effects of beef tenderness on consumer satisfaction with steaks consumed in the home and restaurant. Journal of Animal Science 74, 91–97.

Ingold, K.O., 1962. Metal catalyses. In: Schultz, H.W., Day, E.A., Sinnhuber, R.O. (Eds.), Symposium on Foods: Lipids and Their Oxidation. AVI Publishing, Connecticut.

ISO, 1988. Sensory analysis. General guidance for the design of test rooms. Ref. no. International Organization for Standardization, Genève, Switzerland. ISO 8589, 1988.

Jackson, T.C., Acuff, G.R., Vanderzant, C., Sharp, T.R., Savell, J.W., 1992. Identification and evaluation of the volatile compounds of vacuum and modified atmosphere packaged beef strip loins. Meat Science 31, 175–190.

Jakobsen, M., Bertelsen, G., 2000. Colour stability and lipid oxidation of fresh beef. Development of a response surface model for predicting the effects of temperature, storage time, and modified atmosphere composition. Meat Science 54, 49–57.

Jayasingh, P., Cornforth, D.P., Brennand, C.P., Carpenter, C.E., Whittier, D.R., 2002. Sensory evaluation of ground beef stored in high-oxygen modified atmosphere packaging. Journal of Food Science and Technology 67, 3493–3496.

Jensen, C., Skibsted, L.H., Jakobsen, K., Bertelsen, G., 1995. Supplementation of broiler diets with all-rac-alpha- or a mixture of natural source RRR-alpha-gamma-delta-tocopheryl acetate. 2. Effect of oxidative stability of raw and pre-cooked broiler meat products. Poultry Science 74, 2048–2056.

Jiménez-Colmenero, F., Carballo, J., Cofrades, S., 2001. Healthier meat and meat products: their role as functional foods. Meat Science 59, 5–13.

Jiménez-Colmenero, F., Cofrades, S., López-López, I., Ruiz-Capillas, C., Pintado, T., Solas, M.T., 2010. Technological and sensory characteristics of reduced/low-fat, low-salt frankfurters as affected by the addition of konjac and seaweed. Meat Science 84 (3), 356–363. http://doi.org/10.1016/j.meatsci.2009.09.002.

Johnson, P.B., Civille, G.V., 1986. A standardized lexicon of meat WOF descriptors. Journal of Sensory Studies 1, 99–104.

Kanner, J., 1994. Oxidative processes in meat and meat products: quality implications. Meat Science 36, 169–189.

Khan, M.I., Jo, C., Tariq, M.R., 2015. Meat flavor precursors and factors influencing flavor precursors—A systematic review. Meat Science 110, 278–284.

Kilcast, D., Subramaniam, P., 2004. Ch. 1-Introduction. In: Kilcast, D., Subramaniam, P. (Eds.), The Stability and Shelf-Life of Food. Woodhead Publishing Limited, Cambridge, UK.

Kramlich, W.E., Pearson, A.M., Tauber, F.W., 1973. Processed Meats. AVI Publishing Co. Inc., Westport.

Kropf, D.H., 1993. Colour stability: factors affecting the colour of fresh meat. Meat Focus International 1, 269—275.

Lanari, M.C., Schaefer, D.M., Scheller, K.K., 1995. Dietary vitamin E supplementation and discoloration of pork bone and muscle following modified atmosphere packaging. Meat Science 41, 237—250.

Liu, Q., Lanari, M.C., Schaefer, D.M., 1995. A review of dietary vitamin E supplementation for improvement of beef quality. Journal of Animal Science 73, 3131—3140.

López-López, I., Cofrades, S., Yakan, A., Sola, M.T., Jiménez-Colmenero, F., 2010. Frozen storage characteristics of low-salt and low-fat beef patties as affected by Wakame addition and replacing pork backfat with olive oil-in-water emulsion. Food Research International 43 (5), 1244—1254.

Lund, M.N., Lametsch, R., Hviid, M.S., Jensen, O.N., Skibsted, L.H., 2007. High oxygen packaging atmosphere influences protein oxidation and tenderness of porcine *longissimus dorsi* during chill storage. Meat Science 77, 295—303.

Lynch, M.P., Kerry, J.P., Buckley, D.J., Faustman, C., Morrissey, P.A., 1999. Effect of dietary vitamin E supplementation on the colour and lipid stability of fresh, frozen and vacuum-packed beef. Meat Science 52, 95—99.

Lyon, B.G., 1987. Development of chicken flavour descriptive attribute terms aided by multivariate statistical procedures. Journal of Sensory Studies 2, 55—67.

MacLeod, G., 1994. The flavour of beef. In: Shahidi, F. (Ed.), Flavour of Meat and Meat Products. Blackie Academic and Professional, Glasgow, pp. 4—37.

Mancini, R.A., Hunt, M.C., 2005. Current research in meat colour. Meat Science 71, 100—121.

Martens, H., Martens, M., 1999. Validation of PLS regression models in sensory science by extended cross-validation. In: Tenenhause, M., Monineau, A. (Eds.), Les Methodes PLS. CISIA-CERESTA, France, pp. 149—182.

Martens, H., Martens, M., 2000. Modified jack-knife estimation of parameter uncertainty in bilinear modelling by partial least squares regression (PLSR). Food Quality and Preference 11, 5—16.

Martens, H., Martens, M., 2001. Multivariate analysis of quality. An introduction (1975). In: Marusich, W.L., De Ritter, E., Ogrinz, E.F., Keating, J., Mitrovic, M., Bunnell, R.H. (Eds.), Effect of Supplemental Vitamin E in Control of Rancidity in Poultry Meat, Poultry Science, vol. 54. J. Wiley and Sons Ltd, Chichester, pp. 831—844.

Martens, H., Dijkerhuis, G., Byrne, D.V., 2000. Power of experimental designs, estimated by Monte Carlo simulation. Journal of Chemometrics 14, 441—462.

Marusich, W.L., De Ritter, E., Ogrinz, E.F., Keating, J., Mitrovic, M., Bunnell, R.H., 1975. Effect of supplemental vitamin E in control of rancidity in poultry meat. Poultry Science 54, 831—844.

Maughan, C., Tansawat, R., Cornforth, D., Ward, R., Martini, S., 2012. Development of a beef flavor lexicon and its application to compare the flavor profile and consumer acceptance of rib steaks from grass- or grain-fed cattle. Meat Science 90, 116—121.

McMillin, K.W., 2008. Where is MAP going? A review and future potential of modified atmosphere packaging for meat. Meat Science 80, 43—65.

Meilgaard, M.C., Civille, G.V., Carr, B.T., 1999. In: Sensory Evaluation Techniques, third ed. Academic Press, Florida, pp. 54—55. Chapter 5.

Miller, M.F., Carr, M.A., Ramsey, C.B., Crockett, K.L., Hoover, L.C., 2001. Consumer thresholds for establishing the value of beef tenderness. Journal of Animal Science 79, 3062—3068.

Miller, R.K., 2010. Differentiation of Beef Flavour Across Muscles and Quality Grades (Phase II). National Cattlemens' Beef Association, Centennial, CO.

Morrissey, P.A., Buckley, D.J., Sheehy, P.J.A., Monaghan, F.J., 1994. Vitamin E and meat quality. Proceedings of the Nutrition Society 53, 289–295.

Mottram, D.S., 1998. Flavour formation in meat and meat: a review. Food Chemistry 62, 415–424.

Nattress, F.M., Jeremiah, L.E., 2000. Bacterial mediated off-flavours in retail-ready beef after storage in controlled atmospheres. Food Research International 33, 743–748.

Nottingham, P.M., 1982. Microbiology of carcass meats. In: Brown, M.H. (Ed.), Meat Microbiology. Applied Science Publishers, London, pp. 13–65.

O'Sullivan, M.G., Kerry, J.P., 2009. Chapter 13. Meat packaging. In: Toldrá, F. (Ed.), Handbook of Meat Processing. John Wiley & Sons, Chichester, West Sussex, UK, pp. 211–230.

O'Sullivan, M.G., Kerry, J.P., 2011. Sensory and quality properties of packaged fresh and processed meats. In: Kerry, J.P., Ledward, D.A. (Eds.), Improving the Sensory and Nutritional Quality of Fresh and Processed Meats. Woodhead Publishing Limited, Cambridge, UK.

O'Sullivan, M.G., Kerry, J.P., 2012. Chapter 4. Packaging of (fresh and frozen) pork. In: Nollet, L.M.L. (Ed.), Handbook of Meat, Poultry and Seafood Quality. Wiley-Blackwell Publishing Ltd, Oxford, UK.

O'Sullivan, M.G., Kerry, J.P., Buckley, D.J., Lynch, P.B., Morrissey, P.A., 1997. The distribution of dietary vitamin E in the muscle of the porcine carcass. Meat Science 45, 297–305.

O'Sullivan, M.G., Kerry, J.P., Buckley, D.J., Lynch, P.B., Morrissey, P.A., 1998. The effect of dietary vitamin E supplementation on quality aspects of porcine muscles. Irish Journal of Agricultural and Food Research 37, 227–235.

O'Sullivan, M.G., Byrne, D.V., Martens, M., 2002a. Data analytical methodologies in the development of a vocabulary for evaluation of meat quality. Journal of Sensory Studies 17, 539–558.

O'Sullivan, M.G., Byrne, D.V., Stagsted, J., Andersen, H.J., Martens, M., 2002b. Sensory colour assessment of fresh meat from pigs supplemented with iron and vitamin E. Meat Science 60, 253–265.

O'Sullivan, M.G., Byrne, D.V., Martens, M., 2003a. Evaluation of pork colour: sensory colour assessment using trained and untrained sensory panellists. Meat Science 63, 119–129.

O'Sullivan, M.G., Byrne, D.V., Nielsen, J.H., Andersen, H.J., Martens, M., 2003b. Sensory and chemical assessment of pork supplemented with iron and vitamin E. Meat Science 64, 175–189.

O'Sullivan, M.G., Byrne, D.V., Jensen, M.T., Andersen, H.J., Vestergaard, J., 2003c. A comparison of warmed-over flavour in pork by sensory analysis, GC/MS and the electronic nose. Meat Science 65, 1125–1138.

O'Sullivan, M.G., Byrne, D.V., Martens, H., Gidskehaug, L.H., Andersen, H.J., Martens, M., 2003d. Evaluation of pork meat colour: prediction of visual sensory quality of meat from instrumental and computer vision methods of colour analysis. Meat Science 65, 909–918.

O'Sullivan, M.G., Cruz, M., Kerry, J.P., 2010. Evaluation of carbon dioxide flavour taint in modified atmosphere packed beef steaks. LWT – Food Technology 44, 2193–2198.

O'Sullivan, M.G., 2011. Chapter 25. Case studies: meat and poultry. In: Kilcast, D., Subramaniam, P. (Eds.), Food and Beverage Shelf-Life and Stability. Woodhead Publishing Limited, Cambridge, UK.

O'Sullivan, M.G., 2016. Chapter 18. The stability and shelf life of meat and poultry. In: Subramaniam, P. (Ed.), The Stability and Shelf Life of Food. Elsevier Academic Press Ltd, Oxford, UK.

O'Sullivan, M.G., Kerry, J.P., 2008. Ch 30, sensory and quality properties of packaged meat. In: Kerry, J.P., Ledward, D.A. (Eds.), Improving the Sensory and Nutritional Quality of Fresh Meat. Woodhead Publishing Limited, Cambridge, UK.

Ozdemir, M., Floros, J.D., 2004. Active food packaging technologies. Critical Reviews in Food Science and Nutrition 44, 185–193.

Paik, D.C., Wendel, T.D., Freeman, H.P., 2005. Cured meat consumption and hypertension: an analysis from NHANES III (1988–94). Nutrition Research 25, 1049–1060.

Pearson, A.M., Gillett, T.A., 1995. In: Pearson, A.M., Gillett, T.A. (Eds.), Processed Meats, third ed. Chapman and Hall, London, pp. 53–78.

Pearson, A.M., Tauber, F.W., 1984. Processed Meats. Chapter 3 Curing. AVI Publishing Co. Inc., Westport, pp. 46–68.

Pearson, A.M., 1987. Muscle function and post-mortem changes. Ch. 4. In: Third, J.F.P., Schweigert, B.S. (Eds.), The Science of Meat and Meat Products. Food and Nutrition Press, Inc., Westport, Connecticut, pp. 307–327.

Pennington, J.A.T., 1998. Dietary exposure models for nitrates and nitrites. Food Control 9, 385–395.

Piggott, J.R., 1991. Selection of terms for descriptive analysis. In: Lawless, H.T., Klein, B.P. (Eds.), Sensory Science Theory and Applications in Foods. Marcel Dekker, New York, pp. 339–351.

PROSSLOW, 2016. Development of Consumer Accepted Low Salt and Low Fat Irish Traditional Processed Meats. Project Coordinator: Dr Maurice O' Sullivan. Ref:11F026. https://www.agriculture.gov.ie/media/migration/research/firmreports/CALL2011ProjectAbstracts240216.pdf.

Rason, J., Lebecque, A., leger, L., Dufour, E., 2003. Delineation of the sensory characteristics of traditional dry sausage. I – Typology of the traditional workshops in Massif Central. In: The 5th Pangborn Sensory Science Symposium, July 21–24, Boston, USA.

Resconi, V.C., Escudero, A., Campo, M.M., 2013. The development of aroma in ruminant meat. Molecules 18, 6748–6781.

Rhee, K.I., Ziprin, Y.A., 1987. Lipid oxidation in retail beef, pork and chicken muscles as affected by concentrations of heme pigments and nonheme iron and microsomal enzymic lipid peroxidation activity. Journal of Food Biochemistry 11, 1–15.

Robbins, K., Jensen, J., Ryan, K.J., Homco-Ryan, C., McKeith, F.K., Brewer, M.S., 2003. Effects of dietary vitamin E supplementation on textural and aroma attributes of enhanced beef clod roasts in a cook/hot hold situation. Meat Science 64, 317–322.

Rowe, L.J., Maddock, K.R., Lonergan, S.M., Huff-Lonergan, E., 2004. Influence of early post-mortem protein oxidation on beef quality. Journal of Animal Science 82, 785–793.

Ruusunen, M., Särkkä-Tirkkonen, M., Puolanne, E., 1999. The effect of salt reduction on taste pleasantness in cooked bologna type sausages. Journal of Sensory Studies 14, 263–270.

Ruusunen, M., Särkkä-Tirkkonen, M., Puolanne, E., 2001. Saltiness of coarsely ground cooked ham with reduced salt content. Agricultural and Food Science in Finland 10, 27–32.

Ryan, Y.M., 1997. Meat avoidance and body weight concerns: nutritional implications for teenage girls. Proceedings of the Nutrition Society 56, 519–524.

Sato, K., Hegarty, G.R., 1971. Warmed-over flavour in cooked meat. Journal of Food Science 36, 1098–1102.

Savell, J.W., Branson, R.E., Cross, H.R., Stiffler, D.M., Wise, J.W., Griffin, D.B., Smith, G.C., 1987. National Consumer Retail Beef Study: palatability evaluations of beef loin steaks that differed in marbling. Journal of Food Science 52, 517–519, 532.

Savell, J.W., Cross, H.R., Francis, J.J., Wise, J.W., Hale, D.S., Wilkes, D.L., Smith, G.C., 1989. National Consumer Retail Beef Study: interaction of trim level, price and grade on consumer acceptance of beef steaks and roasts. Journal of Food Quality 12, 251–274.

Schuddeboom, L.J., 1993. Nitrates and Nitrites in Foodstuffs. Council of Europe Press, Publishing and Documentation Service. ISBN:92-871-2424-6.

Schweinsberg, F., Burkle, V., 1985. Nitrite: a co-carcinogen? Journal of Cancer Research and Clinical Oncology 109, 200—202.

Sebranek, J.G., Bacus, J.N., 2007. Cured meat products without direct addition of nitrate or nitrite: what are the issues? Meat Science 77, 136—147.

Seman, D.L., Decker, E.A., Crum, A.D., 1991. Factors affecting catalysis of lipid oxidation by a ferritin containing extract of beef muscle. Journal of food Science 56, 356—358.

Seyfert, M., Hunt, M.C., Mancini, R.A., Hachmeister, K.A., Kropf, D.H., Unruh, J.A., Loughin, M., 2005. Beef quadriceps hot boning and modified atmosphere packaging influence properties of injection-enhanced beef round muscles. Journal of Animal Science 83, 686—693.

Shahidi, F., Pegg, R.B., 1991. Novel synthesis of cooked cured-meat pigment. Journal of Food Science 56, 1205—1208.

Sherbeck, J.A., Wulf, D.M., Morgan, J.B., Tatum, J.D., Smith, G.C., Williams, S.N., 1995. Dietary supplementation of vitamin E to feedlot cattle affects retail display properties. Journal of Food Science 60, 250—252.

Shorland, F.B., Igene, J.O., Pearson, A.M., Thomas, J.W., McGuffey, R.K., Aldridge, A.E., 1981. Effects of dietary fat and vitamin E on the lipid composition and stability of veal during frozen storage. Journal of Agricultural and Food Chemistry 29, 863—871.

Sindelar, J.J., Cordray, J.C., Olson, D.G., Sebranek, J.G., Love, J.A., 2007. Investigating quality attributes and consumer acceptance of uncured, no-nitrate/nitrite-added commercial hams, bacons, and frankfurters. Journal of Food Science 72 (8), S551—S559.

Smiddy, M., Papkovskaia, N., Papkovsky, D.B., Kerry, J.P., 2002. Use of oxygen sensors for the non-destructive measurement of the oxygen content in modified atmosphere and vacuum packs of cooked chicken patties: impact of oxygen content on lipid oxidation. Food Research International 35, 577—584.

Smith, G.C., Savell, J.W., Cross, H.R., Carpenter, Z.L., Murphey, C.E., Davis, G.W., Abraham, H.C., Parrish, F.C., Berry, B.W., 1987. Relationship of USDA quality grades to palatability of cooked beef. Journal of Food Quality 10, 269—287.

Smith, J.P., Ramaswamy, H.S., Simpson, B.K., 1990. Developments in food packaging technology. Part II. Storage aspects. Trends in Food Science and Technology 1 (5), 111—118.

Snyder, H.E., Skrydlant, H.B., 1966. The influence of metallic ions on the autoxidation of oxy-myoglobin. Journal of Food Science 31, 468—479.

Sorheim, O., Nissen, H., Nesbakken, T., 1999. The storage life of beef and pork packaged in an atmosphere with low carbon monoxide and high carbon dioxide. Meat Science 52, 157—164.

Tappel, A.L., 1962. Heme compounds and lipoxidase as biocatalysts. In: Schultz, H.W., Day, E.A., Sinnhuber, R.O. (Eds.), Symposium on Foods: Lipids and Their Oxidation. AVI Publishing, Connecticut, pp. 122—126.

Tichivangana, A.J., Morrissey, P.A., 1985. Metmyoglobin and inorganic metals as pro-oxidants in raw and cooked muscle systems. Meat Science 15, 107—116.

Tobin, B.D., O'Sullivan, M.G., Hamill, R.M., Kerry, J.P., 2012a. Effect of varying salt and fat levels on the sensory quality of beef patties. Meat Science 4, 460—465.

Tobin, B.D., O'Sullivan, M.G., Hamill, R.M., Kerry, J.P., 2012b. Effect of varying salt and fat levels on the sensory and physiochemical quality of frankfurters. Meat Science 92, 659—666.

Tobin, B.D., O'Sullivan, M.G., Hamill, R.M., Kerry, J.P., 2013. The impact of salt and fat level variation on the physiochemical properties and sensory quality of pork breakfast sausages. Meat Science 93, 145—152.

Tobin, B.D., O'Sullivan, M.G., Hamill, R.M., Kerry, J.P., 2014a. European consumer attitudes on the associated health benefits of neutraceutical-containing processed meats using Co-Enzyme Q10 as a sample functional ingredient. Meat Science 97, 207—213.

Tobin, B.D., O'Sullivan, M.G., Hamill, R.M., Kerry, J.P., 2014b. Effect of cooking and in vitro digestion on co-enzyme Q10 in processed meat products fortified with co-enzyme Q10. Food Chemistry 150, 187−192.

Torngren, M.A., September 2003. Effect of packaging method on colour and eating quality of beef loin steaks. In: 49th International Congress of Meat Science and Technology. Brazil, 495−496.

Verbeke, W., Pérez-Cueto, F.J.A., De Barcellos, M.D., Krystallis, A., Grunert, K.G., 2010. European citizen and consumer attitudes and preferences regarding beef and pork. Meat Science 84, 284−292.

WCRF, 2007. World Cancer Research Fund/American Institute for Cancer Research. Food, Nutrition, Physical Activity, and the Prevention of Cancer: A Global Perspective. American Institute for Cancer Research, Washington, DC, p. 517.

Weiss, J., Gibis, M., Schuh, V., Salminen, H., 2010. Advances in ingredient and processing systems for meat and meat products. Meat Science 86, 196−213.

Wheelock, V., Hobbiss, A., 1999. All You Ever Wanted to Know About Salt but Were Afraid to Ask. Verner Wheelock Associates, Skipton, Yorkshire.

WHO, 2003. Diet, Nutrition and the Prevention of Chronic Diseases. Report of a Joint WHO/FAO Expert Consultation. WHO technical report series 919, pp. 148. http://www.who.int/dietphysicalactivity/publications/trs916/summary/en/print.html.

Xiong, X.L., 2000. Protein oxidation and implications for muscle food quality. Antioxidants in Muscle Foods 85−111 (Chapter 4).

Yin, M.C., Faustman, C., Riesen, J.W., Williams, S.N., 1993. The effects of α-tocopherol and ascorbate upon oxymyoglobin and phospholipid oxidation. Journal of Food Science 58, 1273−1276.

Yucel, U., 2016. Intelligent Packaging. Reference Module in Food Science. Elsevier, ISBN 978-0-08-100596-5.

Zakrys, P.I., Hogan, S.A., O'Sullivan, M.G., Allen, P., Kerry, J.P., 2008. Effects of oxygen concentration on sensory evaluation and quality indictors of beef muscle packed under modified atmosphere. Meat Science 79, 648−655.

Zakrys, P.I., O'Sullivan, M.G., Allen, P., Kerry, J.P., 2009. Consumer acceptability and physiochemical characteristics of modified atmosphere packed beef steaks. Meat Science 81, 720−725.

Zakrys-Waliwander, P.I., O'Sullivan, M.G., O'Neill, E.E., Kerry, J.P., 2012. The effects of high oxygen modified atmosphere packaging on protein oxidation of bovine *M. longissimus dorsi* muscle during chilled storage. Food Chemistry 2, 527−532.

Zakrys-Walliwander, P.I., O'Sullivan, M.G., Allen, P., O'Neill, E.E., Kerry, J.P., 2010. Investigation of the effects of commercial carcass suspension (24 and 48 hours) on meat quality in modified atmosphere packed beef steaks during chill storage. Food Research International 43, 277−284.

Zakrys-Walliwander, P.I., O'Sullivan, M.G., Walshe, H., Allen, P., Kerry, J.P., 2011. Sensory comparison of commercial low and high oxygen modified atmosphere packed sirloin beef steaks. Meat Science 88, 198−202.

Zanardi, E., Ghidini, S., Conter, M., Ianieri, A., 2010. Mineral composition of Italian salami and effect of NaCl partial replacement on compositional, physicochemical and sensory parameters. Meat Science 86 (3), 742−747.

Zhang, X., Kong, B., Xiong, Y.L., 2007. Production of cured meat colour in nitrite-free Harbin red sausage by *Lactobacillus fermentum* fermentation. Meat Science 77, 593−598.

Chapter 12

Sensory Properties of Dairy Products

INTRODUCTION

A range of different approaches exist to gain information on the sensory character of dairy products and these can be broadly segmented into four different areas: (1) expert grading (cheese), (2) difference methods, (3) descriptive methods and (4) affective methods. The choice of sensory approach depends upon factors, such as complexity of information required, application, cost and time.

Grading has quite a special and long traditional association with cheese quality assessment by cheese producers. During grading, large numbers of cheeses can be rapidly scored for overall flavour and texture quality, based on an idealized concept of the perfect cheese or graded for a specific market (Kilcawley and O'Sullivan, 2016). Some countries have defined courses and accredited training for graders of cheese, while other countries source graders by simply assessing individuals for their ability to perceive key flavour and aroma attributes. In the latter case, training occurs through mentoring with an experienced grader or graders for a period of time until they can work independently (Kilcawley, 2016). Graders are typically looking for negative attributes and often have a predetermined list of defects that have been developed for that cheese through experience, which can be either formal or informal (Kilcawley and O'Sullivan, 2016; Delahunty and Murray, 1997; Delahunty and Drake, 2004; Drake, 2007; Kilcawley, 2016). Graders can use their expertise to assess over a hundred cheeses in a session and do not always provide a score, just an overall comment as an assessment of quality. For Cheddar cheese, the approach typically involves the grader(s) making rapid visual assessments of cheese blocks, followed by further assessments, such as taking a sample cheese plug using a cheese trier. Graders assess how cleanly the plug emerges, along with its appearance, colour and adhesiveness to the trier. A small sample of the plug is repeatedly manipulated between the index finger and thumb. The aroma of this warm cheese is inhaled and finally the cheese may be rolled within the mouth for aroma and taste (Kilcawley, 2016; Kilcawley and O'Sullivan, 2016). Grading schemes are defect-based judgements for cheese which have been

A Handbook for Sensory and Consumer-Driven New Product Development.
http://dx.doi.org/10.1016/B978-0-08-100352-7.00012-9
259

developed over the years as a tool to determine the reliability and reproducibility of processes; for quality control; as a guide to optimum storage and for marketing strategy. For the latter, the strategy determines what cheeses are selected for specific markets (Muir, 2010). Some experienced graders also use additional compositional data to get a better overall assessment of quality, but also potentially gain a better understanding of predicated ripening characteristics (Kilcawley, 2016). For cheeses with long potential maturation times the grader plays a vital role in monitoring ongoing quality development. Experienced factory graders can diagnose problems and provide feedback to those in production to aid improvement and consistency of manufacture for future reference (Muir, 2010).

As described in Chapter 2 of this book, descriptive sensory methods involve the training of panellists to quantitatively measure the attributes associated with the relevant sensory modalities of 'appearance', 'aroma', 'flavour', 'texture', 'taste' and 'aftertaste'. The language is descriptive and nonhedonic, in that assessors are not asked how much they rate or like the cheese. The different methods for descriptive profiling include the 'flavour profile', 'texture profile', 'free choice profiling', 'spectrum descriptive analysis' and 'quantitative descriptive analysis (QDA) (Kilcawley and O'Sullivan, 2016). With respect to cheese products, descriptive analysis is the most powerful sensory tool in cheese flavour research with QDA and the spectrum method being the most commonly practiced methods employed both by industry and academia. Descriptive profiling can be used to differentiate cheeses based on a full complement of sensory characteristics/attributes/lexicons and to determine a quantitative description of all the sensory aspects that can be identified (Singh et al., 2003). Some of these tests are widely used for product development and for research purposes, while others appear to be of less practical use (Kilcawley and O'Sullivan, 2016). The spectrum method was developed in the 1970s (Civille and Szczesniak, 1973) and is a descriptive profiling method which prescribes the use of a strict technical sensory vocabulary using reference materials. The scales are anchored using extensive reference points which may include a range of foods, which correspond to food reference samples that apparently reduce panel variability. Panellists develop their list of attributes by evaluating a large array of products within the category. Products may be described in terms of only one attribute (e.g., 'appearance' or 'aroma') or they may be trained to evaluate all attributes. This method is pragmatic in that it provides the tools to design a descriptive procedure for a given product category. Its principal characteristic is that the panellist scores the perceived intensities with reference to prelearned 'absolute' intensity scales (Murray et al., 2001). The purpose is to make the resulting profiles universally understandable. The method provides an array of standard attribute names ('lexicons'), each with its set of standards that define a scale of intensity (Muñoz and Civille, 1992; Meilgaard et al., 1999). These descriptive terms have been developed and employed by a number of authors for the sensory evaluation of dairy products, specifically cheese, e.g., Van Hekken et al. (2006) for Mexican cheese; Muir et al. (1996), Murray and Delahunty (2000a,b), and Drake et al. (2005) for Cheddar; Rétiveau et al. (2005) for French cheeses. One of the drawbacks is that extensive training (of panellists) is required when using the spectrum method. Other potential drawbacks include cultural differences of panels and the difficulty of quantifying an attribute over a range of different products (Murray et al., 2001).

In the QDA method, experts with cheese knowledge can evaluate dairy products and suggest descriptive terms that specifically describe them and the sensory dimension to be examined to produce an initial 'meta' sensory list. This meta list is then used to commence sensory training but will be reduced to provide a profiling lexicon for subsequent analysis. These QDA attributes/ lexicons are available for various cheese products: Cheddar cheese (Drake et al., 2005), French cheese (Rétiveau et al., 2005), Camembert (Galli et al., 2016), Pecarino (Torracca et al., 2016) etc.

Descriptive sensory methods, although often the benchmark for quantitative analysis, can be expensive and time consuming because of the requirement to train and profile individual panellists over extended periods of time (O'Sullivan et al., 2011). However, affective methods are restricted to hedonic (liking) assessments, including acceptability or preference, but do not describe the product. Traditionally, this issue was solved by combining descriptive data and hedonic data using data statistical tools involving predominantly chemometrics in a method called 'preference mapping' (Kilcawley and O'Sullivan, 2016). Some sensory scientists consider that using consumers for sensory descriptive tasks is not appropriate as consumers lack consensus and repeatability, or comprehension of the meaning of the sensory attributes (Lawless and Heymann, 1998; Stone and Sidel, 2004). Consumers can only tell you what they 'like' or 'dislike' (Lawless and Heymann, 1998). In contrast, other sensory scientists have shown, through different studies, that consumers can describe the sensory characteristics of products with a precision comparable to experts (Worch et al., 2014). However, only 'simple' sensory attributes/terms can be used (cannot use technical or chemical terms) (Worch et al., 2010, 2014), and larger numbers of consumers are required to make up for a lack of appropriate training. Rapid sensory methods have been designed to provide more cost-effective solutions to these problems and to close the divide between the rigid rules of classic descriptive profiling and the emotional responses involved with affective sensory methods. Rapid sensory evaluation methods can provide quick results with respect to the end user and a reduction in resources required (Kilcawley and O'Sullivan, 2016). The case study presented at the end of this book chapter demonstrates a flash profiling method, ranking descriptive analysis (RDA) combined with sensory acceptance testing for the evaluation of both raw and cooked mozzarella cheese.

SENSORY PROPERTIES OF MILK

Good quality milk has a bland but characteristic flavour with a pleasant mouthfeel, determined by its physical nature, i.e., an emulsion of fat globules in a colloidal aqueous solution, and a slightly salty and sweet taste, due to the presence of salts and lactose (Thomas, 1981). Milk is defined as the secretion of the mammary glands of mammals, its primary natural function being nutrition of the young. Milk of some animals, especially cows, buffaloes, goats and sheep, is also used for human consumption, either as such or in the form of a range of dairy products (Walstra et al., 2006a). Milk is composed of sugars, proteins and fat. Lactose imparts sweetness to milk and is a reducing sugar, also known as milk sugar. Lactose is the distinctive carbohydrate of milk and is a disaccharide composed of glucose and galactose. Milk fat consists of triglycerides ($\sim$98%), the vast majority of which are even-numbered saturated fatty acids esterified on glycerol (Kilcawley and O'Sullivan, 2016). Bovine milk fat is present as emulsified globules surrounded by a thin membrane, called the milk fat globule membrane (MFGM), and imparts the creamy or rich flavour of milk. The MFGM consists of a complex mixture of proteins, phospholipids, glycoproteins, tri-acylglycerides, cholesterol, free fatty acids, monoglycerides, diglycerides and other minor components and acts as a natural emulsifying agent enabling the fat to remain dispersed in the aqueous phase of milk (Wilkinson, 2007). Typically, milk fat contains 66% saturated fatty acids, 30% monounsaturated fatty acids and 4% polyunsaturated fatty acids (Aigster et al., 2000). About four-fifths of the protein in milk consists specifically of casein, a mixture of four proteins: αS1-, αS2-, β- and κ-casein which are to some extent phosphorylated and have little or no secondary structure. The remainder consists of the milk serum proteins, the main one being β-lactoglobulin. Milk contains numerous minor proteins, including many enzymes as well as the minerals K, Na, Ca, Mg, Cl and phosphate (Walstra et al., 2006a). Calcium phosphate in milk and dairy products is absorbed by humans and is important for bone growth and development.

Due to its bland nature, milk can be a carrier of off-odours and flavours. The American Dairy Science Association lists the following flavour criticisms: acid, astringent, barny, bitter, cooked, cowy (acetone), feed, fermented/fruity, flat, foreign, garlic/onion (weedy), lacks freshness (stale), malty, oxidised (metal induced), oxidised (light induced), rancid (lipolyzed), salty and unclean (psychrotrophic) (Azzara and Campbell, 1992). Off-notes may originate from feed (clover, concentrate), microbial spoilage (souring, lactic acid from lactic acid bacteria) or rancid flavours generated by light- or copper (Cu)-catalysed auto-oxidation of dairy fats. A major characteristic of bovine milk fat is the presence of water soluble short-chain fatty acids (SCFAs) with eight carbons or less (Jensen et al., 1962, 1991), which are volatile and highly odour active. Ethyl butanoate and hexanoate (fruity aroma) are the principle ethyl esters in bovine, ovine and caprine milks (Nursten, 1997). The volatile compounds, dimethyl sulphide and also diacetyl, 2-methylbutanol, as well as other alde-hydes are responsible for the characteristic flavour of fresh raw milk (Walstra et al., 2006b). Levels of caproic and caprylic acids are higher in caprine milk, and caprylic acid in ovine milk than in bovine milk (Markiewicz-Keszycka et al., 2013), and are thus important in goat and sheep milk cheeses. Many headspace studies on the volatile profiles of milk or cheese report significant levels of lactones, mainly because of their low volatility rather than their absence. Lactones in general are described as having a buttery type character (Wilkinson, 2007).

When good quality raw milk is pasteurised under minimal conditions, e.g., 72°C for 15 s, the flavour is barely affected. As more stringent conditions are used, the more the flavour gradually moves towards that of UHT milk (Nursten, 1997). More intense heat treatment, e.g., 80−100°C for 20 s, results in a 'cooked' flavour, caused mainly by H_2S (Walstra et al., 2006b). Milk is not just a beverage but a raw ingredient for other dairy products such as butter, yoghurt and cheese.

SENSORY PROPERTIES OF MILK POWDER

Skim milk powder (SMP) and other dried dairy ingredients should ideally have a clean, sweet and pleasant taste free of flavour defects (Bodyfelt et al., 1988). Milk powders are produced, without extensive loss in quality, by drying milk in spray driers (Fig. 12.1) which atomise the milk in a stream of heated air which rapidly dries the milk into a powder in a matter of seconds. Pasteurised milk is firstly separated into skim milk and cream before concentration through evaporation to 45−52% solids. For whole milk powder (WMP) a portion of the cream is added back to the skim milk to produce a milk with a standardised fat content with typically 26−30% fat in the powder (Pearce, 2016). After spray drying, the milk powder is separated from the air in the dryer by a bag filter or cyclone. The bacterial load is dramatically reduced and the low moisture content of the resulting powder inhibits microbial growth producing a long shelf life product ($\sim$ 1 year). A fluidised bed process may also be employed at this stage which involves air being blown up through the powder from below, causing the powder particles to separate and behave like a fluid, as well as to cool and dry the powder further. The powder produced should be free flowing and not stick to containers, vessels, bags etc., and should reconstitute in water readily, without lumps or undissolved particles (Walstra et al., 2006e; Pearce, 2016). Lecithinisation is sometimes used for WMP, because of the higher fat content, which involves covering the powder particles with a thin layer of lecithin to confer better reconstitution properties. This is generally applied before bagging which may involve some form of gas flushing (N_2, CO_2 or both) to prevent oxidation during storage. WMP are particularly more susceptible to oxidation due to their higher fat contents, especially at elevated temperatures (>30°C) (Pearce, 2016). These off-flavours

FIGURE 12.1   GEA-Niro Production Minor spray dryer. Presented is a single stage drying unit (main chamber and cyclone collection) fed by either rotary disc or nozzle. Drying conditions range from 170 to 190°C inlet and 70–90°C outlet with full digital control. Drying capacity is ∼ 15 L/h (30% TS solution) to dry 5 kg powder.

may be described as being 'tallowy', 'rancid', 'oxidised' or 'glue-like'. Drake et al. (2003) undertook descriptive profiling of milk powders and observed that milk fat, fried, fatty/painty flavours were not detected in SMP, but were observed in WMP. For SMP, cooked flavours may be present and vary according to heat treatment (low, medium, high) of the milk prior to evaporation and spray drying (Drake et al., 2003).

SENSORY PROPERTIES OF BUTTER

Butter is a traditional food, which is widely consumed all over the world, directly or as an ingredient in processed foods such as pastries and convenience dishes. Its nutritional value, due to a high content of fats, vitamins and minerals, and its unique and pleasant flavour make butter particularly appreciated by consumers (Mallia et al., 2008). Butter is a water-in-oil (W/O) emulsion in which water forms the dispersed phase and oil forms the continuous phase. Butter is mostly made today by a continuous churn process as opposed to the batch approach. Different types of butter are available on the market such as sweet cream (salted or unsalted), cultured, and whipped butter. Cultured cream butter, as the name suggests, is cultured with lactic bacteria and is usually characterised by its intense diacetyl flavour (Bodyfelt et al., 1988). However, diacetyl is of minor importance in sweet cream butter (Lozano et al., 2007). More than 230 volatile compounds have been identified in butter, however, only a small number of them can be considered as key odorants of butter aroma. Sweet cream butter is characterised by lactones with fruity and creamy notes and by sulphur compounds, having corn-like and garlic odours (Mallia et al., 2008).

The nutritional, organoleptic and rheological (hardness, spreadability, melting) properties of dairy products are largely dependent on the fatty acid composition of milk, particularly polyunsaturated fatty acids (Hurtaud and. Peyraud, 2007). Modifying the fatty acid composition of butter by decreasing the proportions of 12:0, 14:0, 16:0 and stearic acid (18:0) and increasing the proportions of unsaturated and short-chain fatty acids improves its spreadability (Bobe et al., 2007) and also potential health benefits. The natural conditions of the production area directly affect the quality of the dairy products. Thus, the aroma and flavour of butter can vary depending on the season of production as well as the feed the animals are consuming (Gori et al., 2012). For example, a higher amount of fresh grass in the animal diet is reported to significantly increase the relative amount of α-linolenic acid (Dhiman et al., 1999) and CLA (conjugated linoleic acid) (O'Callaghan et al., submitted) in milk. Fresh grass feeding regimens, widely practiced in Ireland and New Zealand, produce a milk fat with higher proportions of unsaturated FA compared to those derived from total mixed ration (TMR) indoor grass/maize/grain silage and concentrate feeding systems (Couvreur et al., 2006), extensively practiced in the United States, Asia and parts of Europe. Additionally, there is evidence that pasture feeding also produces a superior dairy product from a hedonic sensory perspective (O'Callaghan et al., submitted).

The high ratio of saturated to unsaturated fatty acids in milk fat has been a concern because of the link between intake of saturated fatty acids and various biological markers for cardiovascular disease risk, such as elevated blood pressure, insulin resistance, and hyperlipidaemia, particularly of low-density lipoprotein cholesterol (Vessby et al., 2001; Sacks and Katan, 2002; Rasmussen et al., 2006), high cholesterol, atherosclerosis and heart disease (Ulbricht and Southgate, 1991). As a consequence of this, there has been consumer concern regarding the consumption of milk fat compared to spread and margarine alternatives due to high levels of these saturated fatty acids. However, recent reviews and meta-analysis have concluded that there is at least a neutral effect of milk intake on multiple health outcomes, and alternatively, cows' milk consumption may be beneficial in combating osteoporosis, cardiovascular disease, stroke, type 2 diabetes and some cancers (Armas et al., 2016; Lamarche et al., 2016). As well as pasture feeding, a more healthful milk fatty acid composition can be achieved by altering the cow's diet, for example, by feeding supplemental fish oil or roasted soybeans, or by selecting cows with a more unsaturated milk fatty acid composition (Bobe et al., 2007). However, the incorporation of polyunsaturated fatty acids into butter in this fashion will also increase the susceptibility to lipid oxidation. In contrast to short shelf life products, chemical reactions can limit the durability of long shelf life products like butter. The primary oxidation products of unsaturated fatty acids are the hydroperoxides, highly reactive compounds that decompose rapidly, yielding a complex mixture of nonvolatile and volatile compounds such as hydrocarbons, alcohols, acids, aldehydes and ketones which can negatively affect overall quality (Sanches-Silva et al., 2004). Secondary oxidation products cause flavour and taste deterioration in high fat polyunsaturated foods such as butter and are a leading cause of quality deterioration and limit shelf life. Aldehydes are particularly related to flavour alterations because their production leads to the reduction of food shelf life, production of undesirable odours, texture deterioration and reduction of nutritional food value (Lozano et al., 2007).

Dairy manufacturers produce large amounts of butter in the winter months. It is often necessary to store this butter for extended periods until there is a demand for it. During refrigerated or frozen storage, degradation of quality may occur (Krause et al., 2007). Freshly churned salted butter is characterised by an intense cooked/nutty flavour which likely comes from the high heat treatment that the cream receives prior to churning. This flavour is known to rapidly dissipate in butter (Bodyfelt et al., 1988). In general, cooked/nutty flavour decreased more rapidly in butters across storage compared to milk fat flavour while salty taste stays unchanged with storage time and off-flavours described as refrigerator/stale flavours increase to a greater extent in chilled stored as opposed to frozen stored butter (Lozano et al., 2007). Vitamin E (α-tocopherol) is present in pasture and will also be incorporated into milk on consumption by cattle as well as dairy products, including butter, manufactured from that milk. Increasing the concentration of α-tocopherol in milk will improve the oxidative stability of milk (Charmley et al., 1993; Charmley and Nicholson, 1994) and thus milk products. Therefore, butter made from milk rich in vitamin e will be more resistant to oxidative deterioration and the onset of stale flavours.

SENSORY PROPERTIES OF YOGHURT

Yoghurt is a very popular fermented milk that is produced all over the world by acid coagulation of milk without drainage (Sodini et al., 2004). Examples of different yoghurts include: Laban rayeb (Saudi Arabia), Tiaourti (Greece),

Matsoni (Russia), Skyr (Iceland), Jameed (Jordan). Dahi from India is made from Buffalo milk and Zabady from Egypt is traditionally made from ewes' milk. The essential flora of yogurt consists of the thermophiles *Streptococcus thermophilus* and *Lactobacillus delbrueckii* ssp. *bulgaricus*. It is made in a variety of compositions (fat and dry-matter content), either plain or with added substances such as fruits, sugar and gelling agents (Walstra et al., 2006d). Sweeteners (sugar, honey and aspartame), flavourings (vanilla and chocolate) and other ingredients (fruits, preserves and stabilizers such as gelatine to improve the textural property) are added to modify the sensory characteristics of yogurt. The most important aromatic components are acetaldehyde, acetone, acetoin and diacetyl in addition to acetic, formic, butanoic, and propanoic acids. Higher fat content allows longer persistence of volatiles (Routray and Mishra, 2011). Lactic acid itself is suggested to be one of the major compounds significantly contributing to yogurt flavour (Beshkova and others, 1998). Fruit yoghurt is among the most common fermented dairy products consumed around the world (Saint-Eve et al., 2006). The addition of flavours increases consumer appeal with strawberry being the most popular which also increases the functionality and antioxidant capacity of these dairy products (Trigueros et al., 2011) because of the presence of phenolic compounds, in particular anthocyanins (Oliveira et al., 2015).

Probiotic and prebiotic yoghurt products have found great popularity in recent years among consumers. With the emergence of a more health-conscious society, the role of probiotic bacteria in human health has gained considerable attention from both the consumer and the producer (O'Sullivan et al., 1992). The term probiotic refers to live microorganisms which when administered in sufficient quantities confer a health benefit on the host (FAO/WHO, 2001) whereas prebiotics refers to the nondigestible components present in food that helps in the growth of beneficial microorganisms (probiotics) in the digestive system (Routray and Mishra, 2011). Inulin is a widely used prebiotic in these products (Aryana and McGrew, 2007).

SENSORY PROPERTIES OF ICE CREAM

Ice cream is a type of sweetened frozen dessert that is usually flavoured, contains ice and is eaten in the frozen state. Ice cream generally contains dairy ingredients but can also be dairy free made with vegetable fat. Frozen yoghurt can also be classed as ice cream. The Italian variety Gelato is custard based and contains egg yolks (Clarke, 2004a). Ice cream, as we know it today, has been in existence for at least 300 years but the first ice cream making machine that froze the ice cream mix and also whipped in air at the same time was not invented until the 1840s (Underdown et al., 2011). In the United Kingdom, ice cream is defined as a frozen food product containing a minimum of 5% fat which must contain no less than 2.5% milk protein (such as from SMP.) The fat source may be vegetable such as hydrolysed palm kernel oil. The majority of the ice cream in the United Kingdom is made from non-milk-fat ingredients. Ice cream is usually preheated prior to homogenisation, pasteurised ($\sim 90°C$) prior to cooling ($\sim 5°C$) and then undergoes aeration and freezing (continuous or batch) in a scraped surface heat exchanger as an emulsion of fat, milk solids and sugar (or sweetener). Just prior to freezing, it is important the mix is 'aged' (at $5°C$) which involves fat crystallisation or fat destabilisation and is essentially a step that facilitates the formation of the correct microstructure of the mix just prior to freezing. For dairy ice cream the fat content (minimum 5%) must come only from milk fat. Higher value ice creams may contain up to 14% milk fat. In the United States as defined by the FDA (Food and Drug Administration), ice cream must contain at least 10% milk fat, below which it can only be called ice milk, and a minimum of 10% nonfat milk solids. As milk fat percentage increases from 10% the nonfat milk solids may sequentially decrease such that the dairy fat and nonfat dairy ingredients balance at the 20% level (FDA, 2015).

Ice cream products are generally more highly flavoured because of their frozen nature compared to ambient products. Ice cream contains sugar or sucrose as a sweetener which acts as a flavour enhancer as well as counteracting bitter or sour tastes. For texture evaluation a noncomplex vanilla ice cream may be used in order not to complicate the assessment with other flavours (Clarke, 2004b). According to Clarke (2004b) a good way to evaluate ice cream from a sensory perspective is to firstly spoon (e.g., determines hardness, crumbliness etc.), then bite (e.g., resistance) the sample, followed by squashing (e.g., mouthfeel, chewiness, guminess etc.) with the tongue followed by movement around the mouth (e.g., coarseness, smoothness, meltability, thickness, mouth coating) before finally swallowing (e.g., mouth coating, aftertaste). The perception of sandy texture refers to the presence of lactose crystals. A waxy mouthfeel is associated with the presence of high melting point fats that do not quickly melt in the mouth (Clarke, 2004b).

SENSORY PROPERTIES OF CHEESE

There are over 500 different varieties of commercially available cheese which have a wide range of visual, physical and flavour attributes. The flavour of cheese is governed by three main biochemical pathways; glycolysis, lipolysis and proteolysis. In general terms, the extent of each of these processes is characteristic of the individual cheese variety (Kilcawley and O'Sullivan, 2016). Cheese can be made from cows, sheep, goats and even buffalo milk. The milk is acidified typically through the action of an added starter culture of bacteria (lactic acid bacteria) which convert lactose into lactic acid. Rennet is then added which is a complex mixture of enzymes containing chymosin, the main protease, extracted from the stomach of calves which curdles the casein in milk. The curds and whey are then separated with the curds typically formed into moulds and ripened to make cheese. Cheese sensory properties can be categorised as taste, texture and aroma/flavour properties, but colour is also important for some varieties. With regard to taste, salt (NaCl) is of particular importance as it directly impacts on taste and acts as a flavour enhancer and influences structure and rheological properties of cheese. The extent of the impact of salt in cheese flavour depends upon its concentration, cheese composition and the age of the cheese. The biochemical reactions in cheese are primarily initiated by the addition of microbial populations and/or exogenous enzymes during production. Cheese is a dynamic product, with many varieties having up to 100 billion bacteria per gram, all of which metabolise carbohydrates, lipids and/ or proteins to create a myriad of aromatic and sapid compounds that contribute to cheese flavour. These biochemical reactions are in turn influenced by the milk (type, quality and treatment), production equipment/processes, indigenous/exogenous microbial populations (selection and concentration), indigenous/exogenous enzymes (selection and concentration), salting (dry and brine), production processes and ripening regimes (time, temperature and humidity), all of help contribute to the wide variety of cheeses available (Kilcawley and O'Sullivan, 2016).

Lactic acid bacteria are Gram-positive, nonmotile and nonspore forming, which grow anaerobically and obligatory ferment carbohydrates and metabolise lactose to lactic acid. This aids in curd formation (post whey drainage) and in flavour development during the early stages of cheese ripening (Walstra et al., 2006c; Kilcawley and O'Sullivan, 2016). Lactate, diacetyl, acetoin, acetaldehyde and ethanol are all important cheese flavour compounds. Lactic acid has a slightly tart taste, but also more importantly influences the final pH of cheese, thus also influencing overall cheese flavour perception. Acetaldehyde has a characteristic yoghurt aroma, while diacetyl (2,3-butanedione) and acetoin (3-hydroxy-2-butanone) impart a very characteristic creamy buttery aroma, with acetic acid providing a sharp vinegar note (Curionia and Bosset, 2002; Smit et al., 2005). Ethanol likely has a minimal direct contribution to cheese flavour as it is described as having a dry dust aroma (Curioni and Bosset, 2002), but is much more important in relation to the formation of ethyl esters with free fatty acids, via esterification or alcoholysis as these compounds are very odour active and are responsible for fruity flavours in many cheeses. Diacetyl, an important compound in other cheese varieties, such as quark and cottage cheese and acetoin and 2,3-butanediol are commonly found in cheese, but their production is primarily due to amino acid metabolism (McSweeney, 2004). In Swiss-type cheeses, the hot room stage promotes growth of propionibacteria that metabolise L-lactate to propionic acid (contributes to the sweet nutty flavour), acetic acid and CO_2 which contribute to Swiss cheese flavour, texture and visual characteristics (eyes).

Lipolysis is the hydrolysis of free fatty acids from tri-, di- and monoacylglycerides and carried out by two hydrolytic enzymes: esterases and lipases. Both enzymes catalyse the same reaction, the hydrolysis of the ester bond of a glyceride yielding a fatty acid and an alcohol (glycerol), but operate in different environments and have different specificities (Kilcawley and O'Sullivan, 2016). The source of lipolytic enzymes in cheese are relatively widespread and can originate from a number of different sources: (1) the indigenous milk lipase (lipoprotein lipase), (2) pregastric esterases/rennet pastes, (3) starter bacteria, (4) adjunct starter bacteria, (5) NSLAB, (6) yeasts and molds and (7) the addition as exogenous lipases (Deeth and FitzGerald, 1995; Fox and Wallace, 1997; McSweeney and Sousa, 2000). The extent of lipolysis will also vary within each cheese variety, due to differences in manufacture, milk and ripening times. In Camembert and Blue cheeses the flavour perception is reduced and fatty acids are not immediately associated with rancid off-flavours, because of their higher pH, although it can be a problem if the pH is lower (Alewijn, 2006). Camembert and Blue-type cheeses contain extremely high levels of free fatty acids. The metabolism of individual free fatty acids is very important as methyl ketones (major contributors to Blue-veined cheese flavour), secondary alcohols, esters (methyl esters, ethyl esters and propyl esters) and lactones are all odour active and contribute to cheese flavour. Both 2-heptanone and 2-nonanone are important methyl ketones in Blue-veined and hard Italian cheeses and are described as having a Blue cheese, fruity, sweet, and fruity, musty, rose, tea-like aroma, respectively (Qian and Burbank, 2007). The ethyl ketone 1-octen-3-one is thought to be an important component in Parmesan, Grana Padano and Pecorino cheeses and has a mushroom-like earthy aroma (Kubickova and Grosch, 1998; Qian and Burbank, 2007). Secondary alcohols in cheese are formed by the reduction of their corresponding methyl ketones (Engels and Visser, 1997). In Blue-veined cheese *Penicillium* species are directly responsible for the production of 2-pentanol, 2-heptanol and 2-nonanol from acetone, 2-heptanone and 2-nonanone, respectively (Collins et al., 2003). The aroma attributes of these secondary alcohols have been described by Qian and Burbank (2007), Curioni and Bosset (2002), and Kubicikova and Grosch (1998); 2-pentanol (green, fruity, fresh), 2-heptanol (fruity, earthy, green, sweet) and 2-nonanol (fatty, green), but are thought not to be as important as the methyl ketones because of their low aroma activity (Singh et al., 2003). Esterification is the formation of esters from alcohols and carboxylic acids, whereas alcoholysis is the production of esters from alcohols and acylglycerols or from alcohols and fatty acyl-CoAs derived from the metabolism of fatty acids, amino acids and/or carbohydrates. Ethyl acetate, ethyl butyrate, ethyl hexanoate, ethyl octanoate and ethyl decanoate are found in most cheese varieties (Liu et al., 2004). In general they all have fruity aromas, but distinct aroma differences exist. Liu et al. (2004) summarised the aroma of each of these esters as follows: ethyl acetate (solvent, fruity, pineapple), ethyl butanoate (apple, banana, sweet, fruity, fragrant), ethyl hexanoate (banana, pineapple, sweet, fruity, wine-like, brandy, powerful), ethyl octanoate (pear, sweet, fruity, banana, pineapple, apricot, wine, floral), ethyl decanoate (apple, brandy, grape-like, fruity, oily).

Engels and Visser (1997) were able to directly associate fruity notes in Gruyere, Parmesan and Proosdij cheese with ethyl butanoate, as did Lawlor et al. (2003) with Blue-type cheeses, but this could be a defect in Cheddar cheese (McSweeney and Sousa, 2000). In general, ester formation is minimal in cheeses, such as Gouda and Cheddar in comparison to Italian type cheeses, where they are important characteristic aroma compounds. Alewijn et al. (2005) reported that lactones contribute to the flavour of Gouda cheese, although their relatively high flavour threshold may limit their overall contribution to cheese flavour (Qian and Burbank, 2007).

Proteolysis contributes to the softening of cheese texture over ripening due to the hydrolysis of caseins within the curd and through a decrease in water activity (McSweeney, 2004), and directly and indirectly to cheese flavour through the hydrolysis of caseins to small peptides and amino acids, and via the metabolism of amino acids by starter bacteria and secondary cultures (Kilcawley and O'Sullivan, 2016). Milk also contains the indigenous proteolytic enzymes plasmin and cathepsin D that can also play an active role in primary proteolysis. Direct associations between proteolysis and the development of mouthfeel, texture, taste and aroma of maturing cheese have been reported (Pripp et al., 2006). Proteolysis is the most complex of the three biochemical pathways involved in cheese ripening, mainly due to the diversity of potential enzymatic and chemical reactions involved. Proteolysis is initiated primarily by the coagulants action on casein and subsequently by the action of cell-wall proteinases of lactic acid bacteria and intracellular metabolic activity. Bitterness in cheese is due to the accumulation of small peptides that have hydrophobic end sequences due to the presence of certain amino acids at the carboxy or amino terminal. The main bitter amino acids are isoleucine, phenylalanine, leucine, methionine, proline and valine, although lysine, tryptophan, tyrosine, histidine and arginine are also considered bitter (Kilcawley, 2016). Amino acids are also associated with sweetness (proline, lysine, alanine, glycine, serine and threonine), sourness (histidine, asparagine and glutamic acid) and umami flavour (leucine, tyrosine, asparagine and glutamic acid) (McSweeney, 1997). Many cheese aroma compounds are formed from the metabolism of amino acids, a major pathway in cheese flavour development. Possibly the most important aroma compounds resulting from amino acid metabolism are aldehydes, alcohols, carboxylic acids, thiols and thioesters (Kilcawley, 2016).

Some caution must be exercised with regard to relating volatile compounds to cheese flavour, mainly due to the fact that the combined odour effect of these compounds may differ considerably to individual compounds in pure form. Also many studies using advanced gas chromatographic mass spectrometric techniques fail to incorporate odour activity (gas chromatography olfactometry) analysis or descriptive sensory analysis (nor take into account limitations or inherent bias of extraction/concentration techniques, column phases, detector sensitivity) which is important when working with complex fermented foods, such as cheese. Thus it is possible that the influence of some compounds may be overestimated, while other important compounds may be overlooked (Kilcawley and O'Sullivan, 2016).

CASE STUDY: SENSORY QUALITY OF UNHEATED AND HEATED MOZZARELLA

The presented case study evaluated the effects of reducing fat from 23 w/w to 11 w/w and salt from 1.8 to 1.0 w/w on the texture profile attributes, volatile compounds and sensory characteristics of unheated and heated (95°C) mozzarella-style cheese after 15 and 35 days of storage at 4°C. The properties of heated mozzarella cheese are of particular relevance in cooking applications, such as pizza, lasagne and pasta dishes and reductions in fat and salt can potentially improve the nutritional quality of this widely used product. Comparatively, little information is available on the sensory properties of heated cheese and how these are affected by composition or other characteristics of the unheated cheese (Henneberry et al., 2016). Naïve assessors ($n = 25$) evaluated the 12 different Mozzarella cheeses (6 different treatment Mozzarella cheeses described in Table 12.1, each at 15 and 35 days) utilising sensory Hedonic descriptors. Unheated samples (4°C) were coded with a randomly selected three digit code and presented singly to the assessor panel at ambient temperatures (~ 21°C). For the heated samples, ~ 30 g of cheese was placed in a (polyamide/polyethylene) bag, heat-sealed and then placed in water bath at 95°C for 10 min until it reached a core temperature of 95°C (temperature probe); this temperature reflects that obtained on cooking pizza in an electric fan oven at 280°C (Guinee and O'Callaghan, 1997; Henneberry et al., 2015). The cheese was immediately served to assessors. No more than six samples were presented at any individual sensory session. Each assessor was provided with deionised water and instructed to cleanse their palate between tastings. Additionally, each assessor was asked to indicate their degree of liking on a 10-cm line scale ranging from 0 (extremely dislike) at the left to 10 (extremely like) at the right and rating subsequently scored in centimetre from left. An RDA (Dairou and Sieffermann, 2002; Richter et al., 2010) was then carried out after affective testing. All six samples were presented simultaneously and tested in one sitting with short breaks included for the heated cheeses (every three cheeses), so that they could be assessed while still hot. The order of presentation of all test samples was randomised to prevent first order and carryover effects; all samples were presented in duplicate. Texture profile analysis (TPA) and volatile analysis using GCMS were also conducted on cheese samples (Henneberry et al., 2016). Overall, the unheated and heated mozzarella full-fat cheese variants had the highest sensory acceptance of the six treatment cheeses at 15 and 35 days. In contrast, the reduced-fat cheeses generally scored lowest for liking of texture, flavour and overall acceptability at 15 and 35 days. Reducing salt content in the range

TABLE 12.1 Cheese Composition, Biochemical and Textural Attributes

	Cheese Code					
	FFFS	FFRS	RFFS	RFRS	RFFSLC	RFRSLC
Composition at 15 days						
Moisture (%, w/w)	47.6^c	48.0^c	50.4^b	50.1^b	56.7^a	55.8^a
Protein (%, w/w)	25.4^c	23.4^d	32.1^a	32.7^a	28.2^b	29.4^b
M/P (%, w/w)	1.9^b	2.1^a	1.6^c	1.5^d	2.0^b	1.9^b
Fat (%, w/w)	21.5^b	23.1^a	11.4^c	11.0^c	9.0^d	10.6^d
Salt (%, w/w)	1.6^a	1.0^b	1.8^a	1.1^b	1.8^a	1.0^b
S/M (%, w/w)	3.4^a	2.1^{c,d}	3.5^a	2.3^c	3.1^b	1.7^d
Ca (mg/100g)	761^c	655^d	942^b	1009^a	551^e	506^f
Biochemical and Textural Parameters at 15 days						
Protein hydration (g water/g protein)	1.5^c	1.8^b	1.5^c	1.5^c	1.9^a	1.8^{a,b}
WSN (% of total N)	5.4^b	4.4^c	3.4^d	3.5^d	6.4^a	3.1^e
Hardness (N)	71^b	55^c	236^a	141^b	79^b	71^b
Chewiness (N)	79^c	50^e	139^a	106^b	65^d	76^c
Cohesiveness (−)	0.7^b	0.6^c	0.8^a	0.8^a	0.4^a	0.8^a
Adhesiveness (Nmm)	1.8^b	3.9^a	0.1^e	0.5^d	1.0^c	0.6^d
Resilience (−)	0.3^b	0.2^c	0.5^a	0.5^a	0.5^a	0.5^a
Biochemical and Textural Parameters at 35 days						
Protein hydration (g water/g protein)	1.9^b	2.1^a	1.6^c	1.5^c	2.0^{a,b}	1.9^b
WSN (% of total N)	7.7^a	5.3^c	6.4^b	4.1^e	6.6^b	4.9^d
Hardness (N)	93^b	55^d	136^a	135^a	80^c	77^c
Chewiness (N)	75^c	42^e	128^a	100^b	74^c	65^d
Cohesiveness (−)	0.7^c	0.6^c	0.7^b	0.8^b	0.8^a	0.8^b
Adhesiveness (Nmm)	2.9^b	4.1^a	0.3^f	0.6^d	1.1^c	0.5^e
Resilience (−)	0.4^b	0.2^c	0.4^{a,b}	0.5^a	0.5^a	0.4^b

Presented data represent the means of three replicates for each of six treatment mozzarella cheeses: full-fat full salt (FFFS), full-fat reduced salt (FFRS), reduced-fat full salt (RFFS), reduced-fat reduced salt (RFRS), reduced-fat full salt low calcium (RFFSLCa) and reduced-fat reduced salt low calcium (RFRSLCa).
The statistical effects of reducing fat, salt and calcium contents are indicated by lower-case superscripts. Values not sharing a common superscript differ significantly, $P < 0.05$.
Abbreviations: Ca, calcium; M/P, moisture-to-protein ratio; S/M, salt-in-moisture; WSN, water soluble nitrogen (% of total nitrogen).

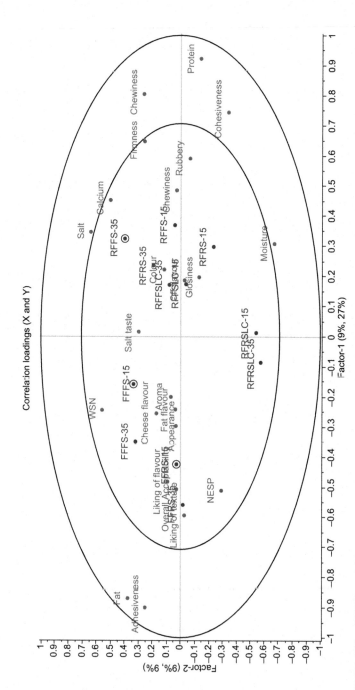

FIGURE 12.2 ANOVA-partial least squares regression (APLSR) plot for unheated mozzarella-style cheeses, showing the affective, descriptive sensory and instrumental data. Where sensory and instrumental terms have the same name, the sensory term is denoted by the suffix S and the instrumental term by I. Full-fat full salt (FFFS), full-fat reduced salt (FFRS), reduced-fat full salt (RFFS), reduced-fat reduced salt (RFRS), reduced-fat full salt low calcium (RFFSLCa) and reduced-fat reduced salt low calcium (RFRSLCa). Reproduced from Henneberry, S., O'Sullivan, M.G., Kilcawley, K.N. Kelly, P.M., Wilkinson, M.G., Guinee, T.P., 2016. Sensory quality of unheated and heated Mozzarella-style cheeses with different fat, salt and calcium levels. International Journal of Dairy Science 69, 38–50.

studied (1.8−1.0%, w/w) had little impact on the sensory properties of the cheese, but reducing fat in the range 21−10% (w/w) had a very definitive negative impact on sensory quality. The overall sensory perception of the reduced-fat cheese was improved by reducing calcium levels but remained less acceptable than the full-fat cheese (Fig. 12.2). The sensory properties and volatile profile of the unheated and heated cheeses were similar for each cheese type. Limonene, cymene, nonanal and toluene correlated with the liking of aroma and flavour and with the attributes cheese flavour and fat flavour in the unheated and heated cheeses. Off-flavour in unheated and heated cheeses was correlated with benzene acetaldehyde, 2-ethyl furan and benzaldehyde. The sensory attributes chewiness, firmness and rubbery generally correlated with instrumental chewiness and hardness. Instrumental heat-induced flow (Schflow) correlated well with sensory attributes fluidity, oiling off and spreadability (Henneberry et al., 2016).

REFERENCES

Aigster, A., Sims, C., Staples, C., Schmidt, R., O'Keefe, S.F., 2000. Comparison of cheeses made from milk having normal and high oleic fatty acid compositions. Journal of Food Science 65, 920−924.

Alewijn, M., 2006. The Formation of Fat-Derived Flavour Compounds During Ripening of Gouda-Type Cheese. Wageningen University, The Netherlands (Ph.D thesis).

Alewijn, M., Sliwinski, E.L., Wouters, J.T.M., 2005. Production of fat-derived (flavour) compounds during the ripening of Gouda cheese. International Dairy Journal 15, 733−740.

Armas, L.A., Frye, C.P., Heaney, R.P., 2016. Effect of cow's milk on human health. In: Beverage Impacts on Health and Nutrition. Springer, pp. 131−150.

Aryana, K.J., McGrew, P., 2007. Quality attributes of yogurt with *Lactobacillus casei* and various prebiotics. LWT − Food Science and Technology 40 (10), 1808−1814.

Azzara, D.C., Campbell, L.B., 1992. Off-flavors of dairy products. In: Charalambous, G. (Ed.), Off-flavors in Foods and Beverages. Elsevier, Amsterdam, pp. 329−374.

Beshkova, D., Simova, E., Frengova, G., Simov, Z., 1998. Production of flavour compounds by yogurt starter cultures. Journal of Industrial Microbiology and Biotechnology 20 (3−4), 180−186.

Bodyfelt, F.W., Tobias, J., Trout, G.M., 1988. Sensory evaluation of butter. In: The Sensory Evaluation of Dairy Products. Van Nostrand Reinhold, New York, NY, pp. 377−417.

Bobe, G., Zimmerman, S., Hammond, E.G., Freeman, A.E., Porter, P.A., Luhman, C.M., Beitz, D.C., 2007. Butter composition and texture from cows with different milk fatty acid compositions fed fish oil or roasted soybeans. Journal of Dairy Science 90, 2596−2603.

Charmley, E., Nicholson, J.W., Zee, J.A., 1993. Effect of supplemental vitamin E and selenium in the diet on vitamin E and selenium levels and control oxidized flavor in milk from Holstein cows. Candian Journal of Animal Science 73, 453−457.

Charmley, E., Nicholson, J.W.G., 1994. Influence of dietary fat source on oxidative stability and fatty acid composition of milk from cows receiving a low or high level of dietary vitamin E. Candian Journal of Animal Science 74, 657−664.

Civille, G.V., Szczesniak, A.S., 1973. Guidelines to training a texture profile panel. Journal of Texture Studies 4, 204−223.

Clarke, C., 2004a. Ch1 the Story of Ice Cream. The Science of Ice Cream. RSC Publishing, pp. 1–12.

Clarke, C., 2004b. Ch6 Measuring Ice Cream. The Science of Ice Cream. RSC Publishing, pp. 104–134.

Collins, Y.F., McSweeney, P.L.H., Wilkinson, M.G., 2003. Lipolysis and free fatty acid catabolism in cheese: a review of current knowledge. International Dairy Journal 13, 841–866.

Couvreur, S., Hurtaud, C., Lopez, C., Delaby, L., Peyraud, J.-L., 2006. The linear relationship between the proportion of fresh grass in the cow diet, milk fatty acid composition, and butter properties. Journal of Dairy Science 89, 1956–1969.

Curioni, P.M.G., Bosset, J.O., 2002. Key odorants in various cheese types as determined by gas chromatography-olfactometry. International Dairy Journal 12, 959–984.

Dairou, V., Sieffermann, J.M., 2002. A comparison of 14 jams characterized by conventional profile and a quick original method, the flash profile. Journal of Food Science 67, 826–834.

Deeth, H.C., Fitz-Gerald, C.H., 1995. Lipolytic enzymes and hydrolytic rancidity in milk and milk products. In: Fox, P.F. (Ed.), Advanced Dairy Chemistry, Lipids, vol. 2. Chapman & Hall, London, pp. 247–308.

Delahunty, C.M., Drake, M.A., 2004. Sensory character of cheese and its evaluation. In: Fox, P.F., McSweeney, P.L.H., Cogan, T.M., Guinee, T.P. (Eds.), Cheese: Chemistry, Physics and Microbiology, General Aspects, vol. 1. Elsevier, Oxford, pp. 455–487.

Delahunty, C.M., Murray, J.M., 1997. Organoleptic evaluation of cheese. In: Cogan, T.M., Fox, P.F., Ross, R.P. (Eds.), Proceedings of the 5th Cheese Symposium. Teagasc, Dublin, pp. 90–97.

Dhiman, T.R., Anand, G.R., Satter, L.D., Pariza, M., 1999. Conjugated linoleic acid content of milk from cows fed different diets. Journal of Dairy Science 82, 2146–2156.

Drake, M.A., Karagul-Yuceer, Y., Cadwallader, K.R., Civille, G.V., Tong, P.S., 2003. Determination of the sensory attributes of dried milk powders and dairy ingredients. Journal of Sensory Studies 18, 199–216.

Drake, M.A., 2007. Invited review: sensory analysis of dairy foods. Journal of Dairy Science 90, 4925–4937.

Drake, M.A., Yates, M.D., Gerard, P.D., Delahunty, C.M., Sheehan, E.M., Turnbull, R.P., Dodds, T.M., 2005. Comparison of differences between lexicons for descriptive analysis of Cheddar cheese flavour in Ireland, New Zealand, and the United States of America. International Dairy Journal 15, 473–483.

Engels, W.J.M., Visser, S., 1997. A comparative study of volatile compounds in the water-soluble fraction of various types of ripened cheese. International Dairy Journal 48, 127–140.

FAO/WHO, 2001. Health and Nutritional Properties of Probiotics in Food Including Powder Milk With Live Lactic Acid Bacteria. Report of a Joint FAO/WHO Expert Consultation on Evaluation of Health and Nutritional Properties of Probiotics in Food Including Powder Milk With Live Lactic Acid Bacteria.

FDA, 2015. 21CFR (Code of Federal Regulations) 135.110. Title 21—Food and Drugs. Chapter I—Food and Drug Administration. Department of Health and Human Services. Subchapter B—Food for Human Consumption. Part 135 — Frozen Desserts. Subpart B—Requirements for Specific Standardized Frozen Desserts Sec. 135.110 Ice Cream and Frozen Custard. http://www.accessdata.fda.gov/scripts/cdrh/cfdocs/cfcfr/cfrsearch.cfm?fr=135.110.

Fox, P.F., Wallace, J.M., 1997. Formation of flavour compounds in cheese. Advances in Applied Microbiology 45, 17–85.

Galli, B.D., Martin, J.G.P., da Silva, P.P.M., Porto, E., Spoto, M.H.F., 2016. Sensory quality of Camembert-type cheese: relationship between starter cultures and ripening molds. International Journal of Food Microbiology 234, 71–75.

Gori, A., Cevoli, C., Fabbri, A., Caboni, M.F., Losi, G., 2012. A rapid method to discriminate season of production and feeding regimen of butters based on infrared spectroscopy and artificial neural networks. Journal of Food Engineering 109, 525–530.

Guinee, T.P., O'Callaghan, D.J., 1997. The use of a simple empirical method for objective quantification of the stretchability of cheese on cooked pizza pies. Journal of Food Engineering 31, 147–161.

Henneberry, S., Kelly, P.M., Kilcawley, K.N., Wilkinson, M.G., Guinee, T.P., 2015. Interactive effects of salt and fat reduction on composition, rheology and functional properties of Mozzarella-style cheese. Dairy Science and Technology 95, 613–638.

Henneberry, S., O'Sullivan, M.G., Kilcawley, K.N., Kelly, P.M., Wilkinson, M.G., Guinee, T.P., 2016. Sensory quality of unheated and heated Mozzarella-style cheeses with different fat, salt and calcium levels. International Journal of Dairy Science 69, 38–50.

Hurtaud, C., Peyraud, J.L., 2007. Effects of feeding Camelina (seeds or meal) on milk fatty acid composition and butter spreadability. Journal of Dairy Science 90, 5134–5145.

Van Hekken, D.L., Drake, M.A., Corral, F.J.M., Prieto, V.M.G., Gardea, A.A., 2006. Mexican Chihuahua cheese: sensory profiles of young cheese. Journal of Dairy Science 89, 3729–3738.

Jensen, R.G., Gander, G.W., Sampugna, J., 1962. Fatty acid composition of the lipids from pooled, raw milk. Journal of Dairy Science 45, 329–331.

Jensen, R.G., Ferris, A.M., Lammi-Keefe, C.J., 1991. The composition of milk fat. Journal of Dairy Science 74, 3228–3243.

Kilcawley, K.N., O'Sullivan, M.G., 2016. Cheese flavour development and sensory characteristics. In: Papademas, P., Bintsis, T. (Eds.), Global Cheesemaking Technology: Cheese Quality and Characteristics. John Wiley & Sons Ltd, Chichester.

Krause, A.J., Miracle, R.E., Sanders, T.H., Dean, L.L., Drake, M.A., 2007. .The effect of refrigerated and frozen storage on butter flavor and texture. Journal of Dairy Science 91, 455–465.

Kubickova, J., Grosch, W., 1998. Evaluation of flavour compounds of Camembert cheese. International Dairy Journal 8, 11–16.

Lamarche, B., Givens, I., Soedamah-Muthu, S., Krauss, R.M., Jakobsen, M.U., Bischoff-Ferrari, H.A., Pan, A., Després, J.-P., 2016. Does milk consumption contribute to cardiometabolic health and overall diet quality? Canadian Journal of Cardiology.

Lawless, H.T., Heymann, H., 1998. Descriptive analysis. In: Lawless, H.T., Heymann, H. (Eds.), Sensory Evaluation of Food, Principles and Practices. Kluwer Academic/Plenum Publishers, New York, London, Dordrecht, Boston, pp. 341–378.

Lawlor, J.B., Delahunty, C.M., Sheehan, J., Wilkinson, M.G., 2003. Relationships between sensory attributes and the volatile compounds, non-volatile and gross compositional constituents of six blue-type cheeses. International Dairy Journal 13, 481–494.

Liu, S.Q., Holland, R., Crow, V., 2004. Esters and their biosynthesis in fermented dairy products: a review. International Dairy Journal 14, 923–945.

Lozano, P.R., Miracle, E.R., Krause, A.J., Drake, M., Cadwallader, K.R., 2007. Effect of cold storage and packaging material on the major aroma components of sweet cream butter. Journal of Agricultural and Food Chemistry 55, 7840–7846.

Mallia, S., Escher, F., Schlichtherle-Cerny, H., 2008. Aroma-active compounds of butter: a review. European Food Research and Technology 226, 315–325.

Markiewicz-Keszycka, M., Czyżak-Runowska, C., Lipińska, P., Wójtowski, J., 2013. Fatty acid profile of milk – a review. Bulletin-Veterinary Institute in Pulawy 57, 135–139.

McSweeney, P.L.H., 1997. The flavour of milk and dairy products: III. Cheese: taste. International Journal of Dairy Technology 50, 123—128.

McSweeney, P.L.H., 2004. Biochemistry of cheese ripening. International Journal of Dairy Technology 57 (2—3), 127—144.

McSweeney, P.L.H., Sousa, M.J., 2000. Biochemical pathways for the production of flavour compounds in cheese during ripening. Le Lait 80, 293—324.

Meilgaard, M.C., Civille, G.V., Carr, B.T., 1999. Sensory Evaluation Techniques, third ed. CRC Press, Boca Raton, New York.

Muir, D.D., 2010. The grading and sensory profiling of cheese. In: Law, B.A., Tamine, A.Y. (Eds.), Technology of Cheesemaking, second ed. Wiley-Blackwell Publication, Oxford, pp. 440—474.

Muir, D.D., Hunter, E.A., Banks, J.M., Home, D.S., 1996. Sensory properties of Cheddar cheese: changes during maturation. Food Research International 28, 561—568.

Muñoz, A.M., Civille, G.V., 1992. The spectrum descriptive analysis method. In: Hootman, R.C. (Ed.), ASTM Manual on Descriptive Analysis. American Society for Testing and Materials, Pennsylvania.

Murray, J., Delahunty, C., Baxter, I., 2001. Descriptive sensory analysis: past, present and future. Food Research International 34, 461—471.

Murray, J.M., Delahunty, C.M., 2000a. Description of Cheddar cheese packaging attributes using an agreed vocabulary. Journal of Sensory Studies 15 (2), 201—218.

Murray, J.M., Delahunty, C.M., 2000b. Selection of standards to reference terms in a Cheddar-type cheese flavor language. Journal of Sensory Studies 15 (2), 179—199.

Nursten, H.E., 1997. The flavour of milk and dairy products: I. Milk of different kinds, milk powder, butter and cream. International Journal of Dairy Technology 50, 48—56.

Oliveira, A., Alexandre, E.M.C., Coelho, M., Lopes, C., Almeida, D.P.F., Pintado, M., 2015. Incorporation of strawberries preparation in yoghurt: impact on phytochemicals and milk proteins. Food Chemistry 171, 370—378.

O'Callaghan, T.F., Faulkner, H., McAuliffe, S., O'Sullivan, M.G., Henness, D., Dillon, P., Kilcawley, K.N., Stanton, C., Ross, P.R., 2016. Characteristics, composition and sensory properties of butter from cows on pasture versus indoor feeding systems. Journal of Dairy Science (submitted for publication).

O'Sullivan, M.G., Thornton, G., O'Sullivan, G.C., Collins, J.K., 1992. Probiotic bacteria: myth or reality? Trends in Food Science and Technology 3, 309—317.

O'Sullivan, M.G., Kerry, J.P., Byrne, D.V., 2011. Use of sensory science as a practical commercial tool in the development of consumer-led processed meat products. In: Kerry, J.P., Kerry, J.F. (Eds.), Processed Meats: Improving, Safety and Nutritional Quality. Woodhead Publishing Ltd, UK, pp. 156—182.

Pearce, K.N., 2016. "Milk Powder", Food Science Section. Dairy Research Institute, New Zealand. In: http://nzic.org.nz/ChemProcesses/dairy/3C.pdf.

Pripp, A.H., Skeie, S., Isaksson, T., Borge, G.I., Sorhaug, T., 2006. Multivariate modelling of relationships between proteolysis and sensory quality of Prast cheese. International Dairy Journal 16, 225—235.

Qian, M.C., Burbank, H.M., 2007. Hard Italian cheeses: Parmigiano-Reggiano and Grana Padano. In: Weimer, B.C. (Ed.), Improving the Flavour of Cheese. Woodhead Publishing Ltd, Cambridge England, pp. 421—443.

Rasmussen, B.M., Vessby, B., Uusitupa, M., Berglund, L., Pedersen, E., Riccardi, G., Rivellese, A.A., Tapsell, L., Hermansen, K., 2006. Effects of dietary saturated, mono-unsaturated, and n-3 fatty acids on blood pressure in healthy subjects. The American Journal of Clinical Nutrition 83 (2), 221—226.

Rétiveau, A., Chambers, D.H., Esteve, E., 2005. Developing a lexicon for the flavor description of French cheeses. Food Quality and Preference 16, 517–527.

Richter, V., Almeida, T., Prudencio, S., Benassi, M., 2010. Proposing a ranking descriptive sensory method. Food Quality and Preference 21, 611–620.

Routray, W., Mishra, H.N., 2011. Scientific and technical aspects of yogurt aroma and taste: a review. Comprehensive Reviews in Food Science and Food Safety 10, 208–220.

Sacks, F.M., Katan, M., 2002. Randomized clinical trials on the effects of dietary fat and carbohydrate on plasma lipoproteins and cardiovascular disease. The American Journal of Medicine 113, 13S–24S.

Saint-Eve, A., Lévy, C., Martin, N., Souchon, I., 2006. Influence of proteins on the perception of flavored stirred yoghurts. Journal of Dairy Science 89 (3), 922–933.

Sanches-Silva, A., Rodriguez-Bernaldo de Quiros, A., Lopez-Hernandez, J., Paseiro- Losada, P., 2004. Determination of hexanal as indicator of lipid oxidation state in potato crisps using gas chromatography and high-performance liquid chromatography. Journal of Chromatography A 1046, 75–81.

Singh, T.K., Drake, M.A., Cadwallader, K.R., 2003. Flavor of Cheddar cheese: a chemical and sensory perspective. Comprehensive Reviews in Food Science and Food Safety 2, 139–162.

Smit, G., Smit, B.A., Engles, W.J.M., 2005. Flavour formation by lactic acid bacteria and biochemical flavour profiling of cheese products. FEMS Microbiology Reviews 29, 591–610.

Sodini, I., Remeuf, F., Haddad, S., Corrieu, G., 2004. The relative effect of milk base, starter, and process on yoghurt texture: a review. Critical Reviews in Food Science and Nutrition 44 (2), 113–137.

Stone, H., Sidel, J.L., 2004. Introduction to sensory evaluation. In: Stone, H., Sidel, J.L. (Eds.), Sensory Evaluation Practices, third ed. Elsevier Academic Press, Oxford, UK, pp. 1–20.

Thomas, E.L., 1981. Trends in milk flavors. Journal of Dairy Science 64, 1023–1027.

Torracca, B., Pedonese, F., Belen Lopez, M., Turchi, B., Fratini, F., Nuvoloni, R., 2016. Effect of milk pasteurisation and of ripening in a cave on biogenic amine content and sensory properties of a pecorino cheese. International Dairy Journal 61, 189–195.

Trigueros, L., Pérez-Alvarez, J.A., Viuda-Martos, M., Sendra, E., 2011. Production of low-fat yoghurt with quince (Cydonia oblonga Mill.) scalding water. LWT — Food Science and Technology 44 (6), 1388–1395.

Ulbricht, T., Southgate, D., 1991. Coronary heart disease: seven dietary factors. The Lancet 338 (8773), 985–992.

Underdown, J., Quail, P.J., Smith, K.W., 2011. Ch17 Saturated fat reduction in ice cream. In: Talbot, G. (Ed.), Reducing Saturated Fats in Foods. Woodhead Publishing Limited.

Vessby, B., Uusitupa, M., Hermansen, K., Riccardi, G., Rivellese, A.A., Tapsell, L.C., Nalsen, C., Berglund, L., Louheranta, A., Rasmussen, B.M., Calvert, G.D., Maffetone, A., Pedersen, E., Gustafsson, I.-B., Storlien, L.H., 2001. Substituting dietary saturated for monounsaturated fat impairs insulin sensitivity in healthy men and women: the KANWU study. Diabetologia 44, 312–319.

Walstra, P., Wouters, J.T.M., Geurts, T.J., 2006a. CH1. Mil: main characteristsics. In: Walstra, P., Wouters, J.T.M., Geurts, T.J. (Eds.), Dairy Science and Technology, second ed. Talyor & Francis Group, Boca Raton, London, New York, pp. 3–16.

Walstra, P., Wouters, J.T.M., Geurts, T.J., 2006b. CH4: milk proteins. In: Walstra, P., Wouters, J.T.M., Geurts, T.J. (Eds.), Dairy Science and Technology, second ed. Talyor & Francis Group, Boca Raton, London, New York, pp. 357–397.

Walstra, P., Wouters, J.T.M., Geurts, T.J., 2006c. Lactic fermentations. In: Walstra, P., Wouters, J.T.M., Geurts, T.J. (Eds.), Dairy Science and Technology, second ed. Talyor & Francis Group, Boca Raton, London, New York, pp. 357–397.

Walstra, P., Wouters, J.T.M., Geurts, T.J., 2006d. CH22: fermented milks. In: Walstra, P., Wouters, J.T.M., Geurts, T.J. (Eds.), Dairy Science and Technology, second ed. Talyor & Francis Group, Boca Raton, London, New York, pp. 357–397.

Walstra, P., Wouters, J.T.M., Geurts, T.J., 2006e. CH20: milk powder. In: Walstra, P., Wouters, J.T.M., Geurts, T.J. (Eds.), Dairy Science and Technology, second ed. Talyor & Francis Group, Boca Raton, London, New York, pp. 513–535.

Wilkinson, M.G., 2007. Lipolysis and cheese flavour development. In: Weimer, B. (Ed.), Improving the Flavour of Cheese. Woodhead Publishing Ltd, Cambridge, UK, pp. 102–120.

Worch, T., Crine, A., Gruel, A., Lê, S., 2014. Analysis and validation of the ideal profile method: application to a skin cream study. Food Quality and Preference 32, 132–144.

Worch, T., Lê, S., Punter, P., 2010. How reliable are the consumers? Comparison of sensory profiles from consumers and experts. Food Quality and Preference 21, 309–318.

Chapter 13

Sensory Properties of Beverage Products (Alcoholic and Nonalcoholic)

INTRODUCTION

Sensory profile changes in a selection of alcoholic and nonalcoholic beverages are discussed in this chapter. Juices, coffee, beers, cream liqueurs, wines and spirits will all have very different sensory profiles and change very significantly from the time of initial production to the extreme end of the respective recommended shelf life. Descriptive sensory profiling methods such as quantitative descriptive analysis (QDA), described in detail in Chapter 2 of this book, have been ubiquitously used to describe beverages. Sensory lexicons have been developed for a number of beverages: coffee (SCAA, 2016), beer (Meilgaard et al., 1979), wine (Noble et al., 1984), whiskey (Lee et al., 2000), brandy (Jolly and Hattingh, 2001) to mention just a few. Usually with QDA, the initial list of sensory terms can be compiled, by experts with product knowledge (Johnson and Civille, 1986; Ordonez et al., 1998), perhaps using lexicons as a starting point along with any additional terms that are deemed relevant or useful. Descriptors may also be incorporated from information from consumers (Sawyer et al., 1988). After a list of appropriate sensory terms has been compiled, the training of panellists can commence. Usually, QDA is performed between 8 and 12 judges (Machado et al., 2010). Machado et al. (2010) used 22 well-trained judges to evaluate 17 red wines using 24 attributes, in 3 repetitions. They found that clearly the standard 8 to 12 judges were sufficient, but it is also important to note, that with well-trained judges and with replication, it may be possible to use as few as 6 judges. This information could be important for timing and cost constrains. A subset of the full complement of samples to be profiled which reflects the sensory variation in these samples is generally used. These could include samples at the beginning or end of the shelf life period along with intermediate samples, depending on the design. This metalist of terms can then be reduced by panel discussion and sensory term reduction protocols, as discussed in the previous chapters.

A Handbook for Sensory and Consumer-Driven New Product Development.
http://dx.doi.org/10.1016/B978-0-08-100352-7.00013-0

The sensory flavour profile of beverage products is dynamic and will change over the course of the respective shelf lives. For nonalcoholic beverages, flavours can fade, whereas products like vintage wine, due to changes in tannin and phenolic compounds, or whiskey, because of flavour concentration in cask evaporation, can actually improve in flavour as they age. Additionally, nonalcoholic beverages like fruit juices or soft drinks can spoil but microorganisms are of less concern in relation to alcoholic beverages because of this antimicrobial effect of ethanol. Additionally, low pH, as is the case with wine or beer, will contribute a hurdle effect along with the ethanol content and increase the antimicrobial properties combined with such components as organic acids in wine or hop bitters in beer.

The limiting factor to beer shelf life is the development of stale or oxidised flavours such as E−2-nonenal with a parallel reduction in fresh fruity or hop notes. Therefore, for alcoholic beverages a finite shelf life is dictated by sensory changes during storage. Cream liqueurs stored at ambient temperatures can have shelf lives of anywhere between 6 months and 2 years depending on the quality of the raw ingredients and the homogenisation technology employed. These beverages tend also to become more 'fruity' with age due to the formation of ethyl esters from the reaction of the alcohol with the cream component of the beverage. This can be viewed as a negative sensory attribute in cream liqueurs.

SENSORY PROPERTIES OF SOFT DRINKS AND FRUIT JUICES

Soft drinks are produced by diluting liquid pasteurised (85°C for 5 min) sugar syrup with potable water with the addition of preservatives, colours and flavourings. Lower calorie variants may be produced by complete replacement of sugar with artificial sweeteners such as sucralose, aspartame, acesulfame K, or saccharin and for reduced calorie drinks by the use of sucrose alternatives such as stevia. They may be flavoured with fruit juices, pulps, vegetable extracts or flavours such as natural or artificial vanilla. The pH (2.5−4.0) is reduced, with generally citric or phosphoric acid and carbonated, both of which have an antimicrobial effect. Soft drinks are quite often evaluated and monitored by manufacturers for appearance, aroma/flavour, taste, aftertaste and off-flavour defects. Carbonation is often described as prickling from a sensory descriptor perspective. QDA may be used for quality control and quality assurance purposes as well as product development and optimisation using trained panels. The degree of difference from a control or standard test is also quite often used in house for routine sensory testing, again using trained panellists (often external), particularly for monitoring of off-flavour defects. Sources of potential defects, which should be rare and infrequent, could be flavour fading but have more often involved cross-contamination issues due to inadequate cleaning in place processes where residues in plant (pipework, containment vessels, balance tanks, etc.) from a previous production run contaminate the following run of a different product.

In general, fruit juices are made by pressing ripe fruit and collecting the juice. Juices can be sold clear as is the case for apple juice or cloudy like orange or grapefruit juices which contain fruit pulp. Fruit juices can be sold either fresh, and refrigerated, where they will have a shelf life of a few days or pasteurised, again refrigerated, where they are processed with either a short flash pasteurisation heat treatment (85–90°C for 15–20 s) or an in-pack pasteurisation process (70°C for 20 min) and can last for several weeks. Fruit juices can also be produced from concentrate which involves aseptically filled cartons of reconstituted juice which will have an ambient shelf life of around 1 year. The concentrate is generally produced through evaporation under heat and vacuum or through freeze concentration (Ashurst, 2011). However, the former process increases the risk of flavour degradation due to thermal destruction of flavour compounds. Before pasteurisation, fruit juices contain a microbial load representative of the organisms normally found on fruits during harvest (bacteria, yeasts and moulds) plus contaminants added postharvest (during transport, storage and processing). Pasteurisation will rid juice of pathogens and other heat-sensitive microbes; therefore, it will reduce the microbial load substantially and extend the shelf life of the product (Tournas et al., 2006). Pasteurisation is used for products that have a pH value of 4.5 or less, where the acidic conditions effectively reduce the risk of growth of pathogenic organisms (Ashurst, 2011). Growth of yeasts can produce cloudiness and off-flavours and off-colours in fruit juices as well as CO_2 which can rupture containers. From a sensory perspective, any departure from the fresh juice product will result in a partial loss of both nutritional and sensory quality due to thermal destruction of flavour components. Juice products packaged in clear containers can undergo bleaching or fading due to the action of sunlight and excessive thermal treatments can produce browning and cooked or off-flavours and losses of flavour volatiles which can diminish overall flavour quality.

SENSORY PROPERTIES OF COFFEE

Commercially available coffee beans come from either one of two species *Coffea canephora* (robusta coffee), or coffee arabic (Drunday and Pacin, 2013; Van der Stegan, 2001). Robusta coffee is cheaper than arabica but has a less desirable flavour and is commonly used in blends or in instant coffee (Banks and McFadden, 2000), whereas arabica coffee has a better aroma (Farah, 2012). The coffee fruit can undergo one of three processes: dry processing, wet processing or semidry processing, to obtain the coffee bean. Generally, dry processing is used for robusta coffee beans and wet processing is used for arabica coffee beans (Chanakya and De Alwis, 2004; Robertson, 2013; Silva et al., 2000). During the drying process a natural microbial fermentation occurs where enzymes, which are naturally present in the coffee fruit, breakdown the mucilage layer and the pulp then the beans are dried to a moisture content of 10−12% and then mechanically hulled (Esquivel and Jiménez, 2012; Joët et al., 2010; Silva et al., 2000). The next step, roasting, results in the beans temperature increasing up to 220/240°C where chemical and physical changes produce hundreds of chemical compound from the various reactions that take place, such as the degradation of sugars, Maillard reaction, Strecker reaction and the breakdown of amino acids (Ku Madihah et al., 2012; Manzocco et al., 2011). The coffee fruit may become contaminated with ochratoxin A (OTA) when the fruit falls to the ground and if they experience inadequate drying during the drying and storage processes. OTA found in coffee is mainly produced by *Aspergillus steynii*, *Aspergillus carbonarius*, *Aspergillus westerdijkiae*, *Aspergillus ochraceus* (Pardo et al., 2004). However, OTA is significantly reduced after roasting (Blanc et al., 1998; Van der Stegan et al., 2001). Additional chemical changes can occur to the coffee bean by roasting and include: oxidation, hydrolysis, polymerisation, reduction and decarboxylation. Above 150°C, large amounts of water, volatile organic compounds and CO_2 are released from the coffee bean (Belitz et al., 2009; Robertson, 2013). The roasting process is a key processing step in coffee manufacture as it allows the development of the coffee bean's aroma, colour and flavour and provides the largest fraction of value added to the product (Chanakya and De Alwis, 2004; Ku Madihah et al., 2012). After roasting, the beans are rapidly cooled with a jet of water or air which prevents the coffee beans burning, which would alter the coffee bean's quality (Boltazzi et al., 2012). Prior to grinding and packaging, the coffee beans must be left to degas for a minimum of 12 h otherwise the resulting flavour will be sour and sharp (Banks and McFadden, 2000). If packed too soon, CO_2, resulting from the decomposition of carbohydrates during the roasting process, can build up and swell and burst the pack (Belitz et al., 2009). To prevent this occurring, a one-way valve or a CO_2 absorber is added to the packaging (Anderson et al., 2003). Generally, North Americans prefer coarse ground coffee (800−1000 µm) and European countries generally prefer a fine ground coffee (500−600 µm) (Perdue, 2009). Different roasts may also be blended together to create a harmonious balance of flavours, acidity and body (Ellis, 2002). A high percentage of robusta coffee is commonly used for instant coffee manufacture as robusta coffee contains a higher amount of soluble solids, in comparison to arabica coffee, which results in a higher yield (Farah, 2012). Instant coffee is packaged in sachets, tin cans or glass jars (Subramaniam, 1993). Instant coffee is obtained by the extraction of volatile aroma/flavour compounds from ground or roasted coffee beans with hot water and high pressure (Farah, 2012). The extracts are then dried by a freeze drying process, which results in better flavour and aroma retention, or by spray drying (Robertson, 2013), to produce a soluble hygroscopic powder with a moisture content of 1−6% (Belitz et al., 2009; Chanakya and De Alwis, 2004).

Caffeine is the main alkaloid found in coffee beans and contributes the essential bitterness component to the brewed product and naturally occurs in coffee, black teas and cocoa (Farah, 2009; Keast, 2008). Caffeine stimulates the nervous system which results in increased alertness, increased energy levels, enhanced cognitive performance, enhanced motor performance and enhanced physical performance (Glade, 2010). Coffee brewed in larger volumes gives higher caffeine content to the beverage (due to more complete caffeine extraction). Finely ground coffee has a larger surface area therefore it has better caffeine extraction in comparison to coarsely ground coffee. The more coffee grounds, the higher the caffeine content (Bell et al., 1996). Decaffeinated coffee (caffeine <0.1%) is produced prior to roasting by placing the raw coffee beans in water or steam at $22-100°C$ and allowed to swell to $30-40\%$ water. The caffeine−potassium−chlorogenate complex is extracted using a water-saturated solvent at $60-150°C$. Then the solvent is removed by applying steam to the coffee beans at $100-110°C$ either under vacuum or with warm air at $40-80°C$ (Belitz et al., 2009). The solvents commonly used during the decaffeination process are ethyl acetate, methylene (Belitz et al., 2009) or supercritical CO_2 (The Institute for scientific research on coffee, 2013). The coffee beans are then dried to a moisture content of $10-12\%$ (Farah et al., 2006).

The diterpenes cafestol and kahweol are present in both arabica and robusta coffee as well as the antioxidants chlorogenic acid (CGA), formed from the esterification of certain trans-cinnamic acids, and the alkaloid trigonelline. CGA and trigonelline are the two main antioxidants found in coffee and also contribute to the bitterness, acidity and astringency of the coffee brew (Farah, 2012). CGA content is generally higher in robusta than that in arabica coffee. Eighty percent of the total CGA content in coffee is made up of caffeoylquinic acids and 60% 5-caffeoylquinic acid dicaffeoylquinic acids with the remainder including caffeoylquinic acids, feruloylquinic acids, and to a lesser extent caffeoyl−feruloylquinic acids and p-coumaroylquinic acids (Clifford, 2000; Farah, 2012).

The main parameters which affect the shelf life of whole and ground coffee beans include oxygen (O_2), carbon dioxide (CO_2) and moisture. Buffo and Cardelli-Freire (2004) defined coffee staleness as 'a sweet but unpleasant flavour and aroma of roasted coffee which reflects the oxidation of many of the pleasant volatiles and the loss of others'. Oxidative reactions can cause an increase in the deterioration rate of ground coffee, instant coffee, coffee concentrate and ready-to-drink coffee beverages. The development of oxidative reactions causes not only the loss of pleasant aroma compounds but also the formations of off-flavours (Nicoli and Savonitto, 2005). There is a large amount of nonvolatile and volatile compounds in roasted coffee, if these compounds react with O_2, oxidative reactions occur (Manzoccoo et al., 2011). The rate of oxidation is largely influenced by temperature and oxygen. The rate of the oxidation reaction increases exponentially with an increase in temperature. Water also influences the oxidation rate; at lower water activities a high oxidation rate occurs (Kong and Singh, 2011). Cardeli and Labuza (2001) measured the sensory acceptability of roasted and ground coffee while varying O_2 partial pressure, water activity and temperature. They found that temperature had less of an effect on the shelf life of coffee than moisture and that O_2 accelerated the deterioration 20-fold as the partial pressure increased from 0.5 to 21.3 kPa. Yeretzian et al. (2012) investigated the free radical contents of whole, halved and ground roasted arabica coffee beans during storage in air. Whole beans were more stable than the ground or halved beans and free radical content increased with storage time. Thus, packaging is essential in maintaining coffee freshness. Roasted coffees can be stored for 8–10 weeks. Green coffees can be stored for 1–3 years. Ground coffee, that is vacuum packaged, can be stored for 6–8 months but once the package is opened the ground coffee will only last for less than two weeks (Belitz et al., 2009). Coffee that is packaged in a modified atmosphere packaging of 100% N_2 will have a shelf life of about 18 months if stored correctly (Parry, 1993).

The main types of packaging that coffee is found in today are pouches (flexible packaging or pillow packs), glass jars (hermetically sealed with a thin layer of foil–polyethylene–paper laminate with a polypropylene lid), metal cans, coffee pods, coffee capsules, 'refill' packs, coffee bags and some instant coffee is packaged into polyethylene terephthalate jars. Metal cans, glass jars and flexible laminates are the main packaging used for instant coffee. Flexible laminates, capsules and pods are common packaging formats for coffee (The Industry council for research on packaging and the environment, 2011). There are two main types of flexible laminated packaging used for coffee, 'hard packs' and 'soft packs'. A 'soft pack' (pillow pack or a pouch) is flushed with N_2 gas prior to sealing. A CO_2 absorber may be used or a one-way valve. A 'hard pack', also made from flexible laminated materials, is filled with coffee and sealed and then a high vacuum is applied to the bag. A CO_2 absorber may be added to the pack. If CO_2 evolves or if O_2 enters the coffee packaging, softening of the package will occur (Robertson, 2013). Resealability is very important for coffee as it is rarely all used in one serving. If coffee is not sealed after it has been opened, it will be exposed to the external environment which will decrease the shelf life of the coffee. Research has shown that if consumers have difficulty opening a pack, they will not purchase it again (Theobald, 2006).

The temperature at which the coffee is brewed is extremely important. The National Coffee Association USA (National Coffee Association USA (NCAUSA, 2015)) recommends the water should be in the range of 90.6–96.1°C (195.08–204.98°F). If coffee is brewed at temperatures below the recommended brewing temperature, it may result in an insufficient extraction process and above the recommended brewing temperature, it may result in an unsatisfactory flavour (NCAUSA, 2015). According to Pipatsattayanuwong et al. (2001), 72.1°C (161.8°F) is the preferred drinking temperature for black coffee. Brochgrevinck et al. (1999) reported that consumers preferred black coffee when it was served in the temperature range of 62.8−68.3°C (145.04−154.94°F). Lee and O' Mahony (2002) reported that consumers preferred to consume black coffee at 61.5°C (142.7°F) and the preferred drinking temperature to consume coffee with added condiments was 59.0°C (138.2°F). Stokes et al. (2016a) evaluated the sensory quality of coffee from paper-based cups and found the optimum temperature to serve black coffee was determined to be 70.8°C (Stokes et al., 2016a).

Consumer coffee consumption patterns have evolved overtime and category coffee groupings like fair-trade, organic and speciality coffees have emerged and become popular. Also, there are numerous coffee types and styles available to the consumer, for example: espresso, cappuccino and latte. The coffee options available to consumers are thus vast. They may choose coffee based on many combinations of origin, type of bean, grinding method, brewing method, packaging and flavouring (Ponte, 2002).

Recent innovations have endeavoured to diversify the coffee sector through the introduction of capsule coffees with bespoke delivery machines. Pod and capsule systems, in particular, have recently gained market share (Parenti et al., 2014). Stokes et al. (2016b,c) found that the consumer sensory preference for an experimental convenient coffee bag (containing 100% arabica ground roasted coffee) compared favourably to instant and filter cafetière (French press) and far better than current coffee bag products coffees currently on the market (UK and Ireland).

SENSORY PROPERTIES OF CREAM LIQUEURS

Cream liqueurs typically contain 10−20% sucrose, thus have a high osmotic potential, and between 10 and 15% ABV ethanol content. This creates a hurdle effect that prevents the growth of pathogenic organisms. Cream liqueurs are typically composed of cream, sodium caseinate, sugar, alcohol, flavours, colours and low-molecular weight surfactants (Banks and Muir, 1988; Lynch and Mulvihill, 1997; O'Sullivan, 2011). Corn syrups, molasses, maltose, ribose, galactose, honey, lactose, sucrose, dextrin, modified starch and glucose have been proposed for use as the carbohydrate source in cream/alcohol containing beverages (Rule, 1983). Typical manufacture of a cream liqueur involves the preparation of a caseinate trisodium−citrate blend at 55−85°C, followed by addition of cream and molten glycerol monostearate with continuous high-speed mixing to give the cream base. An aqueous−ethanol−sucrose solution is then added to the cream base and mixed thoroughly. The blended preemulsion is normally homogenised twice at 45−55°C and 20−30 MPa (two-stage) using a standard radial diffuser homogeniser, and cooled to 20°C (Muir and Banks, 1985, 1986; Dickinson et al., 1989; Lynch and Mulvihill, 1997; O'Sullivan, 2011). Various food grade thickeners may be added to contribute to the mouthfeel of cream liqueurs (Banks et al., 1981a).

Homogenisation is a crucial step in the manufacture of cream liqueurs in order to produce a finely dispersed emulsion with a long shelf life (Banks and Muir, 1988). During homogenisation, the size of the fat globules is decreased and sodium caseinate transfers from the serum phase to the newly exposed fat globule surface to stabilise the product. Droplet size distribution is one of the most important characteristics of an emulsion influencing stability (Heffernan et al., 2009). The more stable the fat globules in the aqueous phase of the emulsion, the less likely they are to coalesce and result in unacceptable sensory quality and creaming or the unsightly neck plug that occurs in the bottle necks as seen in extremely aged cream liqueurs (O'Sullivan, 2011; McClements, 1999). Viscosity of cream liqueurs will increase over time and as such viscosity can be used as an index of shelf life. Eventually the viscosity will become a limiting factor to consumer acceptability. Viscosity can be quantified using a viscometer which will be discussed in more detail in the section covering instrumental analysis (O'Sullivan, 2011).

Fat has a significant effect on the partition of volatile compounds between the food and the air phases with lipophilic aroma compounds being the most affected (Bayarri et al., 2006). The flavour profile of cream liqueurs involves a complex interaction of volatile and nonvolatile flavour compounds which originate from cream, cocoa, vanilla, top notes and whiskey. The way in which these volatiles partition in the continuous phase of the cream liqueur emulsion, within the fat globule itself or across the globule aqueous phase interface, depends on the polarity of these flavour compounds. This flavour partitioning also affects flavour release and ultimately the flavour profile of the liqueur with more polar flavours perceived first and the more nonpolar perceived last perhaps even as aftertaste sensations (O'Sullivan, 2011).

High-quality cream liqueur can have shelf lives of 2 years when stored under ambient conditions and can maintain their sensory quality even further when stored under refrigeration conditions. One of the limiting factors to the sensory quality of cream liqueurs is the formation of ethyl esters which increase in concentration over time. These compounds are formed from the reaction of fatty acids and alcohol and manifest as fruity notes in the cream liqueur. Ethyl esters also contribute to the overall flavour balance of cream liqueurs, but at high concentrations, they become the dominant flavour and thus these fruity notes must be considered and included with respect to sensory quality and descriptive sensory analysis. The level of fruitiness may become a limiting factor to consumer acceptance in aged liqueurs. In beers and wines, ethyl esters contribute to positive sensory attributes, for beer they denote freshness, and decrease during ageing. Sensory descriptors for fruity flavours may include, 'pearlike, bananalike, pineapplelike, applelike, strawberrylike, esterlike, ethereal or just fruity. Ethyl esters form due to the reaction of the alcohol and fatty acids and can result in the formation of excessive levels of short chain fatty acids such as ethyl acetate, ethyl butanoate and ethyl hexanoate, which are the principal causes of this fruity defect (O'Sullivan, 2011).

Ethyl esters can be quantified using GC—MS (gas chromatography—mass spectrometry) and liquid extraction of methylated esters or using solid-phase microextraction. Effectively, flavour partition can be measured under static equilibrium headspace conditions but does not always relate well to the release profile that is observed in vivo during consumption of a product (Doyen et al., 2001).

Cream liqueur manufacturers will also perform regular sensory profiling of their products over the course of their real-time shelf life to monitor flavour profile changes and potentially off-flavour development (ethyl esters). Additionally, accelerated shelf life testing is frequently implemented (O'Sullivan, 2011). Heffernan et al. (2009) measured changes in volume frequency distribution of fat droplet diameter and viscosity to monitor the stability of model cream liqueurs over a 28-day accelerated shelf life storage period at 45°C (Banks et al., 1981b; Muir, 1987; Muir et al., 1991; Power, 1996; Lynch and Mulvihill, 1997). The quantification of ethyl esters, using GCMS can also be measured in an accelerated shelf life context at 45°C. These are directly correlated to the development of fruity notes which could be deleterious sensory quality attributes benchmarked against consumer sensory data. A factor may then be calculated to correlate accelerated storage to real-time storage at ambient temperatures (O'Sullivan, 2011).

SENSORY PROPERTIES OF WINE

Wine flavour is composed of a wide variety of compounds with different aromatic properties where presence and concentration depends on a number of factors including grape cultivar, composition of grape must, yeast strain, fermentation conditions, winemaking practices, wine ageing and storage conditions, among others (Moreira et al., 2016). Wines are inherently stable and long shelf life products but it is their composition that dictates how well they can be cellared in bottles or oak barrelled casks. For the latter woods transfer a series of oak-related compounds into wine, such as ellagitannins, furfural compounds, guaiacol, oak or whisky lactone and eugenol (Tao, 2016). Premium barolos or bordeaux may require many years before developing their desired and traditional characteristics and the development of a complex, rich flavour and bouquet (Jackson, 2011). Some wines maintain their integrity or even improve with cellaring, whilst other wines are produced for consumption when they are young. Beaujolais nouveau, a young wine, is considered to be optimal just after production and can have a fresh fruity and floral fragrance with a nonaggressive flavour. It maintains its character for several months, but often deteriorates quickly thereafter (Jaffré et al., 2009). Essentially, little is known about the assessment of the ageing potential of a wine and the conversion of grape sugars to alcohol and other end products by specific yeast populations may yield wines with distinct organoleptic quality (Dubourdieu, 2005; Jaffré et al., 2009; Romano, 2003). However, during ageing, in addition to changes in colour and structure, the new components of the ageing bouquet develop from primary and secondary aromas or metabolites, under the influence of outside parameters, including ageing in oak barrels and oxygen levels during bottle storage (Picard et al., 2015; Robinson et al., 2014; Styger et al., 2011; Villamor and Ross, 2013).

Colour is a major contributor to wine quality and imparts information to the consumer regarding age, condition, body (concentration of dissolved substances) and possible defects and prepares the consumer for what is to follow with the anticipation of odour, flavour, taste and mouthfeel (Buglass, 2013). Anthocyanins in red wines are progressively transformed into more stable oligomeric and polymeric pigments which give rise to important changes in the colour (from bright red to brick-red hues) and in the astringency of wines (Monagas et al., 2006). For red wines the colour changes over to brown due to degradation and polymerisation of phenolic compounds (Moreira et al., 2016).

Red wine has a complex rich nonvolatile matrix, a quite large amount of different active odorants, and most often no clear impact compounds. In this chemical environment the perception of the different notes is extremely complex (Ferreira et al., 2016). Noble et al. (1984) proposed a list of standardised wine (aroma) terminology. From an initial list composed of virtually all possible wine descriptors, terms were selected which were analytical and free of hedonic or value-judgement connotations. Phenolic compounds constitute one of the most important quality parameters of wines since they contribute to their organoleptic characteristics, particularly colour, astringency and bitterness (Monagas et al., 2006). During wine ageing, these phenolic compounds undergo transformations, such as oxidation, condensation and polymerisation and the modification of these phenolic compounds would lead to a decrease in astringency (Mateus et al., 2004). Volatile esters are only trace compounds in wines; however, they are extremely important for the flavour profile of these drinks. Additionally, fusel alcohols, acids and fatty acid esters contribute significantly to flavour and aroma of wines as well as sulphur and nitrogen compounds, carbonyl compounds, phenolic substances, lactones, acetals and terpenoids (Ribéreau-Gayon et al., 2006; Buglass, 2013).

Atmospheric oxygen permeation through the oak barrel allows certain compounds to be oxidised gently, thus resulting in the changes in colour and the modification of wine sensory properties (Tao et al., 2014). Oxidation is one parameter that determines the shelf life, mainly for table wines. Oxidation will depend on wine resistance and the level of exposure to oxygen. If not properly conditioned, wines losses freshness and new organoleptic impressions are developed (Moreira et al., 2016). It is thus crucial to minimise oxygen uptake to maximise shelf life of wine. Despite all precautions, sufficient oxygen eventually enters to overpower the wine's antioxidant potential, inducing irreversible sensory degradation (Jackson, 2011). Oxidation can be prevented with the addition of or naturally occurring levels of the antioxidant sulphur dioxide. White wines, having lower phenolic contents, need more protection than red wines. Also, red wines typically receive less sulphur dioxide to avoid partial bleaching of the wine's anthocyanin content (Jackson, 2011).

Spontaneous fermentation is still used in numerous wineries, but in the last 20 years, yeast-manufacturing companies have developed selected starter cultures (Marullo et al., 2004). The spread plate technique can be used to isolate and enumerate yeasts with the aim of preventing the growth of all innocent yeast and promote the growth of all spoilage yeasts (Loureiro and Malfeito-Ferreira, 2003). The surface of the grapes contains a large variety of moulds, bacteria and yeasts. In particular, a great variety of yeasts are present on grapes, but only a minor portion can participate in alcoholic fermentation (Romano, 2003). In the wine industry, spoilage yeasts are rarely sought during wine fermentation, but during storage or ageing and during the bottling process (Loureiro and Malfeito-Ferreira, 2003).

SENSORY PROPERTIES OF DISTILLED SPIRITS

Products like vodka (Fig. 13.2B) and flavoured spirits such as gin are based on highly purified neutral alcohol, with very low levels of flavour compounds (Aylott, 2003). Distilled beverages offer some unique challenges for sensory analysis with alcohol content severely limiting the quantity which can be consumed in the course of a tasting session. Traditionally a master taster, blender or cellar master was employed as the main and only assessor of sensory quality within the distillery. Nowadays, trained sensory panellists are additionally employed which frees the master blender for more specialised duties as well as greatly reducing the risk of a master taster making an error on their sole assessment. The majority of quality control and research sensory analysis is by nosing, in a glass with a large headspace and covered to retain volatiles. This allows more samples to be tested than by tasting and facilitates ease of preparation, presentation to the assessors and stability. Samples should be presented coded and in a standardised fashion, as is sensory convention, at room temperature. Samples can also be expectorated if it is necessary to taste them (in order to assess mouthfeel or to detect off-notes) to prevent intoxication. Assessors must never be allowed to drive a vehicle or operate machinery if they are participating in such sensory assessments. For nosing tests the spirit is typically diluted to 20–23% abv with water (Piggott and Macleod, 2010; Jack, 2012). Extensive changes take place in the spirits during ageing, and the final product has a very different flavour and aroma from the newly distilled spirit as is the case with whisky (Nishimura and Matsuyama, 1989; Piggott et al., 2000). For whisky, Lee et al. (2000) reviewed the origins of the flavours in whiskies and developed a revised flavour wheel for Scotch whisky. In whisky, spicy, smooth, vanilla, woody and sweet aroma notes increase with time (Piggott et al., 1993), which is caused primarily by materials such as vanillin, aromatic aldehydes and other materials extracted from the wood

(Conner et al., 1999). The taste and odour of freshly distilled spirits, particularly whisky, is rather raw and unpleasant; desired flavour components develop during years of ageing in wood (Freitas and Costa, 2006). Fig. 13.1 presents two rare bottles of Baileys 'The Whiskey' which was test marketed in Ireland in 1998 but was never fully launched due its legal status (as per the EU regulation) as a whiskey because of the infusion of Baileys flavours during cask finishing (Fig. 13.1).

Rum is a cane spirit obtained by distillation of sugar cane molasses, after fermentation with yeast, and subsequent ageing in oak barrels, where the spirit acquires its special characteristics of flavour and aroma during the time it is in contact with the wood (Pino, 2007). The raw material used to produce cachaça is the juice squeezed from the sugar cane stalks, while rum is mainly obtained from the molasses, a byproduct of the industrial sugar process (Faria, 2012). Dark rum is aged in oak barrels, whereas white rum (Fig. 13.2A) is aged in stainless steel tanks. Some materials reduce in concentration in the spirit, through losses by evaporation through the cask wood; there is a general increase in concentration as water and ethanol are lost by evaporation (Piggott et al., 2000). The sensory profile of cachaça and rum are usually described in terms of the harsh characteristics of new distillates, such as sour, grassy, oily, sulphury, and also to the mature flavours developed during the ageing process,

(A)   **(B)**

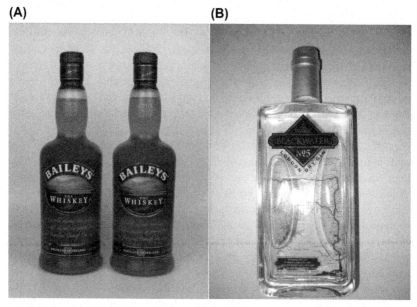

FIGURE 13.1   (A) Baileys 'The Whiskey' had a very limited production run and was test marketed for a short period of time in Dublin, Ireland, around 1998. It never found commercial success due to the ambiguity of its whiskey status because of the incorporation of Baileys flavours during cask finishing. Very few of these rare bottles survive. (B) Blackwater London Dry gin.

(A)   **(B)**

FIGURE 13.2   (A) Rum products popular on the Caribbean island of St. Lucia. Chairman's reserve rum, considered one of the finest on the island and Kweyol spiced rum infused with the bark of the Bois Bande tree, a West Indian aphrodisiac. (B) Vodka produced by the Dingle Distillery, Dingle, Co. Kerry, Ireland.

including vanilla, spicy, floral, wood and smooth flavours (Faria, 2012). Fatty acid ethyl esters are the main components of rum aroma and play an important role in the sensory quality of these distilled alcoholic beverages (Pino et al., 2002). Fusel alcohols, acetic acid and ethyl acetate are present in these beverages at relatively large amounts. Blending may be employed in the production process and caramel may be added to standardise the colour of dark rums.

SENSORY PROPERTIES OF BEER

The assessment of beer quality is a key step for quality control during beer production, while sensory evaluation is one of the necessary means to achieve it (Dong et al., 2014). Beers can have shelf lives of between 6 months and 2 years at ambient temperatures, but this too is increased when stored under refrigerated conditions. Beer quality is known to be affected by storage conditions, as factors such as temperature and light can influence chemical processes related to taste, aroma, mouthfeel and appearance (Stewart, 2004; Vanderhaegen et al., 2006). Many beer manufacturers will perform regular sensory profiling of their beer products over the course of real-time shelf life. Panels are trained as described in the section on descriptor profiling and sensory changes are tracked over time. Producers may also employ triangle tests to differentiate a product at a point in its shelf life from a fresh or gold standard product. Accelerated shelf life storage temperature of 45°C can be used and GCMS analysis to quantify E−2-nonenal as an index of staling, perhaps also correlated to consumer sensory data (O'Sullivan, 2011).

Beers such as lager, stout, bock or wheat beer have their own unique characteristic sensory profiles and sensory attributes which define quality, freshness or the degree to which they have aged (O'Sullivan, 2011). Pale lager is the most widely consumed and commercially available style of beer in the world. The flavour of these products is usually mild and the producers often recommend that the beers be served refrigerated. In general, lagers display less fruitiness and spiciness than ales, simply because the lower fermentation temperatures associated with lager brewing causes the yeast to produce fewer of the esters and phenols associated with those flavours. The beer industry has a standardised terminology wheel of mouthfeel and taste terms (Meilgaard et al., 1979). Meilgaard et al. (1979) showed how the principal flavour notes for beer may be represented as spokes in a 'flavour wheel'. The taste of beer is described in terms of the four basic tastes, from acidic to bitter, plus certain other taste impressions that are grouped under mouthfeel, and fullness. The system divides the more complex part of beer flavour, the odour (or aroma), into eight classes in an approximate order from pleasant to unpleasant (Meilgaard et al., 1982).

The sensory profile changes that occur in the massive and diverse beverage category that is beer are incredibly broad; however, there are some common sensory attributes which are affected by ageing. In contrast to some wines, beer ageing is usually considered negative for flavour quality. Beer typically has a shelf life of between 6 months and 1 year depending on the type and whether it is canned, bottled or kegged. The production of flavour active compounds like carbonyls are associated with extended storage (Hempel et al., 2013). Also, an initial acceleration of sweet aroma development, the formation of caramel, burnt sugar and toffeelike aromas (also called leathery) coincides with the sweet taste increase (Vanderhaegen et al., 2006). Whereas positive flavour attributes of beer, such as fruity/estery and floral aroma tend to decrease in intensity during ageing. The most important flavour-active esters in beer are ethyl acetate (solventlike aroma), isoamyl acetate (fruity, banana acetate (solventlike aroma)), isoamyl acetate (fruity, banana aroma), ethyl caproate and ethyl caprylate (sour apple), and phenyl ethyl acetate (flowery, roses, honey) (Verstrepen, 2003). Volatile compounds, such as aldehydes and sulphur-based compounds, can be major contributors to an aged, stale, flavour profile (Vanderhaegen et al., 2006). For the overall impression, the decrease of positive flavours, such as these esters, may be just as important as development of stale flavours (Bamforth, 1999; Whitear et al., 1979; Vanderhaegen et al., 2006 Bitterness in beer arises when the α-acids in hops are isomerised during boiling with wort. Additionally, a constant decrease in bitterness is also observed during beer ageing which is partly due to sensory masking by an increasing sweet taste (Hough et al., 1982). 'Light struck skunk odour' can also develop which results when light causes riboflavin to react with and break down isohumulones, a bitter compound derived from hops which results in formation of the compound 3-methyl-2-butene-1-thiol. This principally occurs in beers packaged in clear or green bottles; however, brown bottles offer a certain level of protection (O'Sullivan, 2011).

Contact with oxygen causes a rapid deterioration of the flavour of beer and is dependent on the oxygen content of the beer (Clapperton, 1976). Due to oxidation effects the main focus of research on beer ageing has been the study of the cardboard-flavoured component (E)-2-nonenal and its formation by lipid oxidation with other stale compounds (Vanderhaegen et al., 2006) or increases in the compound furfural with beer ageing. Thus the limiting factor to beer shelf life is the development of these stale, oxidised flavours such as E−2-nonenal, with a parallel reduction in fresh fruity or hop notes. A reduction in fresh notes and an increase in stale or oxidised notes can be used as indices to track the change in the consumer flavour profile of beer. Almost all beers taste best immediately after they have been produced and usually start to deteriorate/age/stale over time; this is where best before dates are set. The primary quality issue associated with bottled beer is the change of its chemical composition during storage, which subsequently alters its sensory properties (Vanderhaegen et al., 2006, 2007). Flavour stability in beer depends to a great extent on the lack of oxidation of the beer. If there is a relatively large volume of air in the final container, oxidation inevitably occurs (Wainwright, 1999b). At the end of the fermentation stage, beer is completely free of oxygen. At this point, beer is highly susceptible to oxidation, which has the following effects on the end product; undesirable taste, cloudy/hazy beer, increased beer astringency and darkened beer colour. Fig. 13.3 depicts medium- and small-scale pilot plant brewery systems which can be used for test batch production.

FIGURE 13.3   The picture on the left depicts a 1000-L pilot scale brewery. On the right is a 50-L brew house. This small 50-L brew house consists of a mashing, lauter and wort boiling kettle. The whirlpool, cooler and the three 50-L fermenters ensure a standardised operation and performance in small half technical scale.

Wild yeasts cause less serious spoilage problem than bacteria but are considered a serious nuisance to brewers because of the difficulty to discriminate them from brewing yeasts. The strictly anaerobic Gram-negative beer spoilage bacteria of the genera *Pectinatus* and *Megasphaera* have increased because of the improved technology in modern breweries which results in a significant reduction in O_2 content in the final products (Sakamoto and konings, 2003). Food-poisoning microorganisms are of less concern to alcoholic beverages because of this antimicrobial effect of ethanol. Additionally, low pH, as is the case with wine or beer, will contribute a hurdle effect along with the ethanol content and increase the antimicrobial properties combined with such components as organic acids in wine or hop bitters in beer (O'Sullivan, 2011). Beer is a tough medium for bacterial growth, due to the presence of ethanol (0.5−10% w/w), hop bitter compounds (approximately 17−55 ppm of iso-acids), high CO_2 (approximately 0.5% w/w), low pH (3.8−4.7), extremely reduced content of oxygen (<0.1 ppm) and the presence of only traces of the nutritive substances, glucose, maltose and maltotriose (Sakamoto and konings, 2003). Bacteria of the genera *Lactobacillus* and *Pediococcus* are a significant problem for the brewing industry, and account for the majority of spoilage incidents (Sakamoto and Konings, 2003; Yasui and Yoda, 1997). These bacteria not only produce off-flavour compounds and increase acidity (which alters taste) but they also affect the clarity of the final beer (Fernandez and Simpson, 1995). Aerophillic lactic acid bacteria can grow in beer because of their relative indifference to free oxygen and acquisition of tolerances to hop antiseptics, ethanol up to about 6% and to low pH (Rainbow, 1981). Enrichment cultivation is the only practical quality control method for the presence−absence testing of the strictly anaerobic beer spoilers due to their oxygen sensitivity in packaged beers. In this method, the development of turbidity in a bottled beer mixed with a concentrated growth medium is followed for 2−6 weeks (Haikara and Helander, 2002). Traditional incubation on culture media is used for the detection of beer spoilage bacteria which will form visible colonies on plates or increase the turbidity in nutrients broths after a week or longer. This presents a problem for the brewery as the products are often already released for sale before the microbiological results become available (Sakamoto and konings, 2003). The typical brewing process is finalised by filling, packaging and pasteurisation steps. Depending on the packaging selected, pasteurisation can take place prior to, or postfilling (Hempel et al., 2013). The only successful way to prevent beer spoilage by microorganisms is to thermally destroy organisms via pasteurisation (Wainwright, 1999a).

CASE STUDY: SENSORY EFFECTS OF RESIDUAL OXYGEN IN BOTTLED BEER

Hempel et al. (2013) investigated the use of optical oxygen sensors to monitor residual oxygen in pre- and postpasteurised bottled beer and its effect on sensory attributes and product acceptability during simulated commercial storage. The level of oxygen present prior to pasteurisation was assessed using a bespoke in bottle oxygen sensor in an attempt to ascertain if this oxygen level would have an impact on shelf life and product sensory attributes. A flash profile protocol was utilised which involved brief training of naïve panellists ($n = 26$) in a short session by being presented with fresh (1-month old) and aged beer samples (8-months old) and the relevant sensory descriptors presented (Fig. 13.4). This assisted panellists in identifying the sensory attributes that describe fresh and stale beer attributes and allowed the evaluation of the products on a monthly basis. The descriptors used include those for oxidised flavour, stale flavour, liking of flavour, carbonation and overall acceptability, etc. (Fig. 13.4). Hedonic descriptors were included as intensive sensory training was not undertaken and thus bias was not introduced in to the panel. Flash profiling was carried out in panel booths conforming to international standards (ISO 8589: 2007). Beer samples were categorised by the level of oxygen present in the product pre-pasteurisation and were assigned to the nearest oxygen category outlined; 0%, 0.5%, 1%, 2% and 5%. The level of oxygen present was assessed by panellists to ascertain if beer containing the highest oxygen levels also possessed the most perceived negative attributes. Five samples were taken to

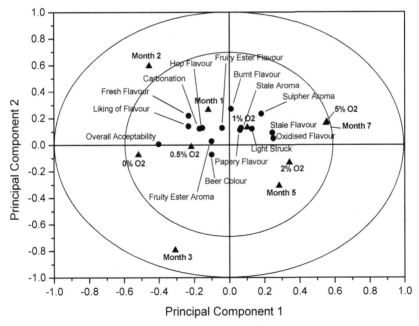

FIGURE 13.4   An overview of the variation found in the mean data from the ANOVA-partial least squares regression correlation loadings plot for each of the five oxygen level groups.

represent varying oxygen contents present in the beer samples prepasteurisation and assessed monthly over the course of the labelled shelf life. Samples were held at refrigeration temperatures (4°C) up to the moment of sampling. One bottle representing each category was relinquished for every test day. Each panellist was presented with five samples (0%, 0.5%, 1% 2% and 5% O_2, respectively) and asked to assess the attributes, according to a 10-point scale. Each panellist was presented with coded (three digit) samples which were assessed blind (Hempel et al., 2013). The order of the presentation of all test samples was randomised to prevent first order and carryover effects (MacFie et al., 1989). As time progressed, these authors (Hempel et al., 2013) found that deteriorative quality attributes became more apparent in some beer samples and it was determined that samples possessing the highest levels of oxygen prepasteurisation were also associated with the most negative sensory attributes (Fig. 13.4). The use of optical oxygen sensors showed a potential quality indicator application to the beer bottling industry. Additionally, sensory evaluation on a progressive month to month basis showed a continued decline in positive beer attributes in samples that were deemed to possess high oxygen prepasteurisation. Increases in prepasteurisation headspace O_2 levels reduced positive fresh beer sensory descriptors, increased negative stale and oxidised descriptors and also reduced the overall acceptability of the beer. The results confirmed that the increased level of oxygen present prepasteurisation has a negative effect on sensory attributes in beer over time. Levels of O_2 in excess of 1% were determined to be unacceptable to the panel (Hempel et al., 2013).

REFERENCES

Anderson, B.A., Simoni, E., Liardon, R., Labuza, T.P., 2003. The diffusion kinetics of carbon dioxide in fresh roasted and ground coffee. Journal of Food Engineering 59, 71–78.

Ashurst, P., 2011. CH19. The stability and shelf life of fruit juices and soft drinks. In: Kilcast, D., Subramaniam, P. (Eds.), The Stability and Shelf-Life of Food. Woodhead Publishing Limited, Cambridge, UK, pp. 571–593.

Aylott, R.I., 2003. In: Vodka, Gin and Other Fl Avoured Spirits', in Lea AGH and Piggott JR, Fermented Beverage Production, second ed. Kluwer Academic/Plenum Publishers, New York, pp. 289–308.

Bayarri, S., Taylor, A.J., Hort, J., 2006. The role of fat in flavor perception: effect of partition and viscosity in model emulsions. Journal of Agriculture and Food Chemistry 54, 8862–8868.

Damforth, C.W., 1999. The science and understanding of the flavour stability of beer: a critical assessment. Brauwelt International 98–110.

Banks, M., McFadden, C., 2000. In: Fox-Davies, F. (Ed.), Coffee Around the World. The Complete Guide to Coffee. Lorenz Books, NY, USA., p. 105

Banks, W., Muir, D.D., Wilson, A.G., 1981a. The formation of cream based liqueurs. The Milk Industry 83, 16–18.

Banks, W., Muir, D.D., Wilson, A.G., 1981b. Extension of the shelf-life of cream-based liqueurs at high ambient temperatures. Journal of Food Technology 16, 587–595.

Banks, W., Muir, D.D., 1988. Stability of alcohol-containing emulsions. In: Dickinson, E., Stainby, G. (Eds.), Advances in Food Emulsions and Foams. Elsevier Science Publishing Co., New York, pp. 257–283.

Belitz, H.-D., Grosch, W., Schieberle, P., 2009. In: Coffee, Tea, Cocoa. Food Chemistry, fourth ed. Springer, Germany, Heidelberg, pp. 938–949 (Chapter 21).

Bell, L.N., Wetzel, C.R., Grand, A.N., 1996. Caffeine content in coffee as influenced by grinding and brewing techniques. Food Research International 29, 785–789.

Blanc, M., Pittet, A., Munoz-Box, R., Viani, R., 1998. Behaviour of Ochratoxin A during green coffee roasting and soluble coffee manufacture. Journal of Agricultural and Food Chemistry 46, 673–675.

Boltazzi, D., Farina, S., Montorsi, 1, 2012. A numerical approach for the analysis of the coffee roasting process. Journal of Food Engineering 112, 243–252.

Borchgrevink, C.P., Susskind, A.M., Tarras, J.M., 1999. Consumer preferred hot beverage temperatures. Food Quality and Preference 10, 117–121.

Buffo, R.A., Cardelli-Freire, C., 2004. Coffee flavor: an overview. Flavor and Fragrance Journal 19, 99–104.

Buglass, A.J., 2013. Instrumental assessment of the sensory quality of wine. In: Kilcast, D. (Ed.), Instrumental Assessment of Food Sensory Quality: A Practical Guide. Woodhead Publishing Limited, Cambridge, UK, pp. 466–546.

Cardelli, C., Labuza, T.P., 2001. Applications of Weibull Hazard Analysis to the determination of the shelf-life of roasted and ground coffee. Lebensmittel-Wissenschaft & Technologie 34, 273–278.

Chanakya, H.N., De Alwis, A.A.P., 2004. Environmental issues and management in primary coffee processing. Process Safety and Environmental Protection 82, 291–300.

Clapperton, J.F., 1976. Ribes flavour in beer. Journal of the Institute of Brewing 82, 175–176.

Clifford, M.N., 2000. Chlorogenic Acids and other cinnamates- nature, occurrence, dietary burden, absorption and metabolism. Journal of Science and Food Agriculture 80, 1033–1043.

Conner, J.M., Paterson, A., Birkmyre, L., Piggott, J.R., 1999. Role of organic acids in maturation of distilled spirits in oak casks. Journal of the Institute of Brewing 105, 287–291.

Dickinson, E., Narhan, S.K., Stainsby, G., 1989. Stability of cream liqueurs containing low-molecular-weight surfactants. Journal of Food Science 54, 77–81.

Dong, J.J., Li, Q.J., Yin, H., Zhong, C., Hao, J.G., Yang, P.F., Tian, Y.H., Jia, R.S., 2014. Predictive analysis of beer quality by correlating sensory evaluation with higher alcohol and ester production using multivariate statistics methods. Food Chemistry 161, 376–382.

Doyen, K., Carey, M., Linforth, R.S.T., Marin, M., Taylor, A.J., 2001. Volatile release from an emulsion: headspace and in-mouth studies. Journal of Agriculture and Food Chemistry 49, 804–810.

Dubourdieu, D., 2005. Comment la garde vient aux vins' Le point. Special Vins 1721, 168–170.

Drunday, V., Pacin, A., 2013. Occurrence of ochratoxin A in coffee beans, ground roasted coffee and soluble coffee and method validation. Food Control 30, 675–678.

Ellis, H., 2002. In: Bevan, S. (Ed.), The Brew Coffee: Discovering, Exploring, Enjoying. Ryland Peters and Small, London, p. 23.

Esquivel, P., Jiménez, V.M., 2012. Functional properties of coffee and coffee by-products. Food Research International 46, 488–495.

Farah, A., de Paulis, T., Turgo, L.C., Martin, P.R., 2006. Chlorogenic acids and lactones in regular and water-decaffeinated Arabica coffee. Journal of Agricultural Food Chemistry 54, 374–381.

Farah, A., 2009. In: Paquin, P. (Ed.), Coffee as a Speciality and Functional Beverage Functional and Speciality Beverage Technology. Woodhead, Cambridge, pp. 371−400 (Chapter 15).

Farah, A., 2012. In: Yi-Fang, C. (Ed.), Coffee Constituents Coffee: Emerging Health Effects and Disease Prevention, first ed. Blackwell Publishing Ltd, pp. 21−39 (Chapter 2).

Faria, J.B., 2012. CH 17. Sugar cane spirits: cachaça and rum production and sensory properties. In: Piggott, J.R. (Ed.), Alcoholic Beverages: Sensory Evaluation and Consumer Research. Woodhead Publishing Limited, Cambridge, UK.

Fernandez, J., Simpson, W.J., 1995. Measurement and prediction of the susceptibility of lager beer to spoilage by lactic acid bacteria. Journal of Applied Microbiology 78, 419−425.

Ferreira, V., Sáenz-Navajas, M.P., Campo, E., Herrero, P., de la Fuente, A., Fernández-Zurbano, P., 2016. Sensory interactions between six common aroma vectors explain four main red wine aroma nuances. Food Chemistry 199, 447−456.

Freitas, M.A., Costa, J.C., 2006. Shelf life determination using sensory evaluation scores: a general Weibull modeling approach. Computers & Industrial Engineering 51, 652−670.

Glade, M.J., 2010. Caffeine- not just a stimulant. Nutrition 26, 932−938.

Haikara, A., Helander, I., 2002. Pectinatus, Megasphaera and Zymophilus. In: Dworkin, M. (Ed.), The Prokaryotes, an Evolving Electronic Database for the Microbiological Community, third ed. Springer Verlag, New York. 3.5.

Heffernan, S.P., Kelly, A.L., Mulvihill, D.M., 2009. High-pressure-homogenised cream liqueurs: emulsification and stabilization efficiency. Journal of Food Engineering 95, 525−531.

Hempel, A., O'Sullivan, M.G., Papkovsky, D., Kerry, J.P., 2013. Use of optical oxygen sensors to monitor residual oxygen in pre- and post-pasteurised bottled beer and its effect on sensory attributes and product acceptability during simulated commercial storage. LWT-Food Science and Technology 50, 226−231.

Hough, J.S., Briggs, D.E., Stevens, R., Young, T.W., 1982. Malting and Brewing Science, second ed., vol. 2. Chapman and Hall, London, p. 389.

Jackson, R.S., 2011. Chapter 18; Shelf life of wine. In: Kilcast, D., Subramaniam, P. (Eds.), The Stability and Shelf-life of Food. Woodhead Publishing Limited, Cambridge, UK, pp. 540−570.

Jaffré, J., Valentin, D., Dacremont, C., Peyron, D., 2009. Burgundy red wines: representation of potential for aging. Food Quality and Preference 20, 505−513.

Johnson, P.B., Civille, G.V., 1986. A standardized lexicon of meat WOF descriptors. Journal of Sensory Studies 1, 99−104.

Joët, T., Laffargue, A., Descroix, F., Doubeau, S., Bertrand, B., de Kochko, A., Dussert, S., 2010. Influence of environmental factors, wet processing and their interactions on the biochemical composition of green Arabic coffee beans. Food Chemistry 118, 693−701.

Jolly, N.P., Hattingh, S., 2001. A brandy aroma wheel for South African brandy. South African Journal of Enology and Viticulture 22, 16−21.

Jack, F.R., 2012. CH 19, Whiskies: composition, sensory properties and sensory analysis. In: Piggott, J.R. (Ed.), Alcoholic Beverages: Sensory Evaluation and Consumer Research. Woodhead Publishing Limited, Cambridge, UK.

Keast, R.S.J., 2008. Modification of the bitterness of caffeine. Food Quality and Preference 19, 465−472.

Kong, F., Singh, R.P., 2011. In: Kilcast, D., Subramaniam, P. (Eds.), Chemical Deterioration and Physical Instability of Food and Beverages Food and Beverage Stability and Shelf Life. Woodhead publishing Limited, pp. 29−44 (Chapter 2).

Ku Madihah, K.Y., Zaibunnisa, A.H., Norashikin, S., Rozita, O., Misnawi, J., 2012. Optimization of roasting conditions for high-quality robusta coffee. APCBEE Procedia 4, 209−214.

Lee, M., Paterson, A., Piggott, J.R., Richardson, G.D., 2000. Origins of flavour in whiskies and a revised flavour wheel: a review. Journal of the Institute of Brewing 107, 287–313.

Lee, H.S., O'Mahony, M., 2002. At what temperatures do consumers like to drink coffee?: Mixing methods. Journal of Food Science 67 (7), 2774–2777.

Loureiro, V., Malfeito-Ferreira, M., 2003. Spoilage yeasts in the wine industry. International Journal of Food Microbiology 86, 23–50.

Lynch, A.G., Mulvihill, D.M., 1997. Effect of sodium caseinate on the stability of cream liqueurs. International Journal of Dairy Technology 50, 1–7.

Macfie, H.J., Bratchell, N., Greenhoff, K., Vallis, L.V., 1989. Designs to balance the effect of order of presentation and first-order carry-over effects in hall tests. Journal of Sensory Studies 4 (2), 129e148.

Machado, B., Heymann, H., Robinson, A.L., Torri, L., 2010. How many judges are required for sensory descriptive analysis. Fourth European Conference on Sensory and Consumer Research: a sense of quality, 5–8 September (Palacio Europa, Vitoria-Gasteiz, Spain).

Manzocco, L., Calligaris, S., Nicoli, M.C., 2011. In: Kilcast, D., Subramaniam, P. (Eds.), The Stability and Shelf Life of Coffee Products Food and Beverage Stability and Shelf Life. Woodhead publishing Limited, pp. 615–637 (Chapter 21).

Marullo, P., Bely, M., Masneuf-Pomarede, I., Aigle, M., Dubourdieu, D., 2004. Inheritable nature of enological quantitative traits is demonstrated by meiotic segregation of industrial wine yeast strains. FEMS Yeast Research 4, 711–719.

Mateus, N., Pinto, R., Ruão, P., De Freitas, V., 2004. Influence of the addition of grape seed procyanidins to port wines in the resulting reactivity with human salivary proteins. Food Chemistry 84, 195–200.

McClements, D.J., 1999. Food Emulsions: Principles, Practices and Techniques. CRC Press, Boca Raton, FL.

Meilgaard, M.C., 1982. Prediction of flavor differences between beers from their chemical composition. Journal of Agriculture and Food Chemistry 30, 1009–1017.

Meilgaard, M.C., Dalgleish, C.E., Clapperton, J.R., 1979. Journal of the Institute of Brewing 85, 38.

Monagas, M., Gomez-Cordoves, C., Bartolomé, B., 2006. Evolution of the phenolic content of red wines from Vitis vinifera L. during ageing in bottle. Food Chemistry 95 (3), 405–412.

Moreira, N., Lopes, P., Ferreira, H., Cabral, M., Guedes de Pinho, P., 2016. Influence of packaging and aging on the red wine volatile composition and sensory attributes. Food Packaging and Shelf Life 8, 14–23.

Muir, D.D., Banks, W., 1985. From Atholl Brose to cream liqueurs: development of alcoholic milk drinks stabilized with trisodium caseinate. In: Galesloot, T.E., Tinbergen, B.J. (Eds.), Proceedings of the International Congress on Milk Proteins, Luxemburg, May 7–11, 1984, pp. 120–128.

Muir, D.D., Banks, W., 1986. Technical note: multiple homogenisation of cream liqueurs. Journal of Food Technology 21, 229–232.

Muir, D.D., 1987. Cream liqueur manufacture: assessment of efficiency of methods using a viscometric technique. Dairy Industries International 52, 38–40.

Muir, D.D., McCrae, C.H., Sweetsur, A.W.M., 1991. Characterization of dairy emulsions by forward lobe laser light-scattering – application to cream liqueurs. Milchwissenschaft 46, 691–694.

Nicoli, M.C., Savonitto, O., 2005. Physical and chemical changes of roasted coffee during storage. In: Viani, R., Illy, A. (Eds.), Espresso Coffee: The Science of Quality, second ed. Elsevier Academic Press, San Diego, CA, pp. 230–245.

Nishimura, K., Matsuyama, R., 1989. Maturation and maturationchemistry. In: Piggott, J.R., Sharp, R., Duncan, R.E.B. (Eds.), The Science and Technology of Whiskies, pp. 235–263 (Harlow: Long-man).

Noble, A.C., Arnold, R.A., Masuda, B.M., Pecore, S.D., Schmidt, J.O., Stern, P.M., 1984. .Progress towards a standardized system of wine aroma terminology. American Society for Enology and Viticulture 35, 107–109.

National Coffee Association USA (NCAUSA, 2015. How to Brew Coffee. http://www.ncausa.org/i4a/pages/index.cfm?pageID=71.

Ordonez, A.I., Ibanez, F.C., Torre, P., Barcina, Y., 1998. Application of multivariate analysis to sensory characterisation of ewes' milk cheese. Journal of Sensory Studies 13, 45–55.

O'Sullivan, M.G., 2011. CH 4, Sensory shelf-life evaluation. In: Piggott, J.R. (Ed.), Alcoholic Beverages: Sensory Evaluation and Consumer Research. Woodhead Publishing Limited, Cambridge, UK.

Pardo, E., Martin, S., Ramos, A.J., Sanchis, V., 2004. Occurrence of ochratoxigenic fungi and ochratoxin A in green coffee from different origins. Food Science and Technology International 10, 45–49.

Parenti, A., Guerrini, L., Masella, P., Spinelli, S., Calamai, L., Spugnoli, P., 2014. Comparison of espresso coffee brewing techniques. Journal of Food Engineering 121, 112–117.

Parry, R.T., 1993. In: Parry, R.T. (Ed.), Introduction Principles and Applications of Modified Atmosphere Packaging of Food. Blackie Academic and Professional, an imprint of Chapman and Hall, Glasgow, UK, pp. 1–4 (Chapter 1).

Piggott, J.R., Macleod, S., 2010. Sensory quality control of distilled beverages. Sensory Analysis for Food and Beverage Quality Control 2010, 262–275.

Piggott, J.R., Conner, J.M., Paterson, A., Clyne, J., 1993. Effects on Scotch whisky composition and flavour of maturation in oak casks with varying histories. International Journal of Food Science and Technology 28, 303–318.

Piggott, J.R., Hunter, E.A., Margomenou, L., 2000. Comparison of methods of analysis of time-intensity data: application to Scotch malt whisky. Food Chemistry 71, 319–326.

Pipatsattayanuwong, S., Lee, H.S., Lau, S., O'Mahony, M., 2001. Hedonic R-index measurement of temperature preferences for drinking black coffee. Journal of Sensory Studies 16, 517–536.

Pino, J., Martí, M.P., Mestres, M., Pérez, J., Busto, O., Guasch, J., 2002. Headspace solid-phase micro extraction of higher fatty acid ethyl esters in white rum aroma. Journal of Chromatography A 954, 51–57.

Pino, J.A., 2007. Characterization of rum using solid-phase microextraction with gas chromatography–mass spectrometry. Food Chemistry 104, 421–428.

Ponte, S., 2002. The 'Latte Revolution'? regulation, markets and consumption in the global coffee chain. World Development 30 (7), 1099–1122.

Perdue, R., 2009. In: Yam, K.L. (Ed.), Vacuum-Bag Coffee Packaging the Wiley Encyclopedia of Packaging Technology, third ed. John Wiley & Sons, Inc., Hobroken, NJ, pp. 1265–1266, Chapter V.

Picard, M., Tempere, S., de Revel, G., Marchand, A., 2015. A sensory study of the ageing bouquet of red Bordeaux wines: a three-step approach for exploring a complex olfactory concept. Food Quality and Preference 42, 110–122.

Power, P.C., 1996. The Formulation, Testing and Stability of 16% Fat Cream Liqueurs. National University of Ireland, Cork (Ph.D. Thesis).

Rainbow, C., 1981. Beer spoilage microorganisms. In: Pollock, J.R.A. (Ed.), Brewing Science, Vol. 2. Academic Press, pp. 491–550.

Ribéreau-Gayon, P., Glories, Y., Maujean, A., Dubourdieu, D., 2006. Part One. The chemistry of wine. In: Handbook of Enology, second ed., vol. 2. John Wiley & Sons Ltd, Chichester.

Robertson, G.L., 2013. Packaging of Beverages Food Packaging and Shelf-life: A Practical Guide, third ed. CRC Press, Boca Raton, FL, pp. 586−588 (Chapter 21).

Robinson, A.L., Boss, P.K., Solomon, P.S., Trengove, R.D., Heymann, H., Ebeler, S.E., 2014. Origins of grape and wine aroma. American Journal of Enology and Viticulture 65 (1), 1−42.

Romano, P., Fiore, C., Paraggio, M., Caruso, M., Capece, A., 2003. Function of yeast species and strains in wine flavour. International Journal of Food Microbiology 86, 169−180.

Rule, C.E., 1983. Cream/Alcohol Containing Beverages. US Patent No. 4,419,378.

Sakamoto, K., Konings, W.N., 2003. Beer spoilage bacteria and hop resistance. International Journal of Food Microbiology 89, 105−124.

Sawyer, F.M., Cardello, A.V., Prell, P.A., 1988. Consumer evaluation of the sensory properties of fish. Journal of Food Science 53, 12−18, 24.

SCAA, 2016. Coffee taster's flavour wheel. Speciality Coffee Association of America. http://www.scaa.org/chronicle/wpcontent/uploads/2016/01/SCAA_FlavorWheel.01.18.15.jpg.

Stewart, G.G., 2004. The chemistry of beer instability. Journal of Chemical Education 81 (7), 963.

Stokes, C., O'Sullivan, M.G., Kerry, J.P., 2016a. Assessment of black coffee temperature profiles consumed from paper-based cups and effect on affective and descriptive product sensory attributes. International Journal of Food Science and Technology (Accepted).

Stokes, C., O'Sullivan, M.G., Kerry, J.P., 2016b. Hedonic and descriptive sensory evaluation for development of novel instant and fresh coffee products. European Food Research and Technology (Accepted).

Stokes, C., O'Sullivan, M.G., Kerry, J.P., 2016c. Hedonic and descriptive sensory evaluation for development of novel coffee products. Journal of Food Processing and Beverages (Submitted).

Styger, G., Prior, B., Bauer, F.F., 2011. Wine flavour and aroma. Journal of Industrial Microbiology and Biotechnology 38, 1145−1159.

Silva, C.F., Schwan, R.F., Dias, Ë.S., Wheals, A.E., 2000. Microbial diversity during maturation and natural processing of coffee cherries of *Coffea Arabica* in Brazil. International Journal of Food Microbiology 60, 251−260.

Subramaniam, P.J., 1993. In: Parry, R.T. (Ed.), Miscellaneous Applications Principles and Applications of Modified Atmosphere Packaging of Food. Blackie Academic and Professional, an imprint of Chapman and Hall, Glasgow, UK, pp. 176−179 (Chapter 8).

Tao, Y., García, J.F., Sun, D.-W., 2014. Advances in wine aging technologies for enhancing wine quality and accelerating wine aging process. Critical Reviews in Food Science and Nutrition 54, 817−835.

Tao, Y., Sun, D., Górecki, A., Błaszczak, W., Lamparski, G., Amarowicz, R., Fornal, J., Jeliński, T., 2016. A preliminary study about the influence of high hydrostatic pressure processing in parallel with oak chip maceration on the physicochemical and sensory properties of a young red wine. Food Chemistry 194, 545−554.

Tournas, V.H., Heeres, J., Burgess, L., 2006. Moulds and yeasts in fruit salads and fruit juices. Food Microbiology 23, 684−688.

The Institute for scientific research on coffee, 2013. Decaffination. http://www.coffeeandhealth.org/all-about-coffee/decaffeination/ (10.12.13.).

The Industry council for research on packaging and the environment, 2011. Why products are packaged. http://www.incpen.org/docs/WPAPTWTAJune%202011.pdf (09.12.13.).

Theobald, N., 2006. In: Theobald, N. (Ed.), Introduction Packaging Closures and Sealing Systems. Winder, Belina. CRC Press: Blackwell publishing, Oxford, U.K, pp. 9−26 (Chapter 1).

Vanderhaegen, B., Delvaux, F., Daenen, L., Verachtert, H., Delvaux, F.R., 2007. Aging characteristics of different beer types. Food Chemistry 103, 404—412.

Vanderhaegen, B., Neven, H., Verachtert, H., Derdelinckx, G., 2006. .The chemistry of beer aging — a critical review. Food Chemistry 95, 357—381.

Van der Stegan, G., Essens, P., van der Lijn, J., 2001. Effect of roasting conditions on reduction of Ochratoxin A in coffee. Journal of Agriculture and Food Chemistry 49, 4713—4715.

Verstrepen, K.J., Derdelinckx, G., Dufour, J.P., Winderickx, J., Thevelein, J.M., Pretorius, I.S., Delvaux, F.R., 2003. Flavor-active esters: adding fruitiness to beer. Journal of Bioscience and Bioengineering 96 (2), 110—118.

Villamor, R.R., Ross, C.F., 2013. Wine matrix compounds affect perception of wine aromas. Annual Review of Food Science and Technology 4, 1—20.

Wainwright, T., 1999a. Basic brewing science. In: Final Processing of Beer. Magic Print Limited, pp. 267—273.

Wainwright, T., 1999b. Basic brewing science. In: Flavour Stability. Magic print Limited, pp. 283—286.

Whitear, A.L., Carr, B.L., Crabb, D., Jacques, D., 1979. The challenge of flavour stability. Proceedings of the European Brewery Convention Congress 13—25.

Yasui, T., Yoda, K., 1997. Imaging of Lactobacillus brevis single cells and microcolonies without a microscope by an ultrasensitive chemiluminescent enzyme immunoassay with a photoncounting television camera. Applied Environmental Microbiology 63, 4528—4533.

Yeretzian, C., Pascual, E.C., Goodmen, B.A., 2012. Effect of roasting conditions and grinding on free radical contents of coffee beans stored in air. Food Chemistry 131, 811—816.

Chapter 14

Sensory Properties of Bakery and Confectionary Products

INTRODUCTION

Bakery goods are an incredibly diverse category of products, which are generally made with a significant proportion of wheat flour. A specific component of this wheat flour, gluten, once hydrated and worked conveys on uncooked doughs a viscoelastic structure that traps air bubbles, which in turn are inflated by the gases produced during leavening. For breads, leavening is usually achieved biologically by the use of baker's yeast during proofing (fermentation). Soda bread is leavened chemically with baking soda (sodium bicarbonate), which reacts with acids in the dough (e.g., buttermilk) to produce gas. In both cases of leavening, either biological or chemical, this gas, carbon dioxide (CO_2), is essential for the formation of the breads light cellular foam like network, which forms a sponge structure on baking. Another class of bakery product, confectionary sponges, muffins and cakes, like breads, are also defined by this cellular sponge structure but are produced quite differently with minimal gluten action, where the foam-stabilising ingredients include egg protein, fat, and emulsifiers. For sponges, leavening is typically achieved physically through the incorporation of air during mixing, e.g., mixing air in to egg whites, which is then folded in to a sponge batter and baked to produce a light airy texture. Again chemical leavening can be also used for some cakes and cookies where baking powder produces CO_2 in the presence of water. Unlike baking soda, baking powder contains sodium bicarbonate and at least one acid salt to produce gas. Essentially cakes, muffins and sponges are fat-in-water emulsions. The high level of liquid from water and the egg ingredients produces a low-viscosity cake batter that is easy to pour and the resulting steam produced in the oven during baking helps produce a light texture in the end product. Biscuit doughs are generally a lot tougher than bread doughs because of a much lower water content, which also inhibits gluten action but also reduces the amount of water that needs to be cooked out during baking and for this reason they have a much harder and dryer finished texture. Pastry products are made from short doughs similar to biscuits. Puff pastry is made by encasing multiple laminated layers of fat and dough to produce a flaky

A Handbook for Sensory and Consumer-Driven New Product Development.
http://dx.doi.org/10.1016/B978-0-08-100352-7.00014-2

texture on baking. Another major difference between breads and confectionary baked goods such as cakes, muffins, pastry and biscuits are their very different sensory properties. The latter are sweet, high-calorie baked products where sucrose contributes to the sweet taste but also acts as a bulking agent in the batter, aids in moisture retention and air entrapment and creates a fine crumb grain in the products. Sucrose in confectionary products promotes fat crystal aggregates, which enhances air entrapment within the batter, leading to a greater stabilization of air bubbles during baking. Sucrose also contributes to colour formation in both the crumb and crust and helps to inhibit microbial spoilage. Sucrose plays a very important role in raising the temperature for starch gelatinization and protein denaturation during the baking process, resulting in the formation of the desired final crumb structure.

SENSORY PROPERTIES OF BISCUITS AND COOKIES

Biscuits are bakery products with a long history. Panis biscoctus is Latin for twice-cooked bread and refers to bread rusks that were made from flour and water for mariners (ship's biscuits) from as long ago as the Middle Ages. These were made by baking in an oven followed by further drying in a cooler oven to produce a low moisture, very long shelf life product that could sustain sailors on long sea voyages (Manley, 2011a). Biscuits today are considerably more appetising and are principally made from flour, water, sugar, fat and salt. A variety of shapes and textures may be produced by varying the proportions of these ingredients (Maache-Rezzoug et al., 1998). Biscuits are characterised by a high fat/sucrose to flour ratio with their respective doughs being tougher than bread doughs because of a much lower water content which inhibits gluten action, but also reduces the amount of water that needs to be cooked out during baking. For this reason, biscuits have a much harder and dryer finished texture. Biscuits are typically baked on trays due to these tougher and dryer doughs, compared to cakes and breads, which require tins, because of their softer doughs. The fat-type content of biscuits is important for providing flavour, texture and eating quality. In the recent past hydrogenated vegetable oils, with their higher melting points were commonly used in biscuit manufacture, but this practice has fallen out of favour due to the presence of trans fats and the restriction of these compounds as enforced by some regulatory bodies. Trans fat consumption has been linked to an increased risk of coronary heart disease by contributing to the buildup of plaque inside the arteries that may cause heart attack. For this reason, Austria, Denmark, Hungary, Iceland, Norway and Switzerland have set limits that virtually ban trans fats from food products while the EU is considering similar legislation for all the nation states. In the United States, the FDA requires the levels of trans fat in foods to appear on nutritional labels. Palm oil has a high melting point (36°C) and has replaced hydrogenated oils in current biscuit manufacture but not without controversy. Palm oil is high in saturated fat and thus is less healthy than more polyunsaturated oils like rapeseed and sunflower oils. Also, the increased subsequent global demand is being linked to deforestation in several countries in Asia such as Indonesia and Malaysia. In biscuit formulations, fat acts as a lubricant and a plasticiser and imparts a desirable flavour to the baked products. By coating the flour particles during mixing, the fat prevents the development of a gluten network, thus resulting in a brittle, crumbly product. In the biscuit manufacturing industry, the presence of fat is vital for machinability and dough moulding properties.

The sucrose present in biscuit and cookie formulations affects the sweet flavour, dimensions/shrinkage, colour, hardness and surface finish. Sucrose inhibits gluten development during dough mixing by competing with the flour for the available water in the formulation. Depending on the amount of water available, the sucrose in biscuit doughs dissolves, or partially dissolves, then recrystallises or forms an amorphous glass (a supercooled liquid) after baking. Essentially, sucrose shifts the starch gelatinisation point to a higher temperature thus allowing the dough more time to rise in the oven (Manley, 2011c). Sugar also acts as a hardening agent by crystallising as the biscuit cools, thus making the product crisp. As sugar is present in such significant quantities in biscuits and cookies, altering the level used will greatly affect dough consistency, rheology, baking regimes employed and final product characteristics.

Modifications to the quantities of fat used in biscuit formulation, particularly in conjunction with a concurrent lowering of sugar, will affect significant aspects of the product, e.g., increased dough stiffness, less crispy/crumbly/brittle texture (i.e., more chewy). Of equal importance is the effect of the distribution and migration on the flavour of the products as the presence of fat and sugar plays an integral part in the development of flavour intensity and release with fat containing the more nonpolar flavours. Therefore, achieving the optimum fat/sugar combinations is crucial. Thus, it is also very important for biscuit formulations that an optimised fat level is determined through sensory and physico–chemical (TPA, texture profile) analysis that can deliver fat-reduced products while maintaining optimal sensory characteristics. Fat replacement technologies are covered in Chapter 9 of the book on 'Nutritionally Optimised Low Fat Foods'.

SENSORY PROPERTIES OF CAKES AND MUFFINS

Cake is a food that is relished by consumers all over the world and it incorporates sugar as one of the basic ingredients (Manisha et al., 2012). In confectionary foods, sugar is responsible for sweetness, while fat contributes to the texture, mouthfeel, flavour and aroma of food (Drewnowski and

Almiron-Roig, 2010). The fat in cake batter not only helps the incorporation of air but it also produces emulsifying properties and holds considerable amounts of liquid to increase and extend cake softness and 'shortens', that is, it interrupts the protein particles to break gluten continuity to tenderise the crumbs (Bennion and Bamford, 1973). Cake batter can be considered to be a complex oil-in-water emulsion with a continuous aqueous phase containing suspended dry ingredients (Barcenilla et al., 2016). Minute air bubbles are trapped in the cake batter by the surface active proteins in the egg, fat, a suitable emulsifier or a combination of all three to form an emulsified foam that forms a sponge on baking (Cauvain, 2011). Muffins, similarly to cakes, are characterised by a typical porous structure and high volume, which confer a spongy texture. To obtain such a final structure, a stable batter retaining many tiny air bubbles is required (Martínez-Cervera et al., 2012). Sugar assists in the incorporation of air cells in cakes and muffins while creaming provides good grain structure, flavour and texture to the product. It aids in retention of moisture, prolongs freshness and promotes good crust colour (Nip, 2007). Sucrose not only contributes to the sweet taste, but acts as a bulking agent, assists products in staying moist, reduces the swelling of starch to produce a finer texture end product and contributes to the browning of the crust. Multiple factors are linked to consumer perceptions of sweetness and fat. Sweetness is mainly due to the sugar content (Drewnowski et al., 1998), but it also depends on the fat content and moisture. However, fat perception is more complex than sweetness. It depends on the fat content, the sugar content, texture, moisture, flavour, the nature of the food (liquid or solid) and mouthfeel (Abdallah et al., 1998). Fat plays vital sensory and functional roles in baked products. The presence of fat contributes to flavour or the combined perception of mouthfeel, taste and aroma. It also contributes to the appearance and lubricity of baked goods and increases the feeling of satiety. For products such as muffins and cakes, which are fat-in-water emulsions, air bubbles form the discontinuous phase, and the egg, sugar, fat and water mixture forms the continuous phase, where the flour is also dispersed. Due to the complex structural, textural and sensory functionality of sugar in baked products, obtaining good quality low-sucrose products is a difficult task. This will be discussed later in the section on calorie reduction strategies.

Fat plays a major structural role in cake and muffin manufacture. Initially, the fat globules entrap air in the batter during mixing, thus aerating the product. Fat also has important emulsifying properties and is able to hold a large amount of liquid, which increases and extends the softness of the cake. Finally, the presence of fat has a 'shortening' effect on the crumb texture of cake and muffins, in that it interferes with the protein matrix, resulting in a 'short', more tender crumb. Therefore, it would be expected that a fat reduction process would induce changes such as lower overall volume (less aeration), increased crumb firmness (because of a greater degree of starch swelling) and a reduced palatability. Optimised fat levels can again be determined through sensory and physico-chemical (TPA, texture profile) analysis. Fat reduction will result in reduced volume and increased crumb firmness (Rodríguez-García et al., 2014). Cake shelf life is between 1 and 4 weeks, but industrial cakes usually have longer shelf lives due to the use of certain additives such as preservatives (propionic, sorbic and benzoic acids) and correct packaging conditions (optimal heat sealing and modified atmosphere) (Barcenilla et al., 2016).

SENSORY PROPERTIES OF CHOCOLATE

Chocolate originated approximately 4000 years ago at the boundary of North and Central America in the regions that now correspond to Mexico and Guatemala. This early version of chocolate was highly regarded by the local population and consumed as a bitter drink and was very different to what we describe as chocolate today. Chocolate is made from a combination of cocoa solids, sugar, cocoa butter, lecithin and, in the case of milk chocolate, milk solids (e.g., milk powder or condensed milk). In some countries, other vegetable fats may also be permitted. According to the European regulation (EC 200/36) vegetable fats other than cocoa butter, up to a maximum of 5%, are permitted in certain Member States. Additionally, milk chocolate must contain not less than 30% total dry cocoa solids and not less than 18% dry milk solids obtained by partly or wholly dehydrating whole milk, semi- or full-skimmed milk, cream, or from partly or wholly dehydrated cream, butter or milk fat, including not less than 4.5% milk fat (EC 200/36). Dark chocolate contains cocoa liquor, sugar and cocoa butter with very dark varieties containing only the butter that is present in the liquor. White chocolate contains just cocoa butter, instead of cocoa solids, as well as sugar and milk powder. There are four types of cocoa bean that originate from trees of the variety Criollo, Forastero, Trinitario and Nacional. All require high rainfall and humidity to successfully grow. Criollo is quite rare, low yielding and an expensive variety grown in Central America and the Caribbean and has a complex flavour. Forastero is the most popular and robust variety, from a flavour and cultivation perspective, which originated in the amazon but is now mostly grown in West Africa (Ivory coast, Ghana, Nigeria), but also in southern and central America as well as Sri Lanka, Malaysia and Indonesia. Trinitario originates from Trinidad and is a hybrid of Forastero and Criollo varieties. Finally, the rarest chocolate variety is Nacional, a member of the Forastero family with up to 40% white beans and once thought extinct, but was recently rediscovered in a mountain valley of the Marañón River in northern Peru. It is said to have an intense rich chocolate flavour but without the bitterness and commands a hefty price.

The cultivation process is essential in the development of the sensory properties of chocolate, particularly flavour. Firstly, the ripened yellow-red pods are cut by hand and then split to extract the white mucus-covered beans. Once heaped in piles or covered with banana leaves a natural fermentation of bacteria and yeasts essentially cleans the beans of this sticky coating and the chocolate flavour starts to develop. Sun drying typically follows to bring the moisture content down to about 7−8%. This is followed by roasting to further develop flavour, reduce microbial loading and facilitate the removal of the bean shell from the cocoa nib in a process called winnowing, where air is used to blow the shell away from the heavier nib. For the manufacture of cocoa powder the nibs are ground to increasing smaller sizes to produce a cocoa liquor, which is then pressed to remove the cocoa butter leaving a hard cake. This is then milled to produce the powder. 'Conching' is a very important mixing step in chocolate flavour development. Small-scale conching uses a melangeur or refiner conche where the initial refining stage involves the grinding of the cocoa powder and ingredients by granite stone rollers to produce a smooth paste of small, grit-free particles. This is followed by the conching stage where the rollers, through friction, heat the mix and after sufficient time allow for off-flavours and water vapour to dissipate as well as producing additional flavour through the production of Maillard reaction and caramelised products. Modern conching occurs in longitudinal cylindrical mixing vessels where the chocolate mix is refined and conched simultaneously by being continually scraped between the internal surfaces with tensioned rotating baffles. Finally a 'tempering' stage is used to modify the six potential varying fat crystal sizes in the product, which can produce a crumbly dull looking product. Firstly, the chocolate is heated to 45°C and then cooled to 27°C, agitated and again heated to 31°C to allow the formation of the correct evenly sized fat crystals and to produce the shiny brittle product that we are all familiar with. Therefore, the sensory properties, especially the flavour of chocolate is determined by variety, fermentation, cocoa solids to cocoa butter ratio, ingredients as well as manufacturing conditions such as conching and tempering.

Chocolate can be drank (drinking, hot chocolate), eaten in block form or used as a coating for biscuits and cakes. As a fat-containing material it acts as a barrier layer preventing moisture migration from cake products and moisture ingress for biscuit products and can thus substantially prolong sensory shelf life. However, the chocolate must be married with the cake so that complementary flavours can be matched (Brown, 2009). This can only be done through sensory evaluation processes.

SENSORY PROPERTIES OF SUGAR

Sweetness has always been a popular characteristic in foods. Honey was probably the first source of sugars used by humans. Many plants have sugars in their tissues but only sugar cane, *Saccharum officinarum*, and sugar beet,

Beta vulgaris, have been used to extract sugar in commercial quantities (Manley, 2011c). There are also many varieties of sugars with different characteristics, colours and crystal sizes and thus different functionalities so it is important to choose the appropriate sugar for the right formulation or baking application. Different sugars and syrups play varied roles in baked goods, such as improving texture crumb and grain. They also provide a tenderising effect. Certain sugars such as fructose, inverted sugar and honey will retain water and thus extend shelf life. The size of the sucrose crystals, and therefore their rate of dissolution as the dough piece warms in the oven, affects the spread of short doughs as they bake and affects the appearance and crunchiness of baked biscuits (Manley, 2011c). Sugars vary in colour from white to dark brown increasing as the amount of molasses is retained during processing. Sugar binds moisture and can increase osmotic potential thus prolonging the shelf life of baked goods. Liquid sugars can retain more moisture than brown sugar, which can hold more moisture than granulated white sugar. Also, the higher the sugar content, the darker the crust.

Sugar plays an important role in cake making by impeding gluten formation by competing with the gluten-forming proteins for water in the batter thus preventing full hydration of the proteins during mixing and preventing the dough becoming too tough. Sugar is not only a sweetener in baking but has a complex role by acting as a preservative and by adding volume, tenderness, texture and colour to the baked product. Sucrose delays starch gelatinisation and protein denaturation temperatures during cake baking so that the air bubbles can be properly expanded by the carbon dioxide and water vapour before the batter sets (Rosenthal, 1995). With the correct level of sugar in the formulation, the gluten maintains the optimum level of elasticity, which in turn traps the right proportion of air in the elastic balloon like pockets of the dough matrix as the gases are formed during leavening and mixing. Thus, the dough will rise as the gases form and the pockets expand under compression by the gluten strands. By competing for the available water the sugar impedes gluten development and conferrs on the final baked product, a good volume with a perfect tender crumb texture. Additionally, the physical act of 'creaming together' the fat and sugar during formulation increases the air in the batter and is not just about the mixing of these two ingredients. The greater the degree of creaming, or working these ingredients together, the greater the amount of air incorporated in to the batter. The sugar granules rub against the fat-producing air bubbles in the fat which subsequently increase in size, causing the batter to rise due to the entrapment of leavening gases and steam produced in the oven. The size of the sugar crystal will also affect the amount of air incorporated during creaming. Granulated sugar, because it is larger in crystal size, will incorporate more air into a batter than confectioner's sugar. Also a confectioner's sugar will dissolve quicker in the batter than granulated sugar because of a smaller particle size.

CALORIE REDUCTION STRATEGIES

High saturated fat and sugar levels present a greater risk of obesity and type 2 diabetes. The prevalence of obesity is increasing around the world and it is a significant public health problem in many countries (International Obesity Task Force, 2002). Overconsumption of fat and sugar is associated with many diseases, such as obesity, high blood cholesterol and coronary heart diseases (Melanson et al., 2009; World Health Organisation, 2003). Several diseases are associated with a high level of sugar consumption, including obesity, insulin resistance, and diabetes mellitus type 2, as well as calories and fatty liver. Therefore, low sugar intake is strongly recommended (FAO/WHO, 2003). This recommendation is addressed not only to the consumer but also requires the food industry to reduce sugar content in the processed food (Barclay et al., 2008). However, as already discussed, fat and sugar play important functional roles in confectionary products and their reduction and removal presents some tough challenges.

Fat reduction and replacement technologies have been discussed in detail in Chapter 9 on Nutritionally Optimised Low Fat Foods. The current chapter will thus concentrate on solely sugar reduction strategies.

Sweetness in confectionary products is mainly due to the sugar content (Drewnowski et al., 1998), but it also depends on the fat content and moisture. However, fat perception is more complex than sweetness. It depends on the fat content, the sugar content, texture, moisture, flavour, the nature of the food (liquid or solid) and mouthfeel (Abdallah et al., 1998). Therefore, the reduction of sucrose levels in a cake system affects structural and sensory properties (Frye and Setser, 1991). Due to the complex structural, textural and sensory functionality of sugar in baked products, obtaining good quality low-sucrose products is also a difficult task. Even though sweetness can be balanced with high-intensity sweeteners while reducing calories, many of the critical functions of sugar in baked products, in particular, aeration, browning, crystallisation, freezing point depression, glass formation, moisture retention, texturisation and whipping are not easy to mimic. Sugar replacement in baked products needs to consider both the sweetness and the bulking effect of sucrose (Martínez-Cervera et al., 2012). Cake and biscuit formulations are complex, containing a wide range of ingredients (flour, fat, egg, raising agents, salt, sugar, water, flavours, etc.). The replacement of sugar with other sources influences the baking regimes of these products. Therefore, during optimisation, it is critical to assess the changes to the mixing and baking processes of the products.

Noncaloric artificial sweeteners (NAS) such as sucralose, aspartame or saccharin are sugar replacers that have found ubiquitous use in the food industry in the production of calorie-reduced food and beverage products. The disadvantages of these sweeteners are that they are often artificial, which are not label friendly, and often carry metallic or bitter taste and aftertastes (Schiffman et al., 1995). Wetzel et al., (1997) investigated the sensory properties of sugar-free cakes sweetened with encapsulated aspartame. Aspartame is usually limited to nonthermal applications such as cooking due to its instability to elevated temperatures, high pH values and high moisture contents (Bell and Hageman, 1994). Wetzel et al. (1997) concluded that a consumer panel found no difference in acceptability between the freshly prepared no-sugar-added and full-sucrose cakes. Martínez-Cervera et al. (2012) investigated the rheological, textural and sensory properties of low-sucrose muffins reformulated with sucralose/polydextrose. These authors found that for a 50% sucrose replacement, the appearance, colour, texture, flavour, sweetness and the general acceptability were similar to those of the control, but significantly less acceptable muffins were obtained with 100% sucrose replacement. Similarly, Aggarwal et al. (2016) demonstrated that highly acceptable reduced-calorie biscuits can be produced by using dairy−multigrain composite flour with maltitol and FOS-sucralose (as sweetener) and polydextrose (as fat replacer). Saccharin, which is 300−400 times sweeter than sugar, has fallen out of favour with product developers mainly due to negative publicity. Bans for use in some jurisdictions, followed by the lifting of bans has left some consumers confused. It also posesses after taste issues (metallic) which limits consumer acceptability. Also, saccharin is associated with the potential risk of bladder cancer when used heavily. Consumption of aspartame must be avoided by people with the metabolic disease, phenylketonuria (Grenby, 1991). The recent articles 'The Weighty Costs of Non-Caloric Sweeteners' (Feehley and Nagler, 2014) and 'Artificial Sweeteners Induce Glucose Intolerance by Altering the Gut Microbiota' (Suez et al., 2014) confirm the potential dangers of NAS ingredients. Suez et al. detected an increase in carbohydrate degradation pathways in the microbiota of NAS-fed mice. Suez et al. (2014) studied around 400 people and found that bacterial populations in the guts of those who consumed NAS were significantly different from those who did not. Moreover, NAS consumption correlated with disease markers linked to obesity, such as elevated fasting blood glucose levels and impaired glucose tolerance. Suez et al. data indicate that NAS consumption may contribute to, rather than alleviate, obesity-related metabolic conditions, by altering the composition and function of bacterial populations in the gut. In summary, their results suggest that NAS consumption in both mice and humans enhances the risk of glucose intolerance and that these adverse metabolic effects are mediated by modulation of the composition and function of the microbiota. Notably, several of the bacterial taxa that changed following NAS consumption were previously associated with type 2 diabetes in humans. In addition, they show that metagenomes of saccharin-consuming mice are enriched with multiple additional pathways previously shown to associate with diabetes mellitus or obesity in mice and humans, including sphingolipid metabolism and lipopolysaccharide biosynthesis. Their findings suggest that NAS may have directly contributed to enhancing the exact epidemic that they themselves were intended to fight.

There is evidence that to a certain extent some consumers also reject the use of these NAS products and embrace a more 'clean label' approach to reduced calorie products. A recent US Mintel survey (Pereira M, 2012; Garder, 2012) found that 64% of respondents indicated they were concerned about the safety of 'artificial' sweeteners. Also, Richardson, O'Sullivan and Kerry (in preparation), in a survey undertaken in Ireland from 1300 respondents on sugar and NAS sugar replacers in food and beverages reported that 57% were wary of the latter NAS. Only 4% of participants said they would drink sugary drinks on a daily basis and 52% of them said they would not buy sugar reduced and NAS-containing sugar-free drinks. Additionally 92% of the people surveyed said there is a greater need for information on the impact of a high sugar and fat diet in general.

Therefore, a clean label approach, where alternatives to NAS compounds are utilised, is of direct interest to the food industry. There are two principal approaches that can be taken with a NAS avoidance strategy for the development of calorie-reduced foods. The first is sugar reduction without replacement or by using alternate sucrose sources. The second approach involves the investigation of sugar reduction via alternate non-NAS and non-sucrose sugar sources and sensory acceptance testing combined with descriptive analysis incorporating sequential sugar reduction similar to the nutritional optimisation strategies discussed in the earlier Chapters 4 and 11 for salt and fat reduction in processed meats. The reduction of sugar in sweet baked products such as cake, muffins and cookies is currently being studied in the SWEETLOW project funded by the Irish Department of Agriculture, Food and the Marine (SWEETLOW, 2016). This project funded as part of the Food Industry Research Measure (FIRM) is investigating sequential sugar reduction without replacement and also the use of alternative sugar sources, such as polyols, steviosides, polydetxrose, inulin/prebiotics and dietary fibres such as arabinoxylan oligosaccharides in sugar reduction strategies. The application of emulsifiers and bulking agents is also being studied, in an effort to reduce the variance of the products being formulated. Polydextrose is widely recognised for its versatility as a bulking agent and texture enhancer and provides only 1 kcal/g. The application of a range of fibres is also being studied as a means to enhance the consistency and texture of the formulations developed. These products, also non-NAS, can then be evaluated and optimised using hedonic sensory affective testing and the descriptive profiling methodologies as described earlier and in previous chapters. Numerous sweet and low calorie compounds are available in nature; some of them are used commercially, e.g., xylitol and stevioside (Nabors and Gelardi, 1986).

GLUTEN-FREE PRODUCTS

Gluten is a protein complex, which functions as the main storage protein in wheat, rye and barley and from a technological perspective is responsible for the viscoelastic properties of dough. This structure consists of thousands of small, balloon like pockets produced through the hydration and working (kneading) of the gluten in the dough which then traps air bubbles as they are inflated by the gases produced during leavening and cooking. Once baked this produces the characteristic aerated structure of bread. Gluten is thus an inherently important compound contributing to the sensory properties of bread products. However, the presence of gluten can produce symptoms from the ∼1% of the population that suffer from coeliac disease. Coeliac disease is an intestinal intolerance to the storage proteins of wheat (i.e., all *Triticum* species), rye and barley (and sometimes oats) (Masure et al., 2016). Coeliac disease is an immune-mediated enteropathy caused by the ingestion of gluten in genetically susceptible individuals and is one of the most common lifelong disorders (Fasano and Catassi, 2012; Biagi et al., 2010). Sensitivity to gluten triggers an immune response that damages the mucosal layer of the small intestine and prevents absorption of nutrients (Pahlavan et al., 2016). The symptoms of gluten intolerance or coeliac disease include diarrhoea, constipation, weight loss, chronic tiredness, anaemia, failure to thrive in children, chronic mouth ulcers, stomach pain and bloating, indigestion, bone pain, moodiness or depression, infertility and recurrent miscarriages (Coeliac Society of Ireland, 2016). As an autoimmune disease, coeliac disease may play a role in some organ-related complications such as autoimmune thyroid disease, hepatitis, type I diabetes, psoriasis, Addison's disease and dermatitis herpetiformis (Nenna et al., 2016).

At present, the only available treatment for coeliac disease is a strict gluten-free diet (Fasano and Catassi, 2012; Masure et al., 2016; Pahlavan et al., 2016). To ensure the safety of gluten-sensitive consumers, Codex Alimentarius, European Commission Regulation and the US Food and Drug Administration have specified 20 mg/kg as the maximum threshold of gluten in foods labelled gluten-free (Pahlavan et al., 2016).

The coeliac population have driven the growth of the 'gluten-free' product sector with companies producing a diverse range of sensory acceptable products from breads to confectionary, pasta and even meat products traditionally produced with cereals such as sausages (Tobin et al., 2013) and black and white puddings (Fellendorf et al., 2015, 2016a,b,c). The growth of this sector has expanded even further with normal consumers choosing to avoid gluten (Capriles et al., 2016) because they believe that gluten-free products (Fig. 14.1) are a 'healthier' option and it is an effective way of weight control, although there is no scientific evidence to support these beliefs (Brouns et al., 2013; Pszczola, 2012).

Many different approaches have been taken for the production of acceptable bread products without using gluten. However, the absence of gluten is well known to show a significant influence on dough rheology resulting in bread with crumbling texture, poor colour, unsatisfying taste and low specific volume with a short shelf life (Houben et al., 2012).

Masure et al. (2016) have extensively reviewed past current and future approaches in gluten-free breadmaking research. Additionally, gluten-free products have been reviewed in detail for alternative ingredients such as cereals (Comino et al., 2013; Huttner and Arendt, 2010) flours (Campo et al., 2016; Paciulli et al., 2016; Turkat et al., 2016), starches (Matos and Rossell, 2013), probiotics (Capriles and Arêas, 2013), hydrocolloids (Anton and Ardfield, 2008), proteins (Shevkani et al., 2015), pseudocereals (Alvarez-Jubete et al., 2010) and sourdough (Arendt et al., 2007; Gobetti et al., 2007). Sourdough has been used to improve dough machinability, as well as nutritional and sensory properties while also prolonging shelf life compared to yeasted breads (Hammes and Ganzel, 1998; De Vuyst and Neysens 2005).

More recently, Campo et al. (2016) investigated the influence of the addition of teff flour (5%, 10% and 20%) and different dried (buckwheat or rice) or fresh (with *Lactobacillus helveticus*) sourdoughs on the sensory quality and consumer preference of gluten-free breads. They found that a

FIGURE 14.1 A selection of gluten-free breads; farmhouse loaf, brown soda, white loaf and poppy seed bagel. Supplying the expanding 'gluten-free' sector.

combination of teff (10%) with cereal sourdough (rice or buckwheat) enhanced bread aroma, increasing the fruity, cereal and toasty notes and bread combining 10% teff and rice sourdough was preferred in terms of flavour.

SHELF LIFE − MICROBIOLOGICAL SPOILAGE, STALING AND OXIDATIVE STABILITY

One of the major differences of bread products compared to cakes and sponges is the much shorter shelf life as a result of spoilage, but more commonly staling or moisture migration. The spoilage of bakery products can be caused by microbial action, unacceptable sensory texture changes due to staling, softening or off-flavour development due to oxidative rancidity. Microbial spoilage, staling or softening will predominantly occur in bread products as the lower sugar content and high water activity as well as the inherent starch structure reduces shelf life considerably in comparison to cakes and other confectionary products and long before oxidation issues can manifest. When all baked products leave the oven their surfaces are sterile and so it is microbial contamination (handling, slicing) of the surface during cooling that leads to product spoilage (Cauvain, 2011). Spoilage is generally due to mould as the water activity of breads supports their growth but yeasts and bacteria may also occur if conditions are right. Bacterial spoilage can occur when the heat-resistant spores of the organism *Bacillus subtilis* contaminate the baker's flour, survive the baking process and then germinate in the bread as it cools producing an unacceptable fruity odour and ropey crumb texture. As *Bacillus* is a soil bacteria it is more likely to contaminate wholemeal or multigrain flours than refined white flours because it exists on the surface of the grains.

Staling is responsible for huge economic losses to the baking and retail sectors every year, as well as to the consumer, due to the development of unacceptable sensory textural properties, primarily crumb firming and to a lesser extent crust softening and is independent of microbial spoilage. Water mobility has an important effect on the structural properties of cereal-based systems, in particular during storage and subsequent staling. Staling of bakery products is a result of moisture migration from the interior to the exterior of the product, in order to achieve equilibrium, resulting in a firming of the crumb and a softening of the crust. From the consumer perspective, crumb firmness during staling is of greater concern than the softening of the crust, which results in a typical tough chewy leathery texture. Although studied for decades, staling still persists in breads and this is due to a lack of complete understanding of the processes involved. Starch is the most abundant component of wheat flour (ca. 70−75%). It occurs as granules and consists mainly of the two glucose polymers: amylose (AM) and amylopectin (AP) (Manley, 2011b). The AP fraction contains ordered regions and is embedded in the noncrystalline matrix of the AM (Schoch, 1945). During baking, wheat starch present in the bread dough (also cake batter) undergoes the transformation known as gelatinisation (Cauvain, 2011; Gray and Bemiller, 2003). The heated water in the dough during baking causes the starch granules to swell and absorb more water and gelatinisation is the point at which the mixture turns from a viscous liquid to a solid foam or sponge. Retrogradation occurs, on cooling, when the AP reverts to a more ordered state after gelatinisation and the starch polymers begin to lose their mobility. This occurs optimally at 4°C and continues during storage contributing to the firming that typically occurs in the staling of the crumb (Cauvain and Young, 2008). Staling is characterised by changes in flavour, taste and aroma, resulting in a loss of perceived sensory freshness as well as changes in firmness, water absorptive capacity, crystallinity, opacity and soluble starch content of the crumb. Additionally, loss of moisture from the product may accelerate reactions leading to staling.

In biscuits and pastries, the moisture contents are so low that moisture may migrate from the atmosphere into the product, rather than from product to atmosphere as with bread and cakes. This is a common mechanism by which cookies and pastries go soft or stale (Cauvain, 2011). Oxidation issues can occur in biscuit products because of their low water activity, high sugar content and resistance to microbial spoilage. They are simply around long enough for off-flavour to potentially form due to oxidation of the fats used in their composition. The oxidation processes for biscuits and the subsequent development of off-flavours can thus be the limiting factor to shelf life in the absence of microbial spoilage and staling. Oxidation has been extensively covered in Chapters 6 and 11.

Packaging is necessary to prevent excessive loss of moisture from bread to the atmosphere and in biscuits to prevent the migration of water in to the package and thus softening of the product. In both cases, moisture barrier layers are necessary to retard this moisture migration. Additionally, MAP (modified atmosphere packaging) can have applications in the shelf life extension of bread and biscuits. These topics have been extensively covered in Chapter 7 of this book on Packaging.

ANTISTALING STRATEGIES

De Vuyst and Neysens (2005) found that sourdough breads had improved technological properties (dough machinability), nutritional and sensory properties as well as longer shelf life compared to yeasted breads. Additionally, flours other than wheat or derived from AM-free wheat flours (waxy) have been proposed in the production of mixed flour breads in order to improve nutritional aspects and bread aging properties (Fadda et al., 2014). Barley flour, due to a high β-glucan content, has been shown to help in reducing starch crystallisation, thus delaying significantly the staling rate of bread when used at the 20% level, but it increased the firmness of the fresh product (Gujral et al., 2003). Soy flour used as a replacement for wheat flour at the 40% level resulted in a significant decrease in AP recrystallisation as well as promoted moisture retention during storage, with respect to control bread, thus leading to decreased staling (Vittadini and Vodovotz, 2003). Additionally, wheat flours mixed with varying quantities of pregerminated brown rice (Watanabe, 2004), flaxseed flour (Mentes et al., 2008), potato paste (Wu et al., 2009), cassava flour (Begum et al., 2010), oat, rye and buckwheat (Angioloni and Collar, 2011), have all demonstrated varying degrees of effectiveness in retarding staling in the resulting bread products.

The antistaling effects of many compounds such as hydrocolloids, modified starches, dextrins and malto-oligosaccharides and other fibres have been extensively reviewed by Fadda et al. (2014). Examples include hydrocolloids such as pectins, hydroxypropyl methylcellulose, guar gum, konjac gum, xanthan gums, chitosan; modified starches such as chemically modified tapioca starches; dextrins, and malto-oligosaccharides, including fructans and exopolysaccharides as well as other fibres and a plethora of other lipid- and enzyme-based ingredients.

Case Study: Use of smart packaging technologies for monitoring and extending the shelf life quality of MAP bread: application of intelligent oxygen sensors and active ethanol emitters (EEs).

The following case study displays a simple storage test for bread samples analysed principally with hedonic analysis and with microbiological assessment in parallel. Total viable counts were carried out in accordance with ISO standards, method 4833:2003. Commercial Ciabatta bread (consisting of wheat flour, yeast, milk, water, sugar, salt and butter) samples were packaged in MAP gas levels consisting of 10% CO_2 and 90% N_2 (Hempel et al., 2013). Each bread pack contained two Ciabatta loafs (2×100 g) and were packaged in high barrier (1 cm^3/m^2, 24 h, bar) OPA/PE laminate films of 38 μm thickness and stored at room temperature (21°C). An optical oxygen sensor (prepared using platinum octaethylporphyrin−ketone, see Hempel et al., 2013) was attached to the inside of all bread packs prior to gas flushing and sealing. Further samples were packaged with air (21% O_2) to serve as a control. The use of ethanol in low density polyethylene-based sachets containing 3 mL of alcohol gel (Hempel et al., 2013) was also incorporated into a number (EE) of packs while an ethanol spray (ES) was applied to other samples.

In order to assess the quality changes of samples and whether hedonic attributes as well as flavour and aroma characteristics were affected, 26 untrained panellists (consumers of speciality bread products) were recruited. Each panellist was presented with samples provided for assessment included control (WOE − without ethanol, EE − ethanol emitter and ES − ethanol spray) and MAP (WOE,

EE and ES) and was assessed on storage days 1, 5 and 10. All samples were assigned with a random three-digit identification code to allow for blind assessment and asked to rate descriptors according to a 10 cm line scale. The list of hedonic descriptors included overall appearance, overall flavour liking, overall acceptability and anchored from extremely dislike or unacceptably to extremely like or acceptable. Additionally the simple descriptors off-aroma, ethanol aroma, fresh aroma, ethanol flavour and astringent taste were also scored and anchored on the line scale from none to extreme (Hempel et al., 2013). The scoring of these attributes cannot be considered as part of a descriptive profile because they were not scored separately to the sensory acceptance. The inclusion of a descriptive component would have been a lot stronger if a ranking descriptive analysis was undertaken using the same attributes.

These authors found that acceptable limits for microbial quality were maintained for 16 days when packaged in air using EEs. Sensory analysis shows that the use of EEs has no negative effect on product quality compared to the ES technique and controls (Fig. 14.2). The addition of ethanol to extend

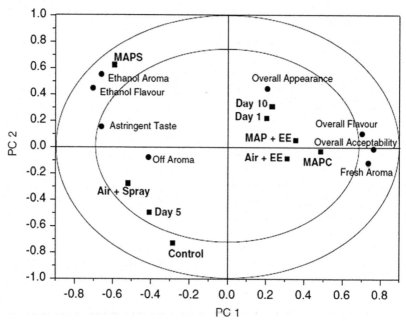

FIGURE 14.2   An overview of the variation found in the mean data from the ANOVA—partial least squares regression (APLSR) correlation loadings plot for each bread packaging treatment (*EE*, Ethanol emitter; *ES*, ethanol spray; *WOE*, without ethanol) and MAP (WOE, EE and ES) and was assessed on storage days 1, 5 and 10. *Reproduced from Hempel, A., O'Sullivan, M.G., Papkovsky, D., Kerry, J.P., 2013. Use of smart packaging technologies for monitoring and extending the shelf-life quality of modified atmosphere packaged (MAP) bread: application of intelligent oxygen sensors and active ethanol emitters. European Food Research and Technology 237, 117–124.*

shelf life shows great promise in delaying the onset of inevitable mould growth. The potential for such smart packaging technology like O_2 sensors and EEs in the bread industry cannot be understated (Hempel et al., 2013).

REFERENCES

Abdallah, L., Chabert, M., Le Roux, B., Louis-Sylvestre, J., 1998. Is pleasantness of biscuits and cakes related to their actual or to their perceived sugar and fat contents? Appetite 30, 309−324.

Aggarwal, D., Sabikhi, L., Sathish Kumar, M.H., 2016. Formulation of reduced-calorie biscuits using artificial sweeteners and fat replacer with dairy−multigrain approach. NFS Journal 2, 1−7.

Alvarez-Jubete, L., Auty, M., Arendt, E.K., Gallagher, E., 2010. Baking properties and microstructure of pseudocereal flours in gluten-free bread formulations. European Food Research and Technology 230, 437−445.

Angioloni, A., Collar, C., 2011. Nutritional and functional added value of oat, KamutR, spelt, rye and buckwheat versus common wheat in bread making. Journal of the Science of Food and Agriculture 91, 1283−1292.

Anton, A.A., Artfield, S.D., 2008. Hydrocolloids in gluten-free breads: a review. International Journal of Food Sciences and Nutrition 59, 11−23.

Arendt, E.K., Ryan, L.A.M., Dal Bello, F., 2007. Impact of sourdough on the texture of bread. Food Microbiology 24, 165−174.

Barcenilla, B., Román, L., Martínez, C., Martínez, M.M., Gómez, M., 2016. Effect of high pressure processing on batters and cakes properties. Innovative Food Science and Emerging Technologies 33, 94−99.

Barclay, A.W., Petocz, P., McMillan-Price, J., Flood, V.M., Prvan, T., Mitchell, P., Brand-Miller, J.C., 2008. Glycemic index, glycemic load, and chronic disease risk—A meta-analysis of observational studies. American Journal of Clinical Nutrition 87, 627−637.

Begum, R., Rakshit, S.K., Mahfuzur Rahman, S.M., 2010. Protein fortification and use of cassava flour for bread formulation. International Journal of Food Properties 14, 185−198.

Bell, L.N., Hageman, M.J., 1994. Differentiating between the effects of water activity and glass transition dependent mobility on a solid state chemical reaction: aspartame degradation. Journal of Agricure and Food Chemistry 42, 2398−2401.

Bennion, E.B., Bamford, G.S.T., 1973. Baking fats. In: Bent, A.J. (Ed.), The Technology of Cake Making, p. 25e47 (Aylesbury: L. Hill).

Biagi, F., Klersy, C., Balducci, D., Corazza, G.R., 2010. Are we not over-estimating the prevalence of celiac disease in the general population? Annals of Medicine 42, 557−561.

Brouns, F., van Buul, V., Shewry, P., 2013. Does wheat make us fat and sick? Journal of Cereal Science 58, 209−215.

Brown, M., 2009. Biscuits and bakery products. In: Talbott, G. (Ed.), Science and Technology of Enrobed and Filled Chocolate, Confectionery and Bakery Products. Woodhead, pp. 152−162.

Campo, E., del Arco, L., Urtasun, L., Oria, R., Ferrer-Mairal, A., 2016. Impact of sourdough on sensory properties and consumers' preference of gluten-free breads enriched with teff flour. Journal of Cereal Science 67, 75−82.

Capriles, V.D., Arêas, J.A.G., 2013. Effects of prebiotic inulin-type fructans on structure, quality, sensory acceptance and glycemic response of gluten-free breads. Food and Function 4, 104−110.

Capriles, V.D., dos Santos, F.G., Arêas, J.A.G., 2016. Gluten-free breadmaking: improving nutritional and bioactive compounds. Journal of Cereal Science 67, 83–91.

Cauvain, S.P., 2011. Chapter 23 The stability and shelf life of bread and other bakery products. In: Kilcast (Ed.), Food and Beverage Stability and Shelf Life. WoodheadPublishingLimited, pp. 657–682.

Cauvain, S.P., Young, L.S., 2008. Bakery Food Manufacture and Quality: Water Control and Effects, second ed. Wiley-Blackwell, Oxford.

Coeliac Society of Ireland, 2016. Symptoms. https://www.coeliac-ireland.com/coeliac-desease/symptoms/.

Comino, I., Moreno, M.D., Real, A., Rodriguez-Herrera, A., Barro, F., Sousa, C., 2013. The gluten-free diet: testing alternative cereals tolerated by celiac patients. Nutrients 5, 4250–4268.

De Vuyst, L., Neysens, P., 2005. The sourdough microflora: biodiversity and metabolic interactions. Trends in Food Science and Technology 16, 43–56.

Drewnowski, A., Almiron-Roig, E., 2010. Human perceptions and preferences for fat-rich foods. In: Montmayeur, J.P., le Coutre, J. (Eds.), Fat Detection: Taste, Texture, and Post Ingestive Effects. CRC Press Taylor & Francis Group, Boca Raton, FL (USA), pp. 243–264.

Drewnowski, A., Nordensten, K., Dwyer, J., 1998. Replacing sugar and fat in cookies: impact on product quality and preference. Food Quality and Preference 9 (1–2), 13–20.

EC 200/36, 2000. Directive 2000/36/EC of the European Parliament and of the Council of 23 June 2000.

Fadda, C., Sanguinetti, A.M., DelCaro, A., Collar, C., Piga, A., 2014. Bread staling:updating the view. Comprehensive Reviews in Food Science and Food Safety 13, 473–492.

FAO/WHO, 2003. Diet, Nutrition and the Prevention of Chronic Diseases. Pages 1–149 in WHO Technical Report Series 916. Food and Agriculture Organization of the United Nations (FAO), Rome, Italy, and World Health Organization (WHO), Geneva, Switzerland.

Fasano, A., Catassi, C., 2012. Celiac disease. The New England Journal of Medicine 367, 2419e2426.

Feehley, T., Nagler, C.T., October 9, 2014. The weighty costs of non-caloric sweeteners. Nature 514.

Fellendorf, S., O'Sullivan, M.G., Kerry, J.P., 2015. Impact of varying salt and fat levels on the physiochemical properties and sensory quality of white pudding sausages. Meat Science 103, 75–82.

Fellendorf, S., O'Sullivan, M.G., Kerry, J.P., 2016a. The reduction of salt and fat levels in black pudding and the effects on physiochemical and sensory properties. International Journal of Food Science and Technology (Online).

Fellendorf, S., O'Sullivan, M.G., Kerry, J.P., 2016b. Effect of using replacers on the physico-chemical properties and sensory quality of low salt and low fat white puddings. European Food Research and Technology (Online).

Fellendorf, S., O'Sullivan, M.G., Kerry, J.P., 2016c. Impact of using replacers on the physico-chemical properties and sensory quality of reduced salt and fat black pudding. Meat Science 113, 17–25.

Frye, A.M., Setser, C.S., 1991. Optimizing texture of reduced calorie sponge cakes. Cereal Chemistry 69, 338e343.

Gardner, C., et al., August 2012. Non-nutritive sweeteners: current use and health perspectives. A Scientific Statement from the American Heart Association and the American Diabetes Association. Diabetes Care 35.

Gobbetti, M., Rizzello, C.G., Di Cagno, R., De Angelis, M., 2007. Sourdough lactobacilli and celiac disease. Food Microbiology 24, 187—196.

Gray, J.A., Bemiller, J.N., 2003. Bread staling: molecular basis and control. Comprehensive Reviews in Food Science and Food Safety 2, 1—21.

Grenby, T.H., 1991. Intense sweeteners for the food industry: an overview. Trends in Food Science and Technology 2, 2—6.

Gujral, H.S., Gaur, S., Rosell, C.M., 2003. Note: effect of barley flour, wet gluten and ascorbic acid on bread crumb texture. Food Science and Technology International 9, 17—25.

Hammes, W.P., Ganzel, M.G., 1998. Sourdough breads and related products. In: Woods, B.J.B. (Ed.), Microbiology of Fermented Foods. Blackie Academic Professional, pp. 199—216.

Hempel, A., O'Sullivan, M.G., Papkovsky, D., Kerry, J.P., 2013. Use of smart packaging technologies for monitoring and extending the shelf-life quality of modified atmosphere packaged (MAP) bread: application of intelligent oxygen sensors and active ethanol emitters. European Food Research and Technology 237, 117—124.

Houben, A., Hochstotter, A., Becker, T., 2012. Possibilities to increase the quality in gluten free bread production: an overview. European Food Research and Technology 235, 195—208.

Huttner, E.K., Arendt, E.K., 2010. Recent advances in gluten-free baking and the current status of oats. Trends in Food Science and Technology 21, 303—312.

International Obesity Task Force, 2002. About Obesity: Incidence, Prevalence & Co-morbidity.

Maache-Rezzoug, Z., Bouvier, J.M., Allaf, K., Patras, C., 1998. Effect of principal ingredients on rheological behaviour of biscuit dough and on quality of biscuits. Journal of Food Engineering 35, 23—42.

Manisha, G., Soumya, C., Indrani, D., 2012. Studies on interaction between stevioside, liquid sorbitol, hydrocolloids and emulsifiers for replacement of sugar in cakes. Food Hydrocolloids 29, 363—373.

Manley, D., 2011a. Chapter 1 Setting the scene: a history and the position of biscuits. In: Manley, D. (Ed.), Manleys's Technology of Biscuits, Crackers and Cookies, pp. 1—9.

Manley, D., 2011b. Chapter 9 Wheat flour and vital wheat gluten as biscuit ingredients. In: Manley, D. (Ed.), Manleys's Technology of Biscuits, Crackers and Cookies, pp. 109—133.

Manley, D., 2011c. Chapter 11 Sugars and syrups as biscuit ingredients. In: Manley, D. (Ed.), Manleys's Technology of Biscuits, Crackers and Cookies, pp. 144—159.

Martínez-Cervera, S., Sanz, T., Salvador, A., Fiszman, S.M., 2012. Rheological, textural and sensorial properties of low-sucrose muffins reformulated with sucralose/polydextrose. LWT — Food Science and Technology 45, 213—220.

Masure, H.G., Fierens, E., Delcour, J.A., 2016. Current and forward looking experimental approaches in gluten-free bread making research. Journal of Cereal Science 67, 92—111.

Matos, M.E., Rosell, C.M., 2013. Quality indicators of rice-based gluten-free breadlike products: relationships between dough rheology and quality characteristics. Food and Bioprocess Technology 6, 2331e2341.

Melanson, E.L., Astrup, A., Donahoo, W.T., 2009. The relationship between dietary fat and fatty acid intake and body weight, diabetes, and the metabolic syndrome. Annals of Nutrition and Metabolism 55 (1—3), 229—243.

Mentes, Ö., Bakkalbassi, E., Ercan, R., 2008. Effect of the use of ground flaxseed on quality and chemical composition of bread. Food Science and Technology International 14, 299—306.

Nabors, O.L., Gelardi, R.C., 1986. Alternative Sweeteners. Marcel Dekker, New York.

Nenna, R., Petrarca, L., Verdecchia, P., Florio, M., Pietropaoli, N., Mastrogiorgio, G., Bavastrelli, M., Bonamico, M., Cucchiara, S., 2016. Celiac disease in a large cohort of children and adolescents with recurrent headache: a retrospective study. Digestive and Liver Disease 48, 495–498.

Nip, W.K., 2007. Sweeteners. In: Hui, H. (Ed.), Bakery Products: Science and Technology. Blackwell Publishing, Chicago, IL, USA, pp. 137–159.

Paciulli, M., Rinaldi, M., Cirlini, M., Scazzina, F., Chiavaro, E., 2016. Chestnut flour addition in commercial gluten-free bread: a shelf-life study. LWT – Food Science and Technology 70, 88–95.

Pahlavan, A., Sharma, G.M., Pereira, M., Williams, K.M., 2016. Effects of grain species and cultivar, thermal processing, and enzymatic hydrolysis on gluten quantitation. Food Chemistry 208, 264–271.

Pereira, M., 2012. Acting Sweet: Drivers, Trends and Forecasts in the Sugar-free Food and Beverage Market Are Explored. CBS Interactive. http://findarticles.com/p/articles/mi_m3289/is_1_175/ai_n26689456/.

Pszczola, D., 2012. The rise of gluten-free. Food Technology 66, 55–66.

Richardson A.M., O'Sullivan M.G.. Kerry J.P., Survey on sugar consumption habit in the Republic of Ireland (in preparation).

Rodríguez-García, J., Sahi, S.S., Hernando, I., 2014. Functionality of lipase and emulsifiers in low-fat cakes with inulin. LWT – Food Science and Technology 58, 173–182.

Rosenthal, A.J., 1995. Application of aged egg in enabling increased substitution of sucrose by litesse (polydextrose) in high-ratio cakes. Journal of the Science of Food and Agriculture 68, 127–131.

Schoch, T.J., 1945. The fractionation of starch. Advances in Carbohydrate Chemistry 1, 247–248.

Schiffman, S.S., Booth, B.J., Losee, M.L., Pecore, S.D., Warwick, Z.S., 1995. Bitterness of sweeteners as a function of concentration. Brain Research Bulletin 36, 505–513.

Suez, J., et al., October 9, 2014. Artificial sweeteners induce glucose intolerance by altering the gut microbiota. Nature 514.

SWEETLOW, 2016. Development of Consumer Optimised Low Carbohydrate Irish Confectionary Products. Project Coordinator: Dr Maurice O'Sullivan. Ref:14/F/812. https://www.agriculture.gov.ie/media/migration/research/firmreports/CALL2014ProjectAbstracts240216.pdf.

Shevkani, K., Kaur, A., Kumar, S., Singh, N., 2015. Cowpea protein isolates: functional properties and application in gluten-free rice muffins. LWT – Food Science and Technology 63, 927–933.

Turkut, G.M., Cakmak, H., Kumcuoglu, S., Tavman, S., 2016. Effect of quinoa flour on gluten-free bread batter rheology and bread quality. Journal of Cereal Science 69, 174–181.

Tobin, B.D., O'Sullivan, M.G., Hamill, R.M., Kerry, J.P., 2013. The impact of salt and fat level variation on the physiochemical properties and sensory quality of pork breakfast sausages. Meat Science 93, 145–152.

Vittadini, E., Vodovotz, Y., 2003. Changes in the physico-chemical properties of wheat and soy containing breads during storage as studied by thermal analyses. Journal of Food Science 68, 2022–2027.

Watanabe, M., Maeda, T., Tsukahara, K., Kayahara, H., Morita, N., 2004. Application of pre-germinated brown rice for breadmaking. Cereal Chemistry 81, 450–455.

Wetzel, C.R., Weese, J.O., Bell, L.N., 1997. Sensory evaluation of no-sugar-added cakes containing encapsulated aspartame. Food Research International 30, 395–399.

WHO, 2003. Diet, Nutrition and the Prevention of Chronic Diseases. Report of a Joint WHO/FAO Expert Consultation. WHO technical report series 919, p. 148. http://www.who.int/dietphysicalactivity/publications/trs916/summary/en/print.html.

Wu, K.L., Sung, W.C., Yang, C.H., 2009. Characteristics of dough and bread as affected by the incorporation of sweet potato paste in the formulation. Journal of Marine Science and Technology 17, 13–22.

Index

Printed in the United States
By Bookmasters